Basic surveyin

Fourth edition

Basic surveying

Fourth edition

W S Whyte
Former Associate Professor,
City University of Hong Kong

and

R E Paul
Former Senior Lecturer,
Leicester Polytechnic

OXFORD AMSTERDAM BOSTON LONDON NEW YORK PARIS
SAN DIEGO SAN FRANCISCO SINGAPORE SYDNEY TOKYO

Butterworth-Heinemann
An imprint of Elsevier Science
Linacre House, Jordan Hill, Oxford OX2 8DP
200 Wheeler Road, Burlington, MA 01803

First published 1969
Second edition 1976
Third edition 1985
Fourth edition 1997
Reprinted 1998, 1999, 2000, 2001, 2003

British Library Cataloguing in Publication Data
Whyte, W.S. (Walter Stuart), 1930-
Basic Surveying - 4th ed.
1. Surveying
I. Title II. Paul, R.E. III. Basic Metric surveying
526.9'024624

Library of Congress Cataloguing in Publication Data
Whyte, W.S. (Walter Stuart.)
Basic surveying/W.S. Whyte and R.E. Paul - 4th ed.
p.cm
Rev. ed. of Basic metric surveying. 3rd ed. 1985.
Includes bibliographical references and index.
ISBN 0 7506 1771 3
1. Surveying. 2. Metric system. I. Paul, R.E.
II. Whyte, W.S. (Walter S.) Basic Metric surveying.
III. Title
TA545 W69 69-29743
526.9-dc21 CIP

ISBN 0 7506 1771 3

For information on all Butterworth-Heinemann publications
visit our website at www.bh.com

Typeset by Genesis Typesetting, Rochester, Kent
Printed and bound in Great Britain by Athenaeum Press Ltd, Gateshead, Tyne & Wea

Contents

Preface to the fourth edition ix

Chapter 1 – Introduction to surveying 1
1.1 Surveying 1
1.2 Objects of surveying 1
1.3 Classifications of surveying 2
1.4 Survey tasks 3
1.5 Scale and scales 4
1.6 Basic survey methods 5
1.7 Principles of survey 9
1.8 Accuracy 10
1.9 Measurement and errors 10
1.10 Numerical values and significant figures 16
1.11 Co-ordinate systems 17
1.12 The use of computers 20

Chapter 2 – National maps 22
2.1 National mapping 22
2.2 Historical background 22
2.3 The Transverse Mercator projection 23
2.4 The National Grid 24
2.5 Maps and map scales 27
2.6 Digital map data 29
2.7 Current OS large-scale mapping products 30
2.8 Interpretation of the 1:1250 and 1:2500 NG scale maps 31
2.9 Other OS services 36
2.10 Copyright 37

Chapter 3 – Site investigations 38
3.1 Purpose 38
3.2 Primary objectives 38
3.3 Initial site appraisal 38
3.4 Initial site appraisal check lists 40
3.5 Detailed site survey 41
3.6 Surveys of existing buildings – condition surveys 42
3.7 Schedule of condition 44
3.8 Sub-surface investigations 45

Chapter 4 – Direct distance measurement 52
4.1 Introduction 52
4.2 Equipment 52
4.3 Direct linear measurement fieldwork 58
4.4 Errors in measurement and corrections 64

Chapter 5 – Height measurement 71
5.1 Introduction 71
5.2 Levelling definitions 73
5.3 Bench marks 74
5.4 Types of levelling 76
5.5 The principle of levelling 77
5.6 Modern surveyor's levels 82
5.7 The levelling staff 88
5.8 Level accessories 90
5.9 Levelling fieldwork 92
5.10 Permanent adjustments to the level 104
5.11 Sources of error in levelling 106
5.12 Computer applications 108

Chapter 6 – Angular measurement 109
6.1 Introduction 109
6.2 The basic construction of the theodolite 110
6.3 Reading systems – optical theodolites 118
6.4 Electronic theodolites 120
6.5 Setting on an angle 122
6.6 Measuring angles 124
6.7 Adjustments 129

Chapter 7 – Indirect distance measurement 132
7.1 Introduction 132
7.2 Optical distance measurement 132
7.3 Electromagnetic distance measurement 136
7.4 Applications of EDM 144

Chapter 8 – Levelling applications 146
8.1 Introduction 146
8.2 Establishing a TBM 146
8.3 Contouring plans by level and staff 147
8.4 Sections and cross-sections 153
8.5 Precise levelling 156
8.6 Reciprocal levelling 159

Chapter 9 – Control surveys 161
9.1 Introduction 161
9.2 Linear survey control 162
9.3 Traverse survey 168
9.4 Triangulation and trilateration 196
9.5 Global positioning systems surveying 199
9.6 Co-ordinate problems 203

Chapter 10 – Detail survey 209
10.1 Introduction 209
10.2 Tape and offset detail survey 209
10.3 EDM radiation detail survey 221
10.4 ODM radiation detail survey 235
10.5 Theodolite and tape radiation detail survey 239
10.6 Digital terrain modelling 239

Chapter 11 – Building surveys 241
11.1 Introduction 241
11.2 The plot survey 242
11.3 The building survey 244
11.4 Office procedure – plotting 249
11.5 Building movement 253

Chapter 12 – Setting out 256
12.1 Introduction 256
12.2 Plan control 259
12.3 Height control 265
12.4 Vertical alignment control 267
12.5 Excavation control 271

Chapter 13 – Curve ranging 275
13.1 Introduction 275
13.2 Horizontal circular curves 275
13.3 Transition curves 289
13.4 Vertical curves 298

Chapter 14 – Areas and volumes 302
14.1 Introduction 302
14.2 Areas of simple figures 302
14.3 Areas from drawings or plans 304
14.4 Areas from survey field notes with no plan 308
14.5 Areas from co-ordinates 309
14.6 Alteration and subdivision of areas 310
14.7 Volume calculations 313
14.8 Volumes from cross-sections 314
14.9 Volumes from contours 316
14.10 Volumes from spot heights 317

Bibliography 319

Index 321

Preface to the fourth edition

In this fourth edition of *Basic Surveying*, the content of the book has been rearranged and almost completely re-written in order to bring it into line with present-day practice. In view of these significant changes to the text, the decision was made to change the title to simply 'Basic Surveying'.

It is intended that the book should continue to provide an up-to-date guide to current practice for non-specialist surveyors in the various professions involved in the construction industry.

The content should generally be suitable for students preparing for degrees and other qualifications in architecture, building, building surveying, quantity surveying, estate management and town planning, and may also be of interest to some engineers.

We are indebted to Ms Valerie Gregg of Leica UK Ltd for permission to use illustrations of survey instruments, and to the Ordnance Survey for information supplied. We are also indebted to our former colleague, Professor Peter Swallow of De Montfort University, for the chapter on Site Investigations.

WSW
REP
1997

Chapter 1

Introduction to surveying

1.1 Surveying

Surveying operations are very often associated with the preparation of *survey drawings*, which may be maps, plans, sections, cross-sections or elevation drawings. The survey of land, together with the natural or man-made features on or adjacent to the surface of the Earth, usually requires two-dimensional drawings of three-dimensional objects. Such drawings are usually produced on either a horizontal or vertical plane.

A drawing on a horizontal plane is known as a plan or map. A *plan* is a true-to-scale representation, while a *map* may contain features which are represented by conventional signs or generalized symbols, these not being true scale representations.

For example, on a map, a feature might be represented by a dot, roads may be drawn to a standard width, or a building with many juts and recesses may be drawn as if it were a simple rectangle on plan.

Drawings in a vertical plane are known as sections, cross-sections or elevations. An *elevation* is a side or end view of an object, such as a view of the end of a building. A *section* is a vertical 'plan' of a line through a building, or a line of a proposed road, etc., as it would appear upon an upright plane cutting through it. A long section such as along a proposed road or rail route is known as a *longitudinal section*, while sections taken at right angles to the longitudinal line are known as *cross-sections*.

1.2 Objects of surveying

Surveying techniques may be considered to be used for three distinct purposes as follows.

1.2.1 Surveying for the preparation of maps, plans, etc.

The determination of the relative positions of natural and artificial features on, or adjacent to, the surface of the Earth, so that they may be correctly represented on maps, plans or sections.

1.2.2 Setting out

The setting out upon the ground of proposed construction or engineering works. The information on the new works is normally found in setting out documents which usually include some of the drawings described above.

1.2.3 Computations such as areas and volumes

The execution of calculations for land areas, for earthworks volumes, etc., either based on 'field measurements' or on measurements abstracted from maps, plans and sections.

1.3 Classifications of surveying

Over the years, surveying operations have been classified or subdivided in a great variety of ways, depending upon the purpose of the work or alternatively the equipment or methods actually used.

1.3.1 Plane surveying or geodetic surveying

When the features of the ground are shown on a map or plan, they are shown on a flat sheet of paper or other medium representing a horizontal plane. Since a horizontal plane touching the Earth's surface at one point will be tangential to the Earth at that point, it is evident that the Earth's surface cannot be accurately represented on a plane. In ordinary geometry, the angles of a triangle always add up to 180°, but on the surface of a sphere the angles of a triangle add up to more than 180°. The Earth, fortunately, is such a large spheroid that for surveys of limited extent there will be no appreciable difference between measurements assumed to be on a plane surface and measurements made on the assumption of a spherical surface.

Where measurements cover such a large part of the Earth's surface that the curvature cannot be ignored, then the operations are termed *geodetic surveying*.

Where surveys cover such a small part of the Earth's surface that curvature can be ignored, then the operations are termed *plane surveying*. The area which can be regarded as a plane will depend upon the accuracy required of the survey, but may be taken as up to 250 km^2 or more. The types of survey dealt with in this book are all aimed at tasks well within this arbitrary limit.

1.3.2 Classification according to purpose or use

1.3.2.1 Geodetic survey

A survey of great accuracy which takes into account the curvature of the Earth and may also provide control for surveys of lower accuracy by creating a three-dimensional framework of very accurately located points on the surface of the earth. This usually involves global position fixing by satellite today.

1.3.2.2 Topographical survey

A survey which results in the production of maps showing the *topography* of an area, i.e. the natural and man-made features on the surface of the earth.

1.3.2.3 Cadastral survey

A survey for the preparation of plans showing and defining legal property boundaries.

1.3.2.4 Engineering survey

A survey preparatory to, or in conjunction with, the execution of engineering works such as roads, railways, dams, tunnels, sewage works and construction works generally.

1.3.2.5 Mining survey

A survey for the control of underground workings for mineral extraction.

1.3.2.6 Hydrographic survey

Surveys to map coastlines, produce nautical navigation charts and control works such as oil exploration, the construction of harbours and waterways, etc.

1.3.3 Classification according to the techniques used

The principal branches of survey technique include:

(i) Linear survey (traditionally termed *chain survey*, occasionally termed *tape and offset survey*)
(ii) Traverse survey
(iii) Tacheometer survey
(iv) Triangulation
(v) Trilateration
(vi) Global positioning systems
(vii) Air survey and photogrammetry, etc.

Some of these will be met later in the text.

1.4 Survey tasks

Dependent upon the object of the particular survey, the surveyor will be required to carry out one or more of the following tasks.

1.4.1 Detail surveys

Surveys for the supply (i.e. the survey or measurement) of detail. *Detail* is the name given to the natural and man-made features of the survey area. Generally, detail is taken to exclude *relief* (heights). Buildings, roads and similar man-made features may be termed *hard* detail, while natural features are *soft*. *Overhead* detail would include power lines, and *underground* detail would include sewers and similar detail.

1.4.2 Heighting

Heighting denotes the supply of height measurements. *Height* is the vertical distance of a feature above or below a datum or reference surface.

1.4.3 Control

A large survey site requires many sets of measurements and calls for methods of 'tying' these measurements together so as to produce an accurate survey. These methods are known as *control*, and they typically involve establishing the relative position of a number of control points with comparatively high accuracy. *Horizontal control* fixes the points in the horizontal plane, while *vertical control* governs height measurements.

Some survey methods establish both horizontal position and vertical height, thus both horizontal and vertical control may be combined and three-dimensional co-ordinates allocated to the control points. Some survey methods are concerned only with plan position, and some only with height.

1.5 Scale and scales

Reference has already been made to *scale* and in the next section the word is used again but with a different meaning. It is essential, at this stage, that the reader understands clearly what is meant by the word *scale* with respect to survey.

Maps, plans, etc., are all proportional representations on paper or other drawing material of actual features on the ground. The ratio of a dimension on a map, plan or other drawing to the same dimension on the ground is known as the *scale* of the map, plan, etc.

Scale may be indicated in several ways, e.g. by a statement that 1 mm on the paper represents 0.5 m on the ground (often abbreviated to 1 mm represents 0.5 m, or 1 mm represents 500 mm). This ratio of 1 to 500, traditionally written 1:500 or 1/500, is termed the *representative fraction*, often abbreviated to RF. Again, the scale of a map or plan may be indicated by a *scale line* (or *linear scale*). Scales vary enormously – the International Map of the World is at a scale of 1:1 000 000, town maps may be at 1:10 000 scale, some cadastral plans are at a scale of 1:1250, and occasionally component and assembly drawings may be as large as 1:1.

Scales may be said to be large or small, but there is no definite dividing point. One scale may be said to be larger than another, on the basis that the larger the denominator the smaller the scale. (An RF is expressed as numerator/denominator.) A user may obtain distances from a map with the aid of a pair of dividers and the scale line. The surveyor can do likewise but the surveyor may also have to produce the map or plan from the survey measurements. To enable these tasks to be carried out to an acceptable accuracy, the surveyor makes use of a piece of draughting (drawing) equipment known as a *scale* or *scale rule*. The scale looks similar in shape and size to a rule. It should be noted that the media on which maps are produced may change its dimensions with varying moisture content, age, etc., leading to inaccuracy in scaled measurements.

1.6 Basic survey methods

The following sections outline the basic survey methods used in the supply of detail, heighting and control.

1.6.1 Methods of supplying detail

In *Figure 1.1*, the line AB represents a straight hedge, and point C represents a tree at some distance from the hedge. If it is required to draw a plan of the area, then the hedge can be represented by a single line drawn to the appropriate scale length on the paper. How can the position of the tree be located and plotted on the plan? A variety of methods are available, as in the following sections.

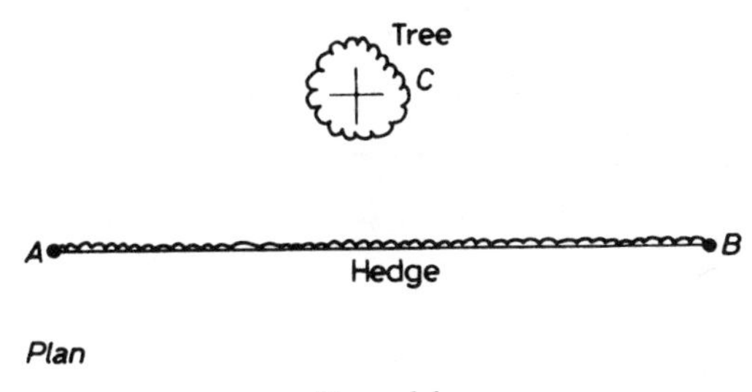

Figure 1.1

1.6.1.1 Offsets (rectangular co-ordinates)

Measure the perpendicular distance (*offset*) from the tree to the hedge line at point D (*Figure 1.2*(*a*)), and the distance from one end of the hedge, say A, to point D. On the paper, scale off distance AD and mark D. Set up a right angle at D, and set out the scale length of DC to locate C.

Offsets are short measurements at right angles to a measured straight survey line. The technique of locating detail by offsets from a survey line is widely used and covered in Chapter 10.

1.6.1.2 Radiation (polar co-ordinates) or bearing and distance

Measure the horizontal angle BAC (a *bearing*) on the ground and measure the distance from hedge end A to the tree. On the paper, use a protractor to set out the angle BAC, then set out the scale length of AC to locate C. The method is described in Chapter 10.

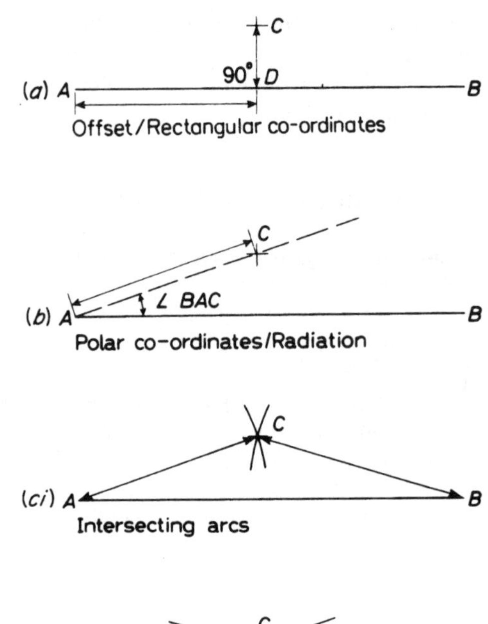

Figure 1.2

1.6.1.3 Intersection

A point of detail may be located by the intersection of a minimum of two lines, one from either end of a base of known length. The lengths of the intersecting lines can be measured from either end of the base line (see (i) below) or alternatively the bearings relative to the base line could be measured as in (ii).

(i) Measure the distances AC and BC on the ground. On the paper, using compasses, swing an arc from A with radius equal in scale length to AC, and a similar arc from B, with radius equal in scale length to BC. The intersection of the arcs locates the point C. This simple method is often used to measure to points of detail in a small area, single-handed. It is also used in the location of detail in tape and offset survey.

(ii) Measure the horizontal angles BAC and ABC on the ground. On the paper, set out these angles by protractor, and their intersection locates point C.

All four methods, shown in *Figure 1.2*, may be used for setting out detail and in filling in map detail. The choice of method, or combination of methods, will depend upon the size and shape of the land, how it is currently used, whether or not heighting and/or control are required, the staff, equipment and time available, and usually also the cost.

1.6.2 Methods of supplying height (spot heights, contours, gradients)

1.6.2.1 Direct heighting – levelling

The commonest method of supplying heights on a site plan (*spot heights* and *contours*) or in setting out (including gradients) is by the use of a *surveyor's level* and a *levelling staff*. The staff, which might be described as a giant rule, is available in a variety of lengths. In use, the staff is held vertically by an assistant and read by the surveyor through a *level* which is basically a tripod-mounted telescope which may be set up in such a way that all lines of sight through the telescope are horizontal (*Figures 1.3* and *1.4*).

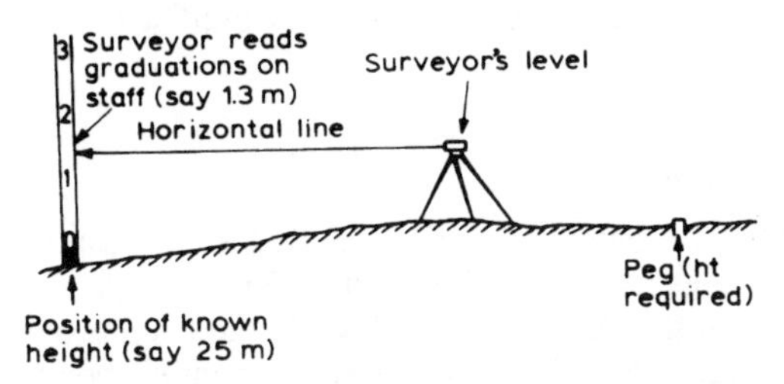

Figure 1.3

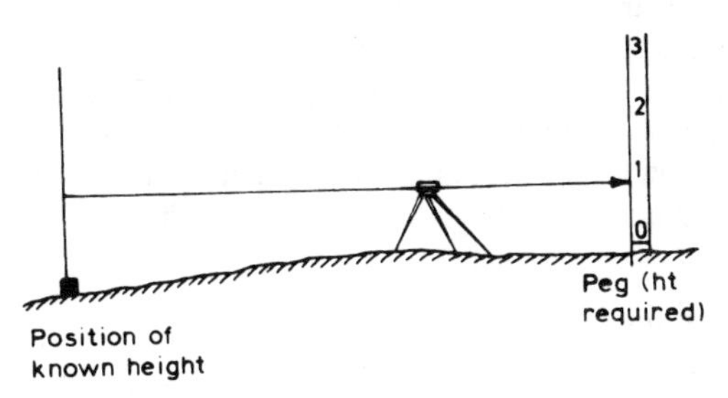

Figure 1.4

It will be observed from *Figure 1.3* that the telescope height is 25 + 1.3 = 26.3 m. The surveyor now rotates the telescope in the horizontal plane and an assistant moves the staff on to a peg (*Figure 1.4*).

If the second staff reading is 0.9 m, then the height to the top of the peg is 0.9 m below the horizontal line of sight, that is, 26.3 – 0.9 = 25.4 m.

There are limitations on the length of sight which may be used, so if it is necessary to transfer heights over a large horizontal distance then the operation has to be repeated. The method is described in Chapters 5 and 8.

1.6.2.2 Indirect heighting

Alternative methods of what might be described as *indirect heighting* include trigonometric heighting, barometric heighting, hydrostatic (water) levelling and tacheometry. These are outlined very briefly in Chapter 5.

1.6.3 Methods of supplying plan control

1.6.3.1 Use of a base or base line

A *base* or *base line* is simply a straight line of known length extending through, adjacent to, or within the area of the survey. Its length should preferably approximate to the maximum length of the site, but this depends upon the site conditions and the methods used to supply detail. The base controls the scale of the survey, that is to say it ensures that the completed survey is neither too small nor too large. Hence the base must be measured with great care, using carefully checked measuring equipment. The base may also serve to prevent distortion, that is to say to ensure that the completed survey is true to shape. Both the methods of supplying detail by intersection make use of a base, and the base principle is used with the other methods of supplying control.

1.6.3.2 Triangulation

Triangulation means a network of triangles tied to a measured base line, all the *angles* of the triangles having been measured. Knowing the base line length, and the magnitude of all the angles in the network, the lengths of all the lines in the net may be calculated and co-ordinates of control point triangle vertices computed. The method was used by the Ordnance Survey to supply the original geodetic control for the mapping of Great Britain, and it involved thousands of triangles, an immense computational feat in those early days.

The introduction of electromagnetic distance measurement has rendered triangulation obsolete as a means of geodetic control, but it may occasionally be useful in conjunction with traversing, using a small number of triangles, to assist in maintaining accuracy (see Section 1.6.3.4 and Chapter 9).

1.6.3.3 Trilateration

This again involves a network of triangles, but in this case all the *side* lengths of the triangles are measured and not the angles, and again co-ordinates are computed for the various control points at the triangle vertices.

Linear surveys in which the control framework line lengths are measured by steel tapes and no angles are measured, are also a form of trilateration, but the triangles are not computed and co-ordinates are not calculated. The term 'trilateration' is normally only applied to the computed higher-order work.

1.6.3.4 Traversing

This is the most flexible method of providing control for surveys, particularly in urban areas where the construction of triangles is difficult and sometimes impossible.

A *traverse* consists of a series of connected straight lines whose lengths and bearings can be determined (*Figure 1.5*). The lines are known as *legs* and the end-points as *stations*. The legs may be used either as survey lines for detail supply or for holding a network of lines. Alternatively, detail may be supplied by radiation from the traverse stations. Traverses may also commence from and close at either end of a base line (see Chapter 9).

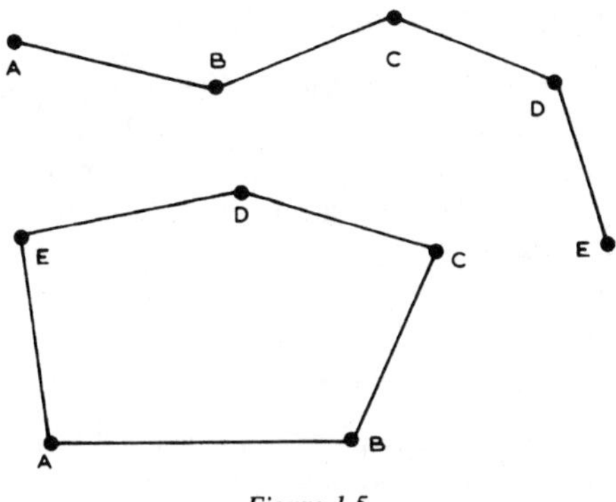

Figure 1.5

1.6.3.5 Global positioning systems (GPS)

This is a system of determining position on the Earth's surface by processing signals from artificial satellites circling the Earth. The system currently in use is NAVSTAR, managed by the US Department of Defense. Given appropriate hardware and software, point co-ordinates can be obtained with high accuracy (see Chapter 9).

1.7 Principles of survey

The following principles should be followed in every survey.

1.7.1 Control

Each survey should be provided with an accurate control framework, the lower-order work (detail and heighting) being fitted and adjusted to the control framework. The traditional expression is *to work from the whole to the part.*

1.7.2 Economy of accuracy

The standard of accuracy aimed at should be appropriate to the needs of the particular task. As a general rule, the higher the standards of accuracy then the higher the cost in time and money.

1.7.3 Consistency

The relative standards of accuracy of the various classes of work (control, detail, heighting) should be consistent throughout a survey task. Thus generally detail will be measured to a lower standard of accuracy than the control framework to which it is fitted.

1.7.4 Independent check

It is desirable that every survey operation (fieldwork, computations, plotting) should either be self-checking or be provided with an independent check. As simple examples, the independent check could mean repeating the task using a different method, such as checking an addition by subtraction, or checking a series of separate measurements by taking an overall measurement and comparing the results.

1.7.5 Revision

Surveys should, where possible, be planned in such a way that their later revision or extension may be carried out without the necessity of having to carry out a complete new survey.

1.7.6 Safeguarding

The results of survey work (survey markers, field and office documents) should, so far as possible, be kept intact for possible use at a later date, e.g. in a revision survey.

1.8 Accuracy

Since no measurement is ever completely accurate, surveys are usually described as being to a certain standard of accuracy. *Accuracy* may be defined as the conformity of a measurement to its true value. The accuracy of a measurement is often quoted as a representative fraction, that is to say as the ratio of the magnitude of the error to the magnitude of the measured quantity. *Error*, here, means the difference between the true value and the measured value of a quantity. If, in measuring a line 1 km long, the expected total error is ±0.01 m, then the error is 1 part in 100 000 (0.01 m/1000 m = 1/100 000). This is an extremely high accuracy, not often attained in plane surveying. On the other hand, if an error of 0.01 m were made in measuring a line 2 m in length, then the standard of accuracy is 0.01 m in 2.0 m, or 1/200, and this is usually too low for any type of work. The greater the degree of accuracy required in a survey, the greater will be the expenditure in effort, time and money. The standard of accuracy must be in keeping with the size and purpose of the survey.

Precision denotes the degree of agreement between several measures of a quantity. If a quantity is measured several times, then the degree of agreement between the measures is the precision of that set of measures. It must be noted that a high degree of precision does not guarantee great accuracy. The classic example is the darts player who, while aiming for the 'bull', places all three darts in 'double one' – the player is throwing with some precision (all darts close together) but the result is not very accurate (should have landed in the bull).

1.9 Measurement and errors

It should be understood that no measure in a survey is ever exact – every measurement, whether linear or angular, contains errors. The types of error and their relative importance must be appreciated, and care taken to keep them to a minimum appropriate to the task in hand.

1.9.1 Mistakes

Mistakes, also known as *gross errors* or *blunders*, are due to carelessness by the surveyor, such as reading a level staff as 2.415 m instead of 3.415 m, or noting a distance measured with the tape as 15.45 m instead of 15.54 m. These can only be eliminated by the use of methods of observing, booking, computing and plotting designed to show up any mistakes.

1.9.2 Systematic errors

These are errors which always recur in the same instrument or operation, they follow some mathematical rule, and they are cumulative, that is to say their effect will increase throughout the survey. As an example, if a nominal 30 m tape has been stretched (such as by hard usage) by an amount of 0.05 m, then every time it is laid upon the ground there will be an error in distance measurement of 0.05 m. If, in measuring the length of a line, the tape is laid down ten times, the length of the line will be noted as 300 m, but the true length will be 300.5 m. The error will have accumulated to $10 \times 0.05\,\text{m} = 0.5\,\text{m}$. This error would be regarded as being *negative*, since its effect is to make the measured length appear less than the true length. Conversely, if the tape is *shortened*, say due to the air temperature being lower than the standard temperature of 20°C, then it will cause cumulative *positive* error, since it will measure the line as being longer than it actually is.

These errors may be guarded against by using suitable operational methods, by standardizing equipment, and by applying appropriate corrections to the actual measurements.

1.9.3 Random errors

If multiple measures are made of the same quantity, with care taken to avoid all mistakes and with all systematic errors eliminated, the results will still show variations. As an example, consider the determination of the length of a line with a steel tape, the line being measured several times, with the same care to eliminate mistakes and systematic error in each measurement. The differences between each of these measures and the true value of the quantity are *random errors*. These errors are small, they are subject to chance, they cannot be eliminated, and their occurrence is assumed to follow the laws of probability.

For repeated measurement of a quantity, all measures being of equal reliability, these laws indicate that:

(i) small errors occur more often than large ones;
(ii) positive and negative errors are equally likely;
(iii) very large errors seldom occur; and
(iv) the distribution of the errors approximates closely to the *normal* or *Gaussian probability distribution*.

1.9.3.1 Most probable value

True random errors cannot be calculated, since the true value of the quantity can never be known. An estimate of the true value must be obtained, and this is given by the *principle of least squares*:

> The sum of the squares of the residuals derived from the arithmetic mean is a minimum

or more simply

> The *most probable value* (MPV) which can be obtained from a set of equally precise measurements is that value for which the sum of the squares of the residuals is a minimum.

This principle shows that if a measurement, say x, is made n times, then the MPV of the measurement is the *arithmetic mean* of the set of n measurements, provided that they are all made independently but under similar conditions. The *arithmetic mean of a set of observations* is the simple average of the set, $\bar{x} = \Sigma x/n$, where $\bar{x}$ is the arithmetic mean, x the individual observation, and n is the total number of observations in the set. Thus:

$$\text{MPV} = \bar{x} = (\Sigma x)/n$$

A *residual* is the difference between a measured value of a quantity and the MPV of the quantity, and since these residuals can be calculated they can be used to analyse a set of measurements instead of the true but unknown errors. If the values of the residuals from a set of measurements are plotted against their frequency of occurrence, a figure will be obtained such as that in *Figure 1.6*, a normal distribution curve.

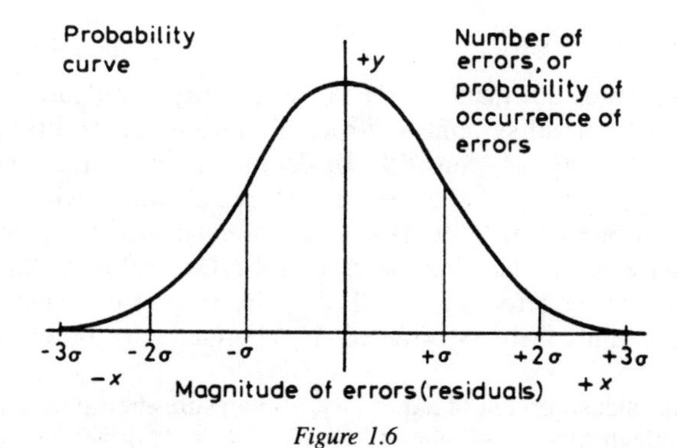

Figure 1.6

The central vertical line indicates the MPV, and the horizontal axis shows positive and negative residual values, i.e. residuals from measurements greater, and lesser, respectively, than the MPV. The frequency of occurrence of each residual is plotted vertically, and thus the area under the curve represents the probability of all errors occurring. If there is no systematic error in the measurement the curve will be symmetrical about the y axis, i.e. about the arithmetic mean.

1.9.3.2 Standard deviation

A normal distribution curve is defined by two parameters, its MPV and its *standard deviation*. The standard deviation, σ, is a measure of the spread of the measurements, and hence of the precision of the work. The larger the standard deviation the greater the spread of results and the lower the precision of the results. The formula for the standard deviation of the variable x, measured n times, is

$$\sigma = \pm \{\Sigma(x - \bar{x})^2/n\}^{1/2}$$

where $\bar{x}$ is the arithmetic mean and x the observation.

Alternative notation is

$$\sigma = \pm \{[\nu^2]/n\}^{1/2}$$

where n is the number of measures, ν_i represents a residual equal to $(\bar{x} - x_i)$ and $[\nu^2]$ represents the sum of the squares of the residuals. (Note that in this aspect of surveying the summation symbol [ ] is often used instead of the more usual mathematical summation symbol Σ.)

1.9.3.3 Standard error

The statistical theory underlying probability distributions assumes that a very large number of observations of the quantity will be made, but in surveying practice only small numbers of observations are made. Accordingly a better and less biased measure for survey work is the *standard error*, s. This is obtained simply by replacing the n in the above formulae with $(n - 1)$, then standard error of a single observation in a set is given by

$$s_x = \pm \{\Sigma(x - \bar{x})^2/(n - 1)\}^{1/2}$$

or

$$s_x = \pm \{[\nu^2]/(n - 1)\}^{1/2}$$

The *standard error of the arithmetic mean of a set* is given by

$$\sigma_{\text{mean}} = \pm \{[\nu^2]/n(n - 1)\}^{1/2}$$

i.e. the standard error of a single observation divided by the root of the total number of observations. This indicates the reliability of the arithmetic mean, since if error distribution follows the normal distribution curve, then:

68.3% of all observations lie within one standard error of the arithmetic mean,
95.4% lie within two standard errors of the arithmetic mean, and
99.7% lie within three standard errors of the arithmetic mean,

and the probability of a residual exceeding 3 × standard error is so low that such a residual is suspect and should probably be rejected.

The *variance of a set of observations* is the square of the standard error of a single observation, and is an indication of the spread of results:

$$\text{variance} = [v^2]/(n-1)$$

1.9.3.4 Weighted observations

The *weight* of an observation is an estimate of its reliability as compared with other observations; the better the observation the greater its weight. The weight of an observation is inversely proportional to the square of its standard error, thus it is useful in comparing the reliability of several observations.

The *weighted mean* of several varyingly weighted observations is the arithmetic mean of all the observations *reduced to the same standard*. If the n observations of quantity x are, respectively, $x_1, x_2, \ldots x_n$, weights being $p_1, p_2, \ldots p_n$, then:

$$\text{weighted mean} = x_0 = (p_1x_1 + p_2x_2 + \ldots p_nx_n)/(p_1 + p_2 + \ldots p_n) = [px]/[p]$$

$$\text{standard error of weighted mean} = s_0 = \pm\, \{[pv^2]/[p](n-1)\}^{1/2}$$

$$\text{standard error of single observation of unit weight} = s = \pm\, \{[pv^2]/(n-1)\}^{1/2}$$

standard error of single observation of weight p_r is

$$s_r = \pm\, \{[pv^2]/p_r(n-1)\}^{1/2}$$

Observations should preferably be weighted in inverse proportion to the squares of their standard errors. If they must be weighted before the standard errors are known, then the weights should be allotted according to the relative precision of the equipment used or the number of observations made.

1.9.3.5 Summary of standard error formulae

Single observation in a set

$$s_x = \pm\, \{[v^2]/(n-1)\}^{1/2} \qquad (1)$$

Mean of n observations

$$s_{\text{mean}} = \pm\, \{[v^2]/n(n-1)\}^{1/2} \qquad (2)$$

Multiple of a quantity, such as $X = a \times x$, where a is a number and x a quantity having standard error of s, then

$$s_X = \pm a \times s \tag{3}$$

Sum or difference of several independent quantities, such as $X = x_1 + x_2 - x_3 + \ldots x_n$, the quantities having standard errors of $s_1, s_2, s_3, \ldots s_n$, then

$$s_X = \pm (s_1^2 + s_2^2 + s_3^2 + \ldots s_n^2)^{1/2} \tag{4}$$

If $s_1 = s_2 = s_3 = \ldots s_n$, t, then

$$s_X = \pm s \cdot n^{1/2} \tag{5}$$

Sum or difference of multiples of several quantities, such as $X = a_1x_1 + a_2x_2 + a_3x_3 + \ldots a_nx_n$, the quantities $x_1, x_2, x_3 \ldots x_n$ having standard errors of $s_1, s_2, \ldots s_n$, then

$$s_X = \pm \{(a_1s_1)^2 + (a_2s_2)^2 + \ldots (a_ns_n)^2\}^{1/2} \tag{6}$$

Product of several *simple* quantities, such as $X = xyz$, these having standard errors of s_x, s_y, s_z, then

$$s_X = \pm (xyz)\{(s_x/x)^2 + (s_y/y)^2 + (s_z/z)^2\}^{1/2} \tag{7}$$

and if $X = x/y$, then

$$s_X = \pm (x/y)\{(s_x/x)^2 + (s_y/y)^2\}^{1/2}$$

Function of several independent quantities, such as $X = f(x_1, x_2, x_3, \ldots x_n)$, these having standard errors of $s_1, s_2, s_3, \ldots s_n$, then

$$s_X = \pm \{(\partial X \cdot s_1/\partial x_1)^2 + (\partial X \cdot s_2/\partial x_2)^2 + \ldots (\partial X \cdot s_n/\partial x_n)^2\} \tag{8}$$

It should be noted that scientific calculators have keys for the calculation of the arithmetic mean and standard deviation, but for survey purposes it is important to check whether the latter routine uses n or $(n - 1)$ as the divisor.

1.9.3.6 The application of standard error

(i) The magnitude of the standard error of the mean of a set of results is an indicator of the precision of the measurements, a large value suggesting low precision and a small value high precision. The measurement is typically stated as:

MPV (arithmetic mean) ± standard error of the mean;
e.g. 120° 15' 43" ± 5"

(ii) When the residual of an observation exceeds ±3 times the standard error of the mean it is usual to discard that observation, since there is a 99.7% probability that all the residuals will lie within that range.

(iii) If the standard error of a single observation is known, it is possible to determine how many observations are required for a specified standard error of the mean. Example – a Wild T1 theodolite is stated to have a standard deviation of ±3" (i.e. standard error) for a single observation. How many observations must be made to achieve a standard error of the mean of ±1"?

$s_x = 3''$; $s_{\text{mean}} = s_x/\sqrt{n}$; then $\sqrt{n} = s_x/s_{\text{mean}} = 3$;
thus $n = 3^2 = 9$

Theoretically a total of at least nine observations will be required.

(iv) Manufacturers specify the precision of their equipment in terms of standard error to allow comparison of equipment. Thus the T1 theodolite (above) is quoted at ±3" while the T2 is quoted at ±0.8". Levels are specified in terms of the standard error in a kilometre run of double levelling, e.g. the NA28 at ±1.5 mm, while the N3 is ±0.2 mm. Steel tapes are typically quoted as having a standard error of ±3 mm.

(v) Where a value is obtained from the combination of several simple quantities, by addition, subtraction, multiplication, or as a function of others, each quantity possessing a standard error, the standard error of the resultant value is obtained from one of the formulae given above, numbered (3) to (8).

1.10 Numerical values and significant figures

1.10.1 Measurement notation

Field notes of measurements such as taped distances are normally expressed in metres and decimals, the number of figures after the decimal point indicating the precision of measurement. If a measurement is to the nearest metre only, the figures should be followed by the letter *m* for metres, but if decimals are shown the *m* should be omitted. Consider an actual distance of 25.643 metres:

Measured to the nearest metre only, it would be written as 26 *m* (note the rounding up).
Measured to nearest decimetre, 0.1 m, written as 25.6, no letter *m*.
Measured to the nearest centimetre, 0.01 m, written as 25.64, no letter *m*.
Measured to the nearest millimetre, written as 25.643, again no *m*.

An exception should be noted in the case of architectural and building drawings, where there is a convention to express all measurements in millimetres. Thus the measurement of 25.643 becomes 25 643, no *m* and no decimal, simply the value written in millimetres.

1.10.2 Significant figures

If readings are estimated to the nearest division on a graduated scale, the maximum estimation error will be ± ½ division, the final recorded digit being subject to a maximum error of half its value.

The number of significant figures in a value is important, and it is not the same as the number of decimals. Thus 25.643 has five significant figures, but 0.00643 has only three significant figures. The number of significant figures in the result of a multiplication or division should be the same as the number of significant figures in the least significant factor. For an addition or subtraction the number of figures after the decimal place in the answer should not be more than that of the number with the least significant decimal place.

1.10.3 Calculation rules

(i) Record all values in ink.
(ii) Make alterations by crossing out and re-writing above – never erase numeric values.
(iii) Group figures in threes in either direction from the decimal point, do not use commas.
(iv) Arrange figures in columns where appropriate.
(v) Use self-checking forms of calculation, or an independent check calculation from the original data.
(vi) Arrange calculations so that they 'read' like a piece of written matter, with punctuation and explanation, e.g. ..but, ...therefore, etc.
(vii) Work generally to one more place of decimals than is required in the answer, round-off as required after completion.

1.11 Co-ordinate systems

In tape and offset linear surveying the results of survey work are produced as purely graphic representations of the plan details of the area surveyed. In all higher survey work, however, the positions of control points are fixed by reference to co-ordinate systems. These systems may involve two or three dimensions.

1.11.1 Two-dimensional systems

The location of a point on a surface may be specified by two measurements, termed *co-ordinates*, but there are several types of co-ordinate systems.

(i) Traditional *mathematical plane rectangular co-ordinates* or *Cartesian co-ordinates* specify the position of a point by two perpendicular distances, measured from given x and y axes, the axes being at right-angles to one another.
(ii) The *geographical co-ordinates* of a point on the surface of the earth are the sexagesimal latitude and longitude of the point as measured from the Equator and the Greenwich meridian, respectively.
(iii) *Surveying rectangular co-ordinates* specify the position of a point by its distance east of a north–south axis (*easting*), and its distance north of an east–west axis (*northing*). The north–south axis is termed the *reference meridian*, the east–west axis is the *reference latitude*. This method is the same as that used on the Ordnance Survey National Grid, see Chapter 2.
(iv) The *polar co-ordinates* of a point are its bearing (angle from a specified direction) and its distance from a given pole point.

1.11.2 Three-dimensional systems

A point in space may be fixed by three co-ordinates. In mathematical work the third co-ordinate is the z value, the co-ordinates then being x, y, z. In the geographical system the third co-ordinate will be the point's height above the particular Earth surface concerned. In the surveying context here, the third co-ordinate is the *reduced level* (RL) of the point above the datum in use, normally the OS Datum in UK practice (see Chapter 5).

1.11.3 Bearings

The *bearing* of an observed distant point is the angle between a specified reference direction from the observer's position and the line from the observer's position to the distant point. A bearing, however, is stated not merely as the angle between the two directions, but as the amount of angular rotation made in turning from the (generally north) reference direction to the direction of the observed point. The *true bearing* of an observed distant point is the angle, measured clockwise, between true north and the line from the observer's position to the distant point.

Similarly, *magnetic* and *grid bearings* are relative to magnetic and grid north, respectively. All of these are, on occasion, referred to as *whole circle bearings*, since they may take any value between 00° 00' 00" and 360° 00' 00".

It is most important to note that surveyors and navigators require and use angles clockwise from north (equivalent to the traditional y axis), while mathematicians use angles measured anti-clockwise from the x axis. Whole circle bearings may be quoted in the form N 185° E, meaning 'from reference direction north, turn eastwards through an angle of 185°'. The E indicates clockwise rotation, this commencing with an easterly swing from north. *Relative* and *quadrant bearings* are occasionally referred to in surveying. A *relative bearing* is one which is measured relative to some line other than north. An example could be a bearing measured relative to a base line. A *quadrant bearing* (now obsolescent) is the whole circle bearing reduced to an angle of less than 90°, so that the 0° to 90° tables of trigonometrical functions can be used in computations. (Today, of course, if computer reduction is not available, it is easier to use a calculator with trigonometric functions which can handle whole circle bearings.) A quadrant bearing is the angle, measured east or west, between the north or south direction and the whole circle bearing direction, for example:

whole circle bearing	N 50° E	=	quadrant bearing N 50° E
	N 100° E	=	quadrant bearing S 80° E
	N 200° E	=	quadrant bearing S 20° W
	N 300° E	=	quadrant bearing N 60° W

1.11.4 Co-ordinates and partial co-ordinates

1.11.4.1 Calculation of partial co-ordinates

In *Figure 1.7* the point O may be considered to represent a survey *station*, while the line OP represents an arbitrary survey line with length equal to unity (i.e. one). The direction OY is the north direction and the *bearing* of the line OP is the clockwise angle measured in turning from the direction O → Y to the direction O → P.

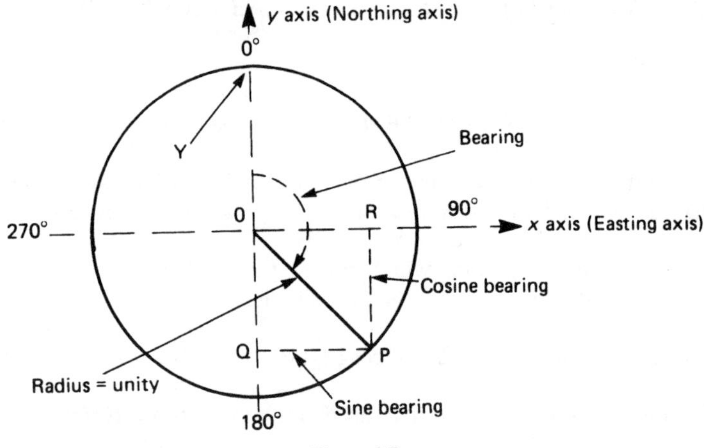

Figure 1.7

If a perpendicular is drawn from P to Q on the y or northing axis, then the distance PQ is equal to the sine of the bearing of OP. Similarly, a perpendicular drawn from P to R on the x or easting axis will give a length PR equal to the cosine of the bearing of OP. In the figure, the sine is positive, since P is to the east of the y axis, and the cosine is negative, since P is south of the x axis.

If OP represented a survey line, running from station O to station P, with length OP, then the distances PQ and PR are its *partial easting*, or difference in easting (ΔE) and *partial northing*, or difference in northing (ΔN), respectively, i.e. the *partial co-ordinates* of line OP. Then

$$\Delta E_{\mathrm{OP}} = \mathrm{OP} \sin \mathrm{brg\,OP}, \quad \text{and} \quad \Delta N_{\mathrm{OP}} = \mathrm{OP} \cos \mathrm{brg\,OP}$$

These partial co-ordinates may be calculated in this manner, preferably, using a calculator with a polar to rectangular key, P → R.

It will be clear that denoting the co-ordinates of point O as E_{O}, N_{O}, and of point P as E_{P}, N_{P}, then

$$\Delta E_{\mathrm{OP}} = E_{\mathrm{P}} - E_{\mathrm{O}}, \quad \text{and} \quad \Delta N_{\mathrm{OP}} = N_{\mathrm{P}} - N_{\mathrm{O}}$$

therefore

$$E_{\mathrm{P}} = E_{\mathrm{O}} + \mathrm{OP} \sin \mathrm{brg\,OP}, \quad \text{and} \quad N_{\mathrm{P}} = N_{\mathrm{O}} + \mathrm{OP} \cos \mathrm{brg\,OP}$$

It is important to note – points have co-ordinates, but lines have partial co-ordinates.

1.11.4.2 Calculation of line bearing and length

It is often necessary to invert the above process and find the *bearing* and *length* of a line from the co-ordinates of its end points (equivalent to using the R → P key). Since for a line AB,

$$\Delta E_{AB} = AB \times \sin \text{brg } AB, \quad \text{and} \quad \Delta N_{AB} = AB \times \cos \text{brg } AB$$

then

$$\Delta E_{AB}/\Delta N_{AB} = (AB \times \sin \text{brg } AB)/(AB \times \cos \text{brg } AB) = \tan \text{brg } AB$$

If the co-ordinates of the end points are given then the partial co-ordinates of the lines may be deduced and the *bearing* is the angle whose tangent equals $\Delta E_{AB}/\Delta N_{AB}$. Thus bearing AB = arctan $\Delta E_{AB}/\Delta N_{AB}$.

Care must be taken regarding which quadrant is involved, of course, and with the signs. Given the line bearing, then the line length is

$$AB = \Delta E_{AB}/\sin \text{brg } AB = \Delta N_{AB}/\cos \text{brg } AB$$

It is often necessary to solve triangles and the most useful formulae then are the *sine* and *cosine rules.*

Sine rule

$$a/\sin A = b/\sin B = c/\sin C, \text{ or } a = b \sin A/\sin B, \sin A = a \sin B/b$$

Cosine rule

$$a^2 = b^2 + c^2 - 2\,bc \cos A, \text{ or } \cos A = (b^2 + c^2 - a^2)/2\,bc$$

1.12 The use of computers

Large mainframe computers have been used for surveying since the early days of computing, usually in batch mode for tasks which entailed handling large volumes of data and high-speed 'number crunching'. Such machines are also used with high-speed plotters for automatic draughting of maps and plans, in national survey organizations and specialist surveying companies. These computer applications, however, are not considered here, since we are concerned with the 'occasional surveyor' having limited resources.

Surveying calculations were formerly carried out using tables of logarithms, tables of trigonometrical functions, and specialist tables such as for stadia reductions, subtense measurements, etc., often with a mechanical calculator. These methods were superseded by the use of electronic scientific calculators and personal computers (PCs). Some models of microcomputer ('laptops', and the even smaller 'notebook' PCs) are small enough to be used in the field. Today every student should have access to a computer.

Like the calculator, the user need have no knowledge of how the computer works, but they must have access to suitable software. Provided the software is robust, easy to use interactively and designed to perform the various survey tasks effectively, the PC may be regarded simply as another tool for the surveyor's use. It is impossible to cover the wide range of software on the market today, and the myriad suppliers, but there is appropriate software available for almost every aspect of surveying.

Many modern survey instruments such as 'total stations' and 'digital levels' actually store observations and compute the results of observations themselves, and either display the results or store them in a *data recorder* for later transfer to the RS232 port of a computer. Increasingly, plug-in PCMCIA cards are used as recording modules, and of course many of these instruments themselves have considerable capacity internal memories. Hand-held PCs may be plugged into survey instruments and can then both store and process the observed data. When survey data has been stored in computer files it may be used to produce survey drawings or it may be input to a computer aided design (CAD) package. The possibility of amending values and re-running the computation in a few seconds allows trial and error of different design parameters in, for example, the design of vertical curves.

The use of standard applications software may encourage the 'occasional surveyor' to use survey methods which he would not have considered worthwhile by hand computation methods, e.g. the radiation methods now popular with electronic distance measuring equipment. It should be noted, however, that the use of a particular application program 'disciplines' the surveyor into executing the survey in a method suited to that program and there is a loss of individuality or freedom in designing the approach to the task.

Developments in technology, allied with the computer, have actually resulted in completely new survey methods being created, such as digital terrain modelling (DTM) and global positioning systems (GPS).

Chapter 2

National maps

2.1 National mapping

The *Ordnance Survey* ('the OS') has been the national mapping agency of Great Britain for over 200 years. Other countries have similar agencies, although in many parts of the world they do not provide large-scale mapping.

2.2 Historical background

During the years 1745 and 1746 the armies of the newly-united United Kingdom were engaged in suppressing the Jacobite rebellion led by Prince Charles Edward Stuart. Troop movements in the Scottish Highlands were considerably handicapped by the absence of any maps of the country. The same problems arose in the 'pacification' of the Highlands, in seeking out and eliminating the rebellious natives, and as a result much of Scotland was mapped by the military, while at the same time a network of military roads was constructed throughout the Highlands. Following these mapping efforts, the military surveyors were employed on a scientific task, triangulation, designed to link the observatories at Greenwich and Paris. This led to the formation of the Ordnance Survey in 1791 as a military organization. An early task was the creation of a 1 inch to 1 mile scale map of southern England, a military (defence in this case) requirement during the Napoleonic wars.

After these wars, the need for maps was better appreciated, for example in the management and transfer of land, in civil engineering (railways, etc.), in geological surveys, and in the survey of archaeological sites.

The Ordnance Survey then assumed its modern role of providing, as a national service, the surveys and maps required by the nation for military, scientific, commercial and industrial purposes. The whole of Great Britain was mapped between 1844 and 1896, providing *County Series* maps at a variety of large and small scales, and various revisions were made up to 1939, the outbreak of World War II. In time it became evident that a new triangulation and new surveys were needed and these tasks were commenced in the 1930s, the resultant triangulation being known as OSGB36.

After the war, several new set of maps were produced, on a single projection, the *Transverse Mercator* (see Section 2.3) and with the new *National Grid* (see Section 2.4) superimposed on the map sheets.

The original maps were produced on engraved plates for printing in standard size sheets, then later offset lithography was used. Today map data can be held in computer-readable digital data form on compact discs, and large-scale maps can be computer printed as desired, for any particular area, with a range of sheet sizes and scales.

2.3 The Transverse Mercator projection

If the curved surface of the Earth is to be projected on to a flat sheet of paper some distortion must occur in one or more of the following: *shape* (orthomorphism), *direction* (bearings), *size* (areas as a quantity).

Clearly, then, scale cannot be consistent in all directions from all points on a map. Nevertheless, for many practical purposes the inconsistencies may be ignored, as is usual in plane surveying.

The early OS maps were a series of *county surveys*, or combined counties, on a series of different projection data, the large-scale maps being known as the *County Series*. To the user who had property stretching across county boundaries it could be most disconcerting since adjoining counties did not always fit together. These discontinued plans and maps of the County Series may still be met in some offices.

Today a single national projection is used for all maps and this is a modified version of the *Transverse Mercator.* In order to appreciate a projection with reasonable simplicity the reader should attempt to visualize a sheet of paper touching (or cutting) the surface of a globe at one or more points. The sheet may be flat, or wrapped around the globe to form a cylinder, a cone, part of a cone, or a spiral. If the lines of latitude and longitude are projected outwards (or inwards) from some point, say the centre of the Earth, on to the paper, then when the paper is laid out flat again some particular *map projection* will have been achieved.

Great Britain is a country whose length is greater in the north–south direction than in the east–west. One possible solution, therefore, is a cylindrical projection touching the globe along a meridian of *longitude* (a line on the surface of the Earth running between the North and South Poles). This form is known as a *Transverse Mercator projection.* A Transverse Mercator projection is orthomorphic over any small area, scale is correct along the 'central meridian' (where the cylinder of paper touches the surface of the globe) and increases gradually to the east and west of the central meridian. To reduce the scale error at the east and west extremities of the country, the OS modified the projection by making the diameter of the cylinder a little less than that of the globe. This reduced the scale error at the extremities but introduced a scale error at the central meridian equal to, but of opposite sign to, that at the extremities. This results in there being two north–south lines (approximately 180 km east and west of the central meridian) where there is no scale error. The Earth actually approximates to an *oblate spheroid* (flattened at the poles) and in practice the lines of latitude and longitude are produced mathematically from formulae defining the projection and the shape of the Earth. (For further reading, see the Bibliography.) The '*origin of the projection*' (and of the calculations) is latitude 49° north and longitude 2° west. The longitude line 2° west then forms the *central meridian* of the projection and of the National Grid.

2.4 The National Grid

When the County Series maps were discontinued, a *National Grid* was introduced and superimposed on all maps (subject to a few minor exceptions) to provide both a single reference system for the whole country and a means of numbering the proposed large-scale maps. The unit used for the grid is the *metre*.

The National Grid (NG) is an imaginary network of lines parallel to, and at right angles to, the central meridian of the projection, so forming a series of squares on the maps. The *origin* of the grid (the zero point) is southwest of the Scilly Isles and lies 400 km west and 100 km north of the origin of the projection, and it is therefore usually referred to as the *false origin*. This false origin means that every point in Great Britain has a co-ordinate reference consisting of a positive *easting* and a positive *northing* with respect to the grid origin. (It is a rectangular co-ordinate system in which the mathematician's x and y distances become 'eastings' and 'northings', respectively.)

2.4.1 Recording the position of survey data

The position of instrument stations in the OS triangulation network or in one of their traverses are calculated and presented in NG units and hence other surveyors may plot or tie their own survey work to these OS positions. The values are quoted in metric units east and north of the false origin, e.g.

Manor Farm triangulation pillar
439 133.37 East
120 988.83 North

Note that eastings are always quoted before northings and the unit, m, is usually omitted. Tying traverses to OS control is referred to in Chapter 9.

2.4.2 A means of reference

The grid is sufficient for most people to devise a simple reference system for their own purposes and many users have done so. The preferred system, however, is as described below.

The NG is a systematic breakdown into squares of varying sizes. The largest unit of the grid is a square of 500 km side length, each square being identified by a letter as shown in *Figure 2.1*. Note that the letter 'I' is not used. Great Britain being a small country, most of it lies in the two squares 'N' and 'S', appropriately North and South, while a little, the Orkney and Shetland Islands, lies in square 'H' (Higher still).

A little less, if any, lies in square 'O' (the Open sea), and the mouth of the Thames lies in square 'T'. The false origin of the NG is the southwest corner of square 'S'.

A	B	C	D	E
F	G	H	J	K
L	M	N	O	P
Q	R	S	T	U
V	W	X	Y	Z

Figure 2.1

Each square of 500 km side is subdivided into 25 further squares, each with a side length of 100 km. The new squares within a 500 km square are similarly identified by letters as illustrated in *Figure 2.2*. Thus, each 100 km side square is uniquely identified by a two-letter reference.

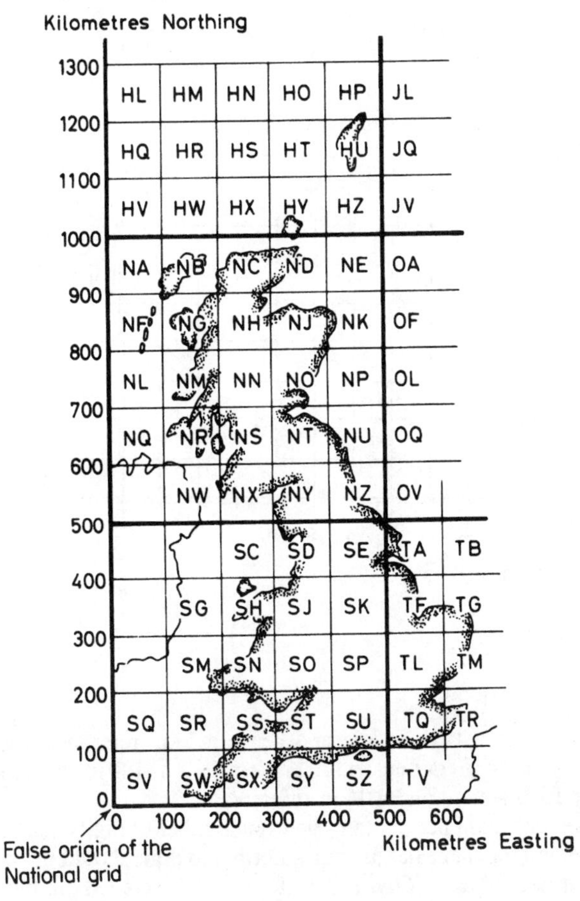

Figure 2.2

Any smaller square within a 100 km square is identified by the distances from the southwest corner of the 100 km square to the southwest corner of the smaller square, expressed in eastings and northings, respectively. For example, SK 91 identifies a 10 km square and hence a 10 km reference, see *Figure 2.3*.

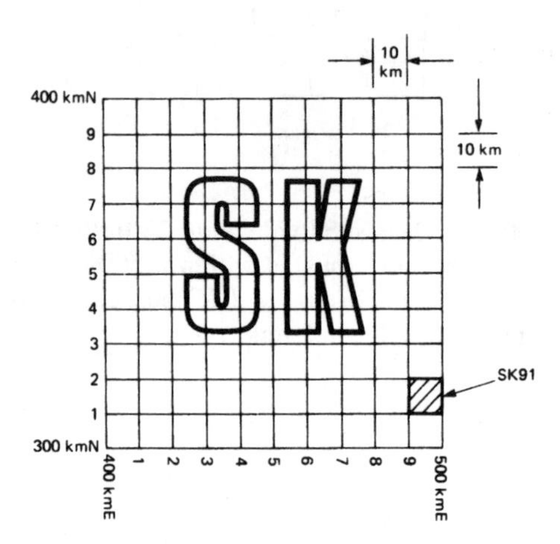

Figure 2.3

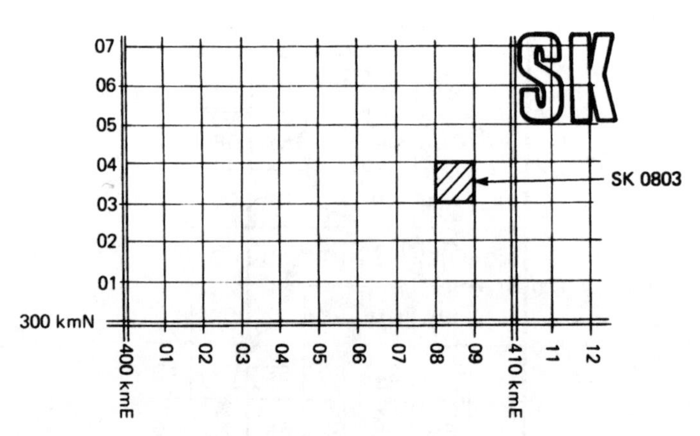

Figure 2.4

SK 0803 defines a 1 km square and hence a 1 km reference, see *Figure 2.4*. Similarly, SK 123456 defines a 100 m square, a 100 m reference, and SK 12345678 a 10 m square and a 10 m reference.

Note that every reference contains an even number of digits, half indicating the distance east and the latter half distance north. No digit should be omitted, a zero being used if needed as in *Figure 2.4*. If the two letters (SK in these examples) are included, the reference is unique and does not recur in the country. If the letters are omitted, the reference repeats in every 100 km square. If all work is confined to the same 100 km square and confusion cannot arise, the letters may be omitted. The grid lines appearing on individual maps depends upon their scale, see Section 2.5. When giving 10 or 100 m references, the final eastings and northings figures are often estimated.

2.5 Maps and map scales

The basic scales for OS large-scale mapping survey are 1:1250, 1:2500 and 1:10 000 and the standard scales for small scale maps are 1:50 000 and 1:25 000. Standard map sheets are no longer published at 1:1250, 1:2500 and 1:10 000, as these have been replaced by *Superplan Plots* and *Landplan Plots* (see later).

2.5.1 1:2500 scale

This is the basic survey scale for rural Britain, except for some mountain and moorland areas. Originally each standard 1:2500 sheet covered an area of 1 square kilometre, its sheet number being the NG reference of the 1 km square, e.g. SZ 3178 (the grid values at the SW corner of the sheet). Later, the maps were printed as double sheets, paired horizontally. *Figure 2.5* shows a portion of a 1:50 000 sheet and how it could be used as an index for the 1:2500 map sheets. The NG lines on both this scale and the 1:1250 maps were at 100 m, forming squares of 1 hectare (1 ha).

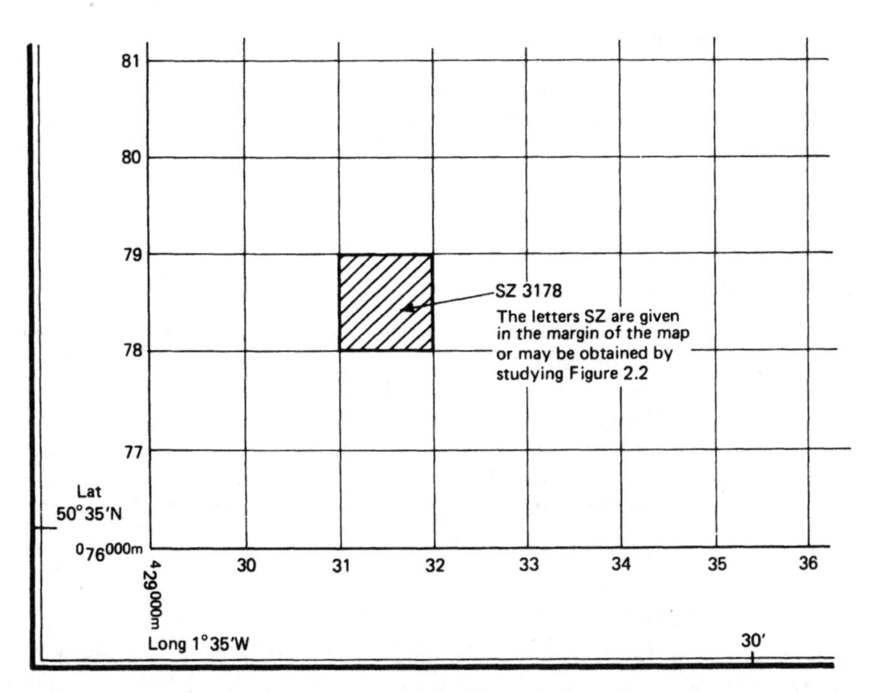

Figure 2.5

These maps covered the whole of Great Britain, the sheets for the larger urban areas being based on the 1:1250 survey of these areas. As stated above, these maps have been superseded by the Superplan system, but facsimiles are available from the OS.

2.5.2 1:1250 scale

Twice the scale of the 1:2500 map, this was the largest scale map sheet in regular production and was the basic scale of survey in the larger urban areas. Each sheet was a perfect square, using the same paper size as the original NG 1:2500 map. Hence a 1:1250 map sheet covered one-quarter of the ground area of the 1:2500 sheet, that is to say 500 m by 500 m.

The 1:1250 scale map was identified by its 1 km reference and its quadrant within the 1 km square, i.e. NW, NE, SW or SE. Again, superseded by Superplan, these are available as Superplan Plots.

2.5.3 1:10 000 scale

Individual sheets of these large-scale maps were identified by their grid reference in a similar manner to the 1:1250 scale map, since each sheet covered exactly one-quarter of a single 1:25 000 scale map (5 km × 5 km). An example reference is SX 88 NW.

The 1:10 000 scale maps covered the whole of Great Britain and this is the basic survey scale in some mountainous and moorland areas, where contours are shown at 10 m vertical interval (VI). In urban and rural areas these maps are derived from the 1:1250 and 1:2500 scale mapping, and the contours are at 5 m VI. The NG on both these scales and the 1:25 000 scale map is at 1 km intervals, hence these scales may also be used as an index to the 1:1250 and 1:2500 scale map sheets.

The 1:10 000 published maps have been replaced by Landplan Plots, a 'plot-on-demand' service for 1:10 000 and 1:5000 scale maps. These may be supplied in any of three formats – Standard National Grid, Pre-defined (i.e. site-centred) and User-defined. They can be supplied either on paper or film, colour or monochrome, with or without contours.

2.5.4 1:25 000 scale

Still published, this small-scale map covers the majority of Great Britain. Originally each sheet covered an area of 10 km by 10 km, the sheet being known by its 10 km reference, e.g. SK 91. The *Pathfinder* series at this scale are generally produced as a double sheet, usually paired horizontally, e.g. SX 88/98. A number of special sheets at the same scale, called the *Outdoor Leisure* and *Explorer* maps are also produced. These special sheets are named and not given a grid reference identification, although the NG is superimposed on them. Contours are at 5 or 10 m VI.

2.5.5 1:50 000 scale

Now termed the *Landranger Series*, this small-scale map is published in sheets which cover an area of 40 by 40 km. One of the uses of this, possibly the most popular of the OS scales, is to serve as an index to the basic large-scale mapping because the National Grid is printed upon the sheets at 1 km intervals, each 10 km line being shown in a broader gauge of line. The sheets of this map are not themselves numbered with respect to the NG. Maps are available in black and white (outline edition) as well as colour. Contours are at 10 m VI.

2.6 Digital map data

The OS has taken the data from the surveys for Urban Areas (1:1250), Rural Areas (1:2500) and Moorland (1:10 000) and stored this in digital form (point co-ordinates and strings) on CD ROMs. The basic system, termed Land-Line, provides detail similar to that originally available on the map sheets at the different scales (see later), while the premium version, Land-Line Plus, gives an extra 27 layers of landscape features, vegetation information and slopes.

The data is held in map files entitled 'tiles', thus a Land-Line.Urban data tile covers an area of 500 by 500 metres, while a Land-Line.Rural tile covers 1 km by 1 km and a Land-Line.Moorland tile covers 5 km by 5 km.

Land-Line data can be displayed on a computer screen, merged with other graphical data, and plotted on paper or film. It may be displayed or plotted at a wide range of scales, and individual feature layers may be coloured or omitted to customize plots. The Land-Line database now holds 228 000 tiles, covering the whole of Great Britain, and it is the basis for a number of OS products, some of which are outlined below. It is available in DXF format for CAD users.

2.6.1 Land-Form PROFILE

This is the definitive National Height Dataset, derived from contoured 1:10 000 mapping, available either as digital contours or Digital Terrain Models (DTMs, see Section 10.6). Contours are at 5 m VI, with some moorland areas being at 10 m VI, together with spot heights from Land-Line and high and low water marks. The OS's DTMs consist of height values at each intersection of a 10 m horizontal grid, interpolated from the contours and air heights on the published 1:10 000 map.

2.6.2 Land-Form PANORAMA

Similar to PROFILE, this is the digital representation of the contours from the OS's 1:50 000 Landranger maps, again available as digital contours or DTMs. Contours are at 10 m VI, while the DTMs are based on a 50 m horizontal grid. Heights are rounded to the nearest metre.

2.6.3 Meridian

This is a vector data product produced from a variety of databases, with selected detail only, it is not a replica of any existing map. The detail includes motorways, major and minor roads, together with County, District and Unitary divisions, the main line railway network, and place and station names.

2.6.4 Raster data

Raster data is available in black and white at 1:10 000 scale, without contours, and in colour at 1:50 000.

2.6.5 Other digital data products

OSCAR is a family of digital road centreline products (Ordnance Survey Centre Alignment of Roads). These give a detailed representation of the road network in the UK, at several levels of complexity. The various OSCAR products are used for managing and producing inventories of the road network, as the basis for in-car navigation applications, or for route planning, e.g. commercial deliveries, RAC rescue planning, etc.

ADDRESS-POINT is a dataset uniquely defining every postal address in Great Britain, aligning with the Royal Mail's Postcode Address File.

Boundary-Line is the definitive data set of electoral and administrative boundaries in England and Wales, derived from the OS's 1:10 000 scale mapping.

Strategi is a regionally based dataset used as a decision-making tool in planning and analytical environment, allowing business users to link business data with up-to-date geographic data.

2.6.6 Revision of digital map data

For *Urban Areas* surveyed at 1:1250, any changes in major features will normally be re-surveyed within six months. If any features are missed in this programme, they will be surveyed within three months if a customer notifies the OS of the omission.

For *Rural Areas*, a five-year revision programme started in 1995, with changes in detail being updated by aerial survey methods and major landscape changes being surveyed as they occur. For *Moorland and Mountain Areas* the revision programme is similar, but over a ten year period.

When digital map files are updated, they are sent out overnight to all agents who hold a particular map in their local cover area, using BT's Integrated Services Digital Network.

2.7 Current OS large-scale mapping products

The map facilities available at large scale today include Superplan Plots, Superplan Data Service and Landplan Plots.

OS large-scale map products are available only from authorized OS Superplan Agents.

2.7.1 Superplan Plots

These are high accuracy graphic plots produced from digital data recorded from large-scale mapping as described in Section 2.6. The plotting scale may be anything from 1:200 to 1:10 000, according to the user's requirements, the plot size may be from A4 to A0, and the material may be paper or film.

The user is not restricted to the old sheet lines, and the plot can be centred on a National Grid reference, or on a postcode or a map sheet number. The user has a choice of the features to be shown, the orientation of plots, and whether to have areas calculated on plots. Maps may have up to 60 layers of detail, using the BS 1192 Pt. 5 layer names method.

2.7.2 Landplan Plots

This 'plot-on-demand' service by Superplan agents, described in Section 2.5.3, is gradually replacing the old 1:10 000 published maps. National coverage should be complete by 1999.

2.7.3 Superplan Data Service

Until recently, Superplan plots were produced only on request by the OS Superplan Agents. Users may now obtain site-centred data in DXF format on 3.5" 1.44 Mb diskettes. They may then prepare their own plots or use the data in CAD systems. The source database is in effect seamless, thus there are no 'sheet edge' problems, and the latest survey data is included. The user has complete control over the site-centring and the extent of the data file, and the data diskette is produced 'while you wait' at local OS Superplan Agents.

The supply charge includes a copyright licence for one year, which covers all on-screen use and up to 20 hard copies derived from the data.

2.8 Interpretation of the 1:1250 and 1:2500 NG scale maps

2.8.1 Margin and border information

Referring to *Figure 2.6*, the following are shown:

(1) National Grid values.
(2) A scale in tens of metres, completely surrounding the map detail.
(3) Sheet and adjoining sheet numbers.
(4) Names of features whose centre (area or linear feature) falls outside the area of the map although part of the feature is contained within the map.

The OS defines *names* to be: *administrative* (e.g. West Dorset County Constituency); *distinctive*, being a proper name or noun other than administrative (e.g. Gernick Estate); or *descriptive*, such as a common noun (e.g. garage). Some or all of these names may appear within the margins of a map.

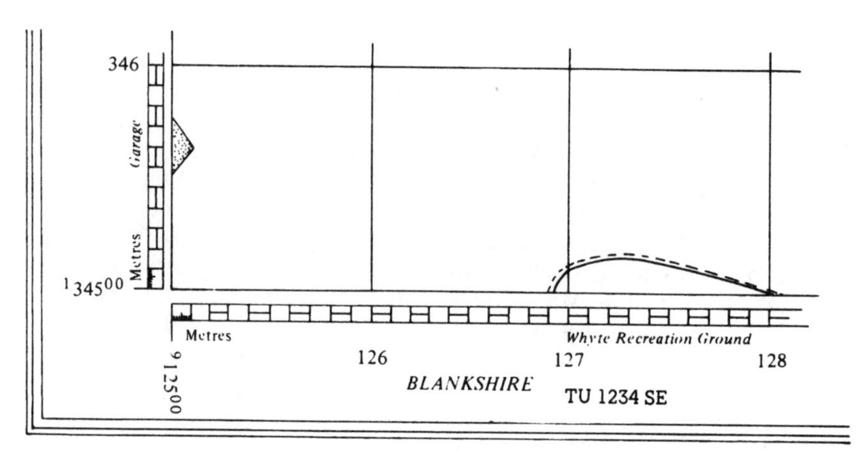

Figure 2.6

2.8.2 Map detail

Detail is the name given in large-scale mapping to the natural and man-made features on or adjacent to the surface of the Earth. One of the principles in OS map preparation was that all 'permanent detail' should be shown in its true plan position as at ground level, although with a few exceptions. This has led to the incorrect belief among a few map users that these map series show all detail and that it is true to scale.

However, a modern built-up area may be extremely complex – consider any city centre area the reader may be familiar with and the problems of any surveyor or draughtsman attempting to map or draw such a place will be self-evident. If the detail is too small to be drawn to scale, or if two parallel features are so close that the draughtsman cannot show both of them with clarity and in their true plan position then the surveyor has the choice of not showing the features, or using conventional signs, or generalizing the detail.

Similarly, some overhead detail is shown, particularly if it is considered to 'constitute a useful feature of the map', e.g. cantilevered bridges and balconies, buildings supported on pillars or columns, etc.

Detail is represented by three classes of conventional signs – *line* symbols, *point* symbols and *area* symbols. All linear detail at ground level which might constitute an obstruction to an able-bodied pedestrian (walls, fences, hedges, water's edge, railway lines, etc.) are shown by firm lines, while other linear detail (overhead and underground detail, edge of footpaths, grass verges, etc.) is shown by pecked lines. A *pecked line* is a form of broken or dashed line (see *Figure 2.7*).

Point symbols are used to represent detail which is too small to show to scale (see *Figure 2.7*).

Area symbols are used to define roofed-over areas, a glazed roof by cross-hatching and other roofs by stippling (dots). Areas of vegetation and surface features depicted by area symbols are shown in *Figure 2.7*.

Names (as described above) whose centre falls within the body of the map are shown, together with postal (house) numbers of buildings, if any. Where numbers

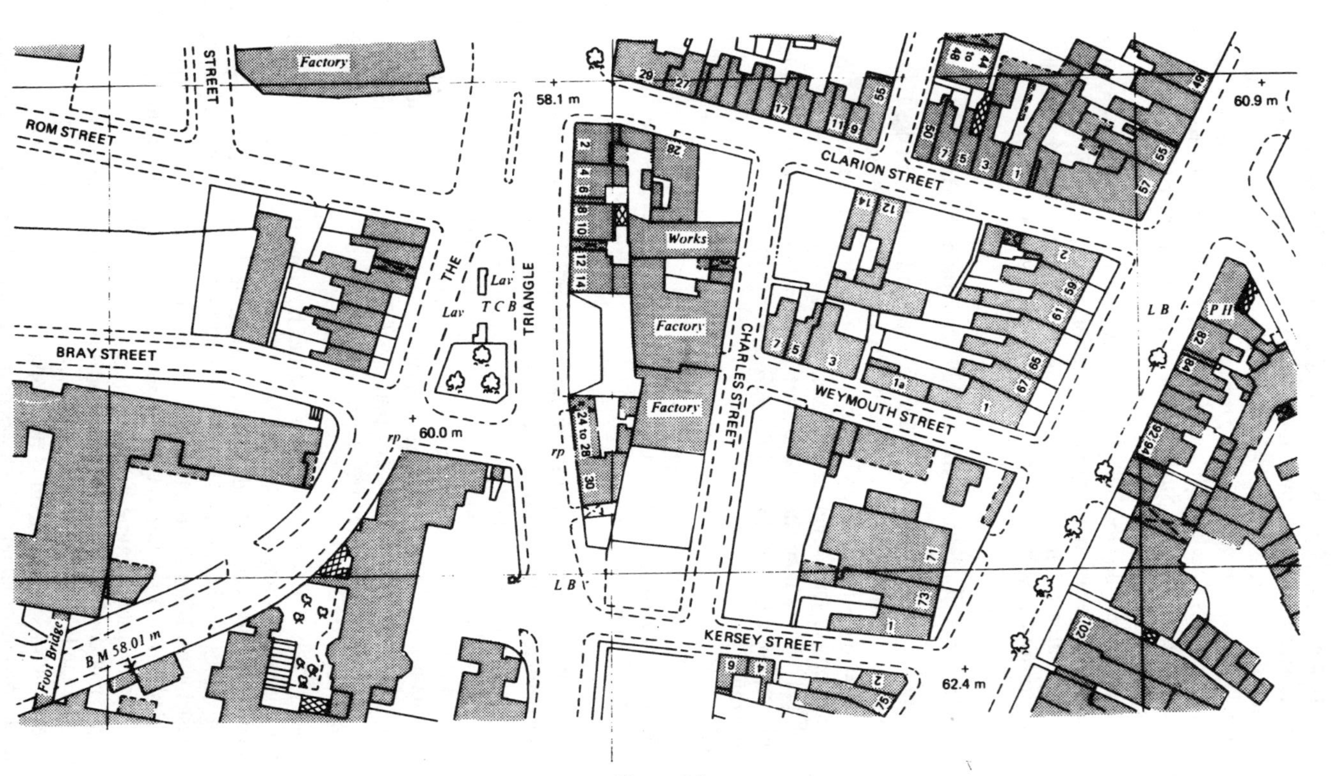
ROM STREET
STREET
Factory
58.1 m
BRAY STREET
THE TRIANGLE
Lav
T C B
60.0 m
rp
Works
Factory
Factory
24 to 28
30
L B
CHARLES STREET
CLARION STREET
WEYMOUTH STREET
KERSEY STREET
44 to 48
50
60.9 m
L B
P H
82
84
92 94
102
62.4 m
B M 58.01 m
Foot Bridge
71
73
75

Figure 2.7

follow a regular sequence, sufficient only are shown in order to enable the user to identify the remainder. The numbers are printed with the base of the number facing the street to which they refer, unless lack of space dictates otherwise. Horizontal and vertical control points are also shown, e.g. minor control points either as .rp (*revision point*) or .ts (*traverse station*). *Triangulation stations* are shown as a dot within a triangle, and *benchmarks* as a 'broad arrow' followed by, for example, BM 56.64 m. Surface spot heights of some roads are indicated, to 0.1 m, e.g. + 27.2 m (see *Figure 2.7*).

A further principle is that all administrative boundaries be shown, these being depicted by a range of line symbols and annotation. A descriptive abbreviation is given of the area defined, e.g. 'Co Const Bdy', that is to say a Parliamentary County Constituency Boundary.

2.8.3 Parcels and areas on the 1:2500 scale map sheets

Apart from the difference in scale, 1:2500 maps differed from the 1:1250 in that in rural localities the area of each 'parcel' of land was shown. A *parcel* of land is defined as a distinct portion of land, although in some cases the OS grouped small portions of land together into one parcel.

Each parcel on a map sheet bore a reference number, together with its area in hectares and/or acres as shown in *Figure 2.8*.

For example

1793	(*reference number*)
1.763 ha	(*area in hectares*)
4.29	(*area in acres*)

Here, 1793 is the parcel reference number, generally a simple 10 m reference to its approximate centre. Exceptions to this rule arose where parcels lay across map edges. Where a parcel lay across a corner of maps, it was allotted a number 0001 to 0006 such that no number is repeated in any one kilometre sheet and the number used was common to the four map sheets upon which the parcel lay. A parcel which lays across map edges but did not cross the corners was given the 10 m reference of the centre of that portion of the map edge passing through the parcel. Where parcels lay across map edges, their areas were quoted to the map edge only.

A parcel was defined above as a distinct portion of land, but this may not always appear so when studying an OS 1:2500 map. A parcel might consist of a group of small portions of land, or on occasion, be subdivided by another parcel and sometimes the boundary of a parcel may be the centre of a double feature. To assist in resolving these situations, the OS used a map symbol known as a *brace*, similar to a long S or the integration symbol. Brace symbols were of four types – single brace, double brace, open brace, or centre brace.

Single and *double braces* symbolized the linking together of two or more portions of land to form a single parcel. An *open brace* tied together the parts of a parcel which itself was divided by another parcel, e.g. a bridge across a river, or an island in an ornamental lake where the island is considered to be part of the adjoining land area. A *centre brace* defined the edge of a parcel which is the centre of a 'double' feature such as adjacent and parallel fences or hedges, see *Figure 2.8*.

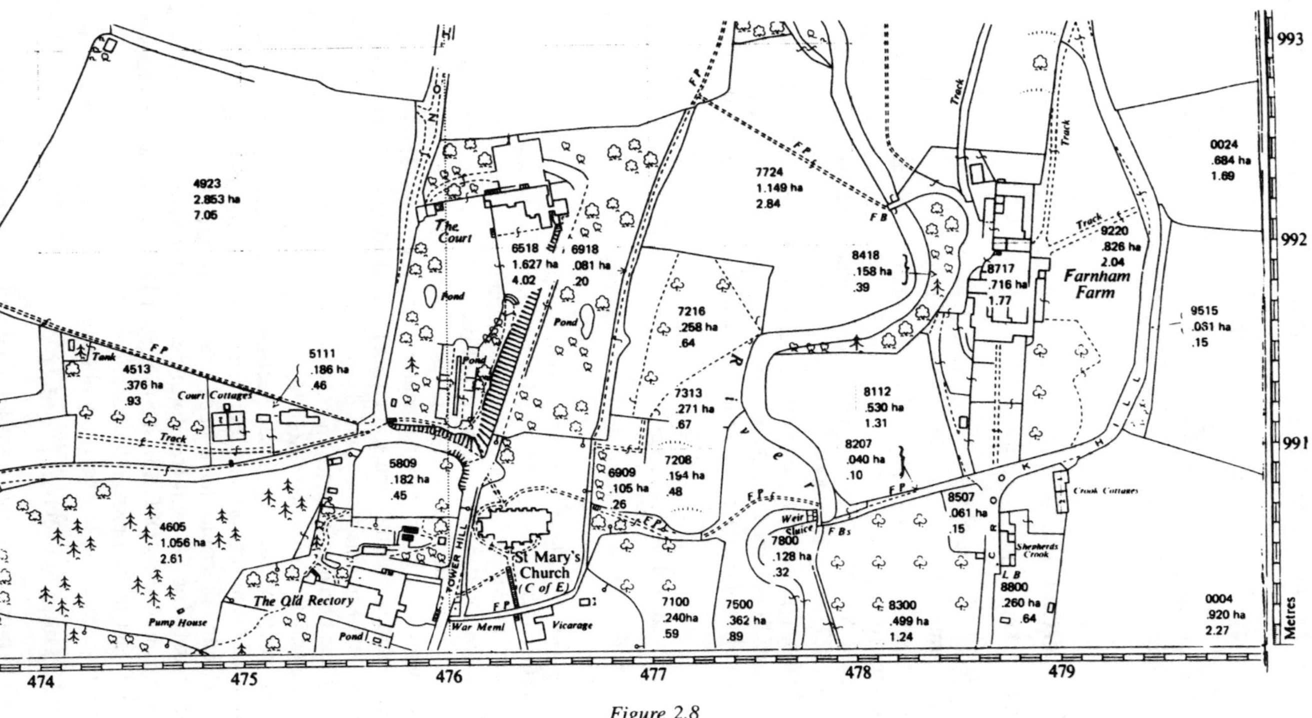
4923
2.853 ha
7.05
The Court
6518
1.627 ha
4.02
6918
.081 ha
.20
Pond
Pond
Pond
7724
1.149 ha
2.84
F P
F B
8418
.158 ha
.39
Track
Track
9220
.826 ha
2.04
Farnham Farm
8717
.716 ha
1.77
0024
.684 ha
1.69
9515
.031 ha
.15
7216
.258 ha
.64
7313
.271 ha
.67
7208
.194 ha
.48
6909
.105 ha
.26
8112
.530 ha
1.31
8207
.040 ha
.10
Tank
4513
.376 ha
.93
Court Cottages
5111
.186 ha
.46
Track
5809
.182 ha
.45
4605
1.056 ha
2.61
The Old Rectory
Pump House
Pond
TOWER HILL
St Mary's Church
(C of E)
War Meml
Vicarage
7100
.240ha
.59
7500
.362 ha
.89
Weir
Sluice
7800
.128 ha
.32
F Bs
8507
.061 ha
.15
Crook Cottages
Shepherds Crook
L B
8800
.260 ha
.64
8300
.499 ha
1.24
0004
.920 ha
2.27
Metres
993
992
991
474
475
476
477
478
479

Figure 2.8

2.8.4 Parcels and Superplan

Superplan plots and the published 1:10 000 map sheets do not, as standard, carry the old 1:2500 map parcel reference numbers and areas, but for both of these users may make use of the OS *Superplan Area Measurement Service* to obtain the ground area and OS parcel references for plots of land. These should be requested when the plot or map is ordered, and they incur an extra charge. A detailed information leaflet on the service is available from Superplan Agents.

2.9 Other OS services

2.9.1 Aerial photography

Some of the mapping of Great Britain has been carried out from aerial photographs, and the revision survey of rural and moorland areas is carried out by air survey. Contact prints, film diapositives, negatives and enlargements of this photography are available for purchase. The photography is black and white 'vertical air photography'. The approximate scale of photography is usually either 1:7500 or 1:24 000.

Photography dating back to 1971 can be obtained through OS Superplan Agents, and the OS-administered Aerial Photography Advisory Service can advise on the availability of both OS and commercial air survey company photography.

2.9.2 Control information

The horizontal and vertical (height) control required to map Great Britain includes *triangulation stations*, points established from traversing (known as *minor control points*) and height control points (known as *benchmarks*). This control information is available for purchase. Horizontal control point information consists of a description, a photograph or sketch and the NG co-ordinates of points. In some cases also, details of ownership of land or the tenant of the land upon which the control point is located. Use of this horizontal control information is referred to in Chapter 9 on Traversing. The use of vertical control benchmarks, possibly the most widely used of the OS services, is covered in some detail in Chapters 5 and 8.

All the horizontal control points used for the OS mapping are based on the OS re-triangulation network OSGB36. A world geodetic reference system for global positioning systems (GPS, see Section 9.5) has been developed, WGS84, and a National GPS Network has been established by the OS for the UK. The National GPS Network points have been specifically selected as being suited to GPS surveying. The OS will transform co-ordinates between any of the systems on request, including OSGB36, OS(SN)80, ED50, ED87, and ETRF89 (realization of WGS84 in Europe).

2.9.3 Other map-based products

The OS produce motoring atlases of Great Britain, road atlases, and street atlases for many cities, the last also available on CD-ROMs.

Recreational maps at 1:25 000 include Outdoor Leisure Maps, Explorer Maps and the Pathfinder Maps mentioned above.

Travelmaster Maps at 1:625 000 are used for route planning, and maps are produced at 1:1 000 000 for communication, physical information, protected environment, rainfall, etc.

A wealth of guides and educational products are also available, including walking guides, waterways guides, a national gazeteer of names, etc.

2.9.4 Old Maps

The OS will supply facsimiles of out-of-print maps, made on order from the originals still held in their archives. The County Series maps produced between 1844 and 1896, and various revisions up to 1939, including Town Maps at a variety of scales, and NG maps produced between 1947 and 1991, are generally obtainable as Old Maps. Some small-scale maps dating back to 1801 may also be obtained as Old Maps.

2.10 Copyright

All users of OS maps, other services, or mapping based upon OS material, must obtain permission before producing copies in whole or in part. Crown copyright of OS publications remains in existence for 50 years. A copyright licence fee may be payable. The reproduction of the NG on maps and other publications is permitted without reference to the OS and without payment provided an acknowledgement is given, for example 'The grid on this map is the National Grid taken from the Ordnance Survey map with the permission of the Controller of Her Majesty's Stationery Office'.

All maps and related products are sold through a network of authorized Ordnance Survey Superplan Agents, who may be located under 'Maps and Charts' in Yellow Pages.

Chapter 3

Site investigations

3.1 Purpose

Site investigations are concerned with the assessment of the suitability of an area of land for the construction of building or civil engineering works. These investigations involve the acquisition of knowledge of the factors which will affect the design and construction of such works and the security of adjacent land and properties. The findings of the site investigation will influence decision making as to whether the proposed development of the site is feasible in technical and economic terms.

3.2 Primary objectives

The primary objectives of a site investigation are set out in BS 5930: 1981 British Standard Code of Practice for Site Investigations. These may be summarized as follows:

(1) To assess the suitability of the site for the proposed works.
(2) To enable an adequate and economic design to be prepared.
(3) To identify factors which may create difficulties during the construction process, in order that they may be planned for and the best method of construction chosen.
(4) To determine the effect of changes, either natural or arising from the proposed works, on adjacent buildings, the local environment and the proposed works themselves.
(5) Where alternatives exist, to evaluate the relative suitability of sites or parts of the same site.

3.3 Initial site appraisal

Much information concerning a site under consideration may be available from existing records and enquiries should be made to determine whether the owner of the site or his agent has any such information.

BRE Digest 318 (February 1987) *Site investigation for low-rise buildings: desk studies* emphasizes the importance of an initial desk study as part of the site investigation process to assess the suitability of a site for development, and lists several sources of published information which might prove helpful. These include OS maps both old and current, geological maps and memoirs, air photography and borehole records. The Digest demonstrates the value of desk studies by reference to eight clearly presented case studies.

The local authority responsible for the area within which the site is situated can often provide useful information, and copies of old maps, etc., are frequently held in the archives of local libraries, county record offices and museums and are available for public inspection.

The companies responsible for providing utilities such as gas, water and electricity should be consulted as to the location and capacity of their mains to service the site. It is helpful if a letter enquiring into these matters is accompanied by a plan of the site and its immediate environs (perhaps a photocopied extract from the relevant OS map) for them to 'mark up' with the relevant information and return.

Every effort should be made by the client's legal advisers to establish any matters such as easements (the rights which one owner of land may acquire over the land of another) or restrictive covenants (an obligation that prevents the person who takes the land with notice of the covenant and his successors in title from doing something on their land, e.g. building more than one house) which might frustrate the proposed development, at the earliest opportunity to avoid further, perhaps abortive investigations. However much information can be collected off-site, an early reconnaissance is advised. BRE Digest 348 (December 1989) *Site investigation for low-rise buildings: walk-over survey* contains helpful advice on information on this aspect of the site investigation process.

Land which has been previously used for industrial purposes may be contaminated. Guidance and information on hazards associated with contaminated land may be found in BS Draft for Development DD 175: 1988 *Code of practice for the identification of potentially contaminated land and its investigation.* Landfill sites may still be producing gas as organic material is broken down by micro-organisms under oxygen-free conditions. Practical guidance is available in the BRE report (1991) *Construction of buildings on contaminated land.* In certain parts of the country naturally occurring radon gas may be present in sufficient quantities to require protective measures to be taken in the design of new buildings. Sound advice is available in BRE report (1991) *Radon: guidance on protective measures for new dwellings.*

During the initial inspection it is invaluable to have a camera handy in order to prepare a photographic record which can then be studied later, together with other data collected in the preparation of a feasibility report. The time available on site is often limited and the camera offers a very quick and accurate method of recording information which will be readily assimilated by others who have not visited the site. A camera with a 'data back', i.e. a camera back containing a quartz clock which will automatically record time and date by exposing these by means of a tiny internal flash at the bottom of the frame each time a picture is taken, is a most useful device.

3.4 Initial site appraisal check lists

The following sections provide check lists for the initial site appraisal. After this has been carried out the extent of the additional detailed information required will be clearer. This additional information may include a detailed site survey, surveys of existing buildings on site and sub-surface investigations.

3.4.1 Site access check list

Classification of roads serving the site, together with their general condition.
Road widths, traffic flow and restrictions, e.g. low bridges, weight limits.
Positions of road intersections and discharge points (e.g. factory gates) which may be potential hazards.
Existing vehicular and pedestrian access points to the site and visibility at junctions.
Footpaths and rights of way.

3.4.2 Services check list

Drainage and sewerage: location and level of existing systems (identifying whether foul, stormwater or combined), pipe sizes, existing flow and ability to take additional flow. Check signs of liability to surcharging or flooding.
Water supply: location, size and depth of main, pressure available.
Electricity supply: location, size and depth of mains, voltage, phases and frequency. Capacity to supply additional requirements. Transformer requirements.
Gas supply: location, size and depth of main. Pressure. Capacity to supply additional requirements.
Telecommunications: location of existing lines and capacity to supply additional lines.
Heating: availability of fuel supplies, district heating, smokeless zones.
Positions of *poles*, *pylons* and *overhead lines* (state safe clearance below).

3.4.3 Topography check list

Site boundaries and how defined, e.g. fence, hedge, ditch.
Positions, types and sizes of trees and hedgerows. Check with Local Planning Authority for Tree Preservation Orders.
Types and differences in vegetation (may indicate changes in soil conditions).
Ground contours and general drainage features.
Mounds or hillocks (natural or man-made?).
Positions of ponds, watercourses and wet patches. Liability to flooding.
Presence of cuttings and excavations or other signs of workings subsequently filled.

3.4.4 Underground hazards check list

Check with the local authority and other statutory bodies for mine workings, tunnels, water table, springs and ground movement, underground services.

3.4.5 Environmental check list

Orientation.
Degree of exposure and local climatic or other hazards by sea air, pollution.
Meteorological information: direction and strength of prevailing winds, annual rainfall and seasonal distribution, temperature range (seasonal and daily), severity and incidence of storms, liability to fogs, etc.
Presence of undesirable features, e.g. noisy or smelly adjacent land uses.
Adequacy of local facilities, e.g. transport, shops, schools, etc.
Photographs: include general views, views from known points, panoramic views, views of specific features (e.g. adjacent buildings, buildings on site, dominant features, particularly good or bad outlooks). Viewpoints should be recorded on a rough sketch or OS map extract and a note made of the date and time of day taken.

3.4.6 Legal and statutory check list

Names and addresses of: vendors; their agents and solicitors, neighbouring landowners; local authority; statutory undertakers for the area (e.g. gas, water, electricity).
The client's actual or potential legal interest in the land – freehold or leasehold.
Ownership of boundary fences and walls.
Easements: such as rights of way, rights of support to neighbouring land and buildings, right to light and air.
Restrictive covenants.
Existence of any valid Town and Country Planning Consents for the site.
Local Planning Authority policy for the area.
Features giving rise to special planning restrictions, e.g. Ancient Monuments, buildings of architectural or historic interest, burial grounds.

3.5 Detailed site survey

The measured survey of the site should be plotted to an appropriate scale and should possibly show the following physical details:

Precise boundaries.
Rights of way across the site.
Position of gates and access roads.
Hedgerows.
Position of trees, noting girth and spread.
Ditches and watercourses.
Ponds and wet areas.
Rock outcrops.
Hillocks.
Ancient monuments.
Benchmarks, levels and contours.
Drain runs, levels and invert levels.
Electricity and telegraph poles, positions of overhead lines including a note of clearance or headroom available.

3.6 Surveys of existing buildings – condition surveys

In addition to preparing the detailed measured drawings as described in Chapter 11, an inspection and report on the condition of the building may also be required. To carry out such an inspection requires not only a sound understanding of construction technology but also a good knowledge of common building defects and their causes. A detailed consideration of these subjects is beyond the scope of this text, but the following sections may provide general guidance.

3.6.1 Scope of the survey

Although the survey should concentrate on the main structure, i.e. roof, walls and floors, some consideration should also be given to the services and the site. Every effort should be made to inspect as much of the building as possible.

Care should be taken to think of the property as a whole, and to consider individual defects as symptoms which may be interrelated and indicative of less obvious but more serious problems. The connection (and the differences) between cause and effect must always be borne in mind.

3.6.2 Equipment

The following basic equipment is recommended:

A4 Millboard and paper, or prepared check list forms.
Moisture meter. Handlamp (plus spare bulb and battery).
Probing instruments (e.g. cheap screwdrivers).
Sectional aluminium ladder (4 m length).
Protective clothing.
Measuring and other equipment as in Chapter 11.

3.6.3 Inspection check lists

The following subsections provide check lists for general guidance as to those points to which the surveyor should pay particular attention.

3.6.3.1 Roofs externally

(i) Pitched roofs
Slopes, ridges and hips for signs of deflection or movement which might indicate failure of the structural framework.
Coverings for missing or defective slates/tiles which could permit water penetration, and for regularity of the course (irregular courses may be due to failure of the battens or nails).
Ridge and hip tiles, noting the condition of mortar pointing and bedding.
Leadwork (or other metals) particularly to valley and parapet gutters.
Eaves and gable details for defects, e.g. rotting fascia, soffit and barge boards, defective verge pointing.

(ii) Flat roofs
Adequacy of falls to dispose of rainwater (ponding may indicate deflection of the roof structure).

Upstands and abutments for cracks and the adequacy of the detailing (e.g. sufficiently high, properly flashed).
Coverings for splits, bubbles, cracks and crumbling asphalt.

(iii) Chimneys
Safety and stability – plumbness, flaunching to pots, etc.
Detailing at the junction with the roof covering, e.g. flashings, aprons, back gutters, etc.
Condition of pointing/rendering.

(iv) Rainwater disposal
Adequacy of design (e.g. size and position of gutters, r.w.p.s. and outlets, correct falls).
Gutters and outlets for freedom from obstruction. Joints for leakage (stains and dampness on adjacent walls).
Rainwater goods for cracking and corrosion.

3.6.3.2 Roofs internally

(i) Roof void (where accessible)
Adequacy of the structural frame (sizes, arrangement and connection of members as related to span/loading; bearings of purlins and rafters).
Coverings, battens and nailing (where not obscured by underslating felt).
Signs of damp penetration, particularly at chimney stacks, soil and vent pipes, abutments and beneath valleys.
Timbers for infestation by wood-boring beetles or wood-rotting fungi.
Thermal insulation and lagging to cisterns and pipework.
Signs of condensation, inadequate insulation.

(ii) Ceiling below
Signs of dampness and staining.

3.6.3.3 Main walls externally

Adequacy of foundations: although foundations are not normally exposed, an opinion as to their adequacy can be based on indirect evidence by examination of the visible superstructure, e.g. leaning, bowing and cracking often indicate foundation problems (where walls lean outwards be sure to check the bearings of structural members resting on the wall).

Walls for signs of structural movement, e.g. cracks, walls out of plumb, distorted window and door openings.

Wall surfaces for defective pointing, porous and soft brickwork, cracked or otherwise defective rendering which may allow water penetration.

Bonding of walls at abutments, particularly extensions to older work.

D.p.c. (if present) for continuity, and the relative levels of the d.p.c. with respect to the surrounding ground and ground floor levels.

Air bricks ventilating suspended timber ground floors for adequacy and freedom from obstruction.

Copings and pointing to parapet walls.

Windows: check for rot in timber windows, particularly cills, and check for corrosion in metal windows.

External decoration.

3.6.3.4 *Main walls internally*

Wall surfaces for signs of dampness, check effectiveness of d.p.c. with a moisture meter, check that detailing around openings in external walls does not permit water penetration.

Walls and openings for signs of structural movement, e.g. cracking in plasterwork (not to confuse shrinkage cracks with cracks of a structural significance), split or stretched wallpapers, distorted door and window openings.

Windows to see that they are operational, open windows at upper levels and lean out to inspect the condition of the window externally and the condition of surrounding brickwork, rainwater goods, roofs at lower levels, etc.

3.6.3.5 *Floors*

(i) Timber floors
Signs of collapse, spongeing, sagging or unevenness, with particular vigilance when inspecting suspended ground floors in older properties without effective d.p.c.s or adequate underfloor ventilation.
Exposed surfaces for signs of timber decay and beetle infestation.

(ii) Solid floors
Surface for signs of dampness evidenced by, for example, presence of salts, detached tiles, moisture visible when covering rolled back.
Surface for signs of movement, e.g. settlement on poorly compacted hardcore (fill deeper than 600 mm is prone to this form of failure), lifting from sulphate attack (note brickwork at the perimeter walls may be displaced laterally at d.p.c. level).

(iii) Ceilings
Surface for signs of cracking, deflection, detachment of lath and plaster.

3.6.3.6 *Joinery*

Exposed surfaces for evidence of wood-boring beetle, wood-rotting fungi and excessive warping and shrinkage.

3.6.3.7 *Services*

A cursory examination should be made, looking for signs of inadequacy.

Old electrical wiring (25 years plus).
Leakage from sanitary plumbing, heating and hot and cold water services.
Blockage or leakage in the drainage system: lift manhole covers, look for soft ground and depressions adjacent to the drain runs.

3.7 Schedule of condition

The information gathered from the inspection of the building may be presented in a report for the client's consideration, but it is often more convenient to present it in the form of a schedule as in the following extract:

Schedule of condition
'Akin Arms', Glass Street, Drinkfield
11th August 1995

Elevation	*Item*	*Description*	*Cond'n*	*Defects*
Front(N)	Roof	Pitched, interlocking conc. tiles with half round ridge	3	Mortar joints between ridge tiles failed 6 No. tiles cracked
	Ch'y	Brick, terra cotta pot	4	Apron flashing loose and torn Mortar joints eroded
	R/W goods	100 mm half round pvc gutters, 65 mm pvc down pipes	1	Nil, recently renewed
	Wall	270 cavity, facing brick to ground storey and conc. vertical tile hanging at 1st floor level	3	Slight settlement cracking below LH gd. floor window (inactive), 20 No. cracked-slipped tiles
	D.p.c.	Bitumen felt	3	Discontinuity to RH of saloon bar entrance, with damp staining extending 150 mm above d.p.c. level

Note: As a time-saving device the condition of the various elements and finishes may be described by reference to a numbered scale, e.g.

1 Very good
2 Good
3 Fair
4 Poor
5 Very poor.

If such a scale is used it must be defined in the schedule.

3.8 Sub-surface investigations

The initial site appraisal should have provided sufficient information to highlight any ground problems which will need to be investigated before the design and construction can proceed. These problems should then be the central considerations of a sub-surface investigation.

3.8.1 Soil investigation methods

For most low-rise small-span structures, e.g. housing, trial pits provide the best method of investigating the subsoil since they allow a very detailed examination of near-surface deposits. Trial pits are cheap, usually being dug using a hydraulic backactor machine such as a JCB or a Hymac as in Figure 3.1.

The pits are normally approximately 1.5 m^2 in plan and excavated to a depth at least one and a half times the width of the proposed base below the level at which the base is to be founded. For example: a typical house wall strip foundation of 600 mm width, which is to have its base set 900 mm below finished ground level, would require a trial pit of 900 + (1.5 × 600) = 1800 mm depth for adequate information to be gathered to design the foundation. Pits should be adequately supported to prevent collapse by means of temporary timber or metal planking and strutting and guarded by a clearly visible barrier at ground level. The maximum depth to which trial pits are dug is generally in the region of 4 m, beyond this depth they become impracticable and expensive. BRE Digest 381 (April 1993) *Site investigation for low rise buildings: trial pits* contains useful advice.

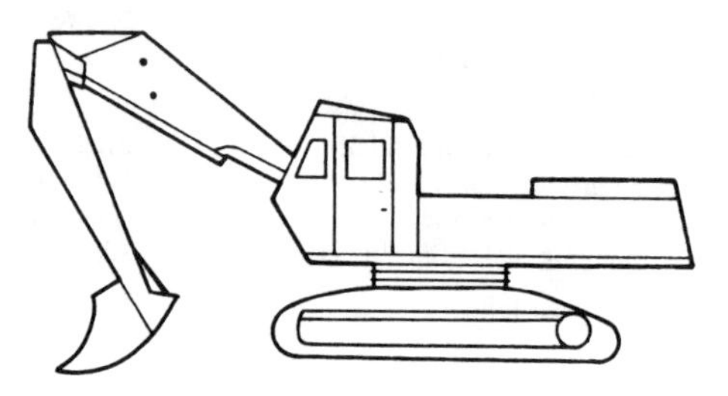

Figure 3.1

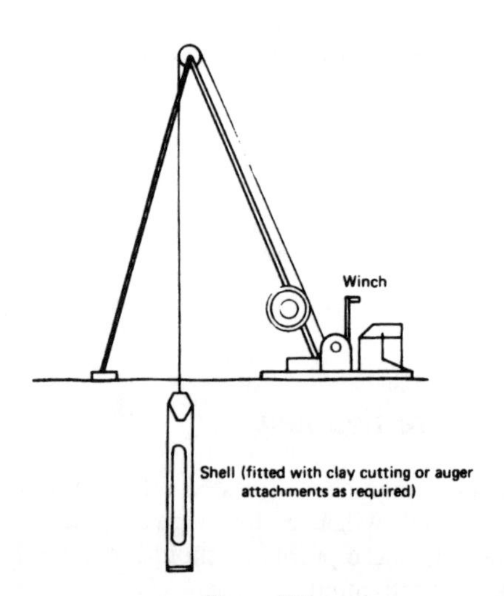

Figure 3.2

Where medium to high-rise developments are proposed, investigations at greater depths than 4 m are usually necessary. This may be done by forming boreholes. These are commonly made using a lightweight, towable shell and auger rig which will mechanically produce holes of 150 to 200 mm diameter to a depth of 45 m in clay. In sandy soils and soft strata the borehole may collapse if the wall of the hole is not supported. The most commonly used method to prevent this is lining the hole with steel tubes or casings. A borehole in sandy soils using 300 mm tools and casings can achieve depths of up to 60 m.

Soil investigations and the interpretation of the findings is a matter best left to geotechnic specialists, therefore the following sections are intended only as a general outline.

3.8.2 Soil profiles

From the trial pits or boreholes a soil profile can be established and a drawing produced showing the depth and composition of the various soil layers.

In describing soils the following standard sequence is recommended in BRE Digest 383 (June 1993) *Site investigation for low-rise building: soil description*:

Moisture condition
Colour
Consistency/strength
Structure
Soil type
Other features
Origin
Ground water (although not strictly part of the soil description it is appropriate to record observations made during borehole drilling).

Example description: Moist, medium brown, firm, closely fissured, slightly silty CLAY. Fissures closed, but show pale grey; fine rootlets present. In situ London Clay.

3.8.3 Soil identification

General advice on soil identification is contained in Building Research Establishment Digest 64, *Soils and Foundations Part 2*, Table 1, Soil Identification. The table is reproduced at Table 3.1.

Table 3.1 Soil identification

Soil type	*Field identification*	*Field assessment of structure and strength*	*Possible foundation difficulties*
Gravels	Retained on No. 7 BS sieve and up to 76.2 mm Some dry strength indicates presence of clay	Loose – easily removed by shovel 50 mm stakes can be driven well in	Loss of fine particles in water-bearing ground
Sands	Pass No. 7 and retained on No. 200 BS sieve Clean sands break down completely when dry. Individual particles visible to the naked eye and gritty to fingers	Compact – requires pick for excavation. Stakes will penetrate only a little way	Frost heave, especially on fine sands Excavation below water table causes runs and local collapse, especially in fine sands

Silts	Pass No. 200 BS sieve. Particles not normally distinguishable with naked eye Slightly gritty; moist lumps can be moulded with the fingers but not rolled into threads Shaking a small moist lump in the hand brings water to the surface Silts dry rapidly; fairly easily powdered	Soft – easily moulded with the fingers Firm – can be moulded with strong finger pressure	As for fine sands
Clays	Smooth, plastic to the touch. Sticky when moist. Hold together when dry. Wet lumps immersed in water soften without disintegrating Soft clays either uniform or show horizontal laminations Harder clays frequently fissured, the fissures opening slightly when the overburden is removed or a vertical surface is revealed by a trial pit	Very soft – exudes between fingers when squeezed Soft – easily moulded with the fingers Firm – can be moulded with strong finger pressure Stiff – cannot be moulded with fingers Hard – brittle or tough	Shrinkage and swelling caused by vegetation Long-term settlement by consolidation Sulphate-bearing clays may attack concrete and corrode pipes Poor drainage Movement down slopes; most soft clays lose strength when disturbed
Peat	Fibrous, black or brown Often smelly Very compressible and water retentive	Soft – very compressible and spongy Firm – compact	Very low bearing capacity; large settlement caused by high compressibility Shrinkage and swelling – foundations should be on firm strata below
Chalk	White – readily identified	Plastic – shattered, damp and slightly compressible or crumbly Solid – needing a pick for removal	Frost heave Floor slabs on chalk fill particularly vulnerable during construction in cold weather Swallow holes
Fill	Miscellaneous material, e.g. rubble, mineral, waste, decaying wood		To be avoided unless carefully compacted in thin layers and well consolidated May ignite or contain injurious chemicals

3.8.4 Bearing capacity

The foundation of a building serves to distribute the loads to be carried over a sufficient area of bearing surface so that the subsoil will be prevented from spreading, and also to avoid excessive or unequal settlement of the structure. To enable the foundation to be designed the bearing capacity of the subsoil must be determined. Methods available include:

Simple field assessment. Reference may be made to Table 12 in Approved Document A of the Building Regulations which tabulates subsoil types, field tests, and suitable minimum foundation widths for residential buildings. Note that the average loading for a two-storey dwelling of traditional construction is of the order of 30–50 kN/m run (3–5 tonnes/m).

Site testing using specialist equipment such as the dynamic penetration test, or vane testing.

Taking samples for analysis and testing off-site in a soils laboratory.

BS 5930 Table 4 provides a schedule of laboratory tests on soil which are described in detail in BS 1377 *Methods of test for soils for civil engineering purposes Part 2: 1990.* Classification tests.

3.8.5 Shrinkable soils

Clay soils are prone to seasonal moisture content changes leading to shrinkage (in summer) and swelling (in winter). In order to avoid such dimensional changes disrupting foundations, they must be placed at a depth below the zone of seasonal movement. A measure of the shrinkability of clay is its plasticity index, which may be determined by laboratory test. Problems arise where clay soils have a plasticity index in excess of 40 and an index of 50 is considered a high risk necessitating deep foundations.

3.8.6 Harmful ground water

If a soil analysis is being carried out it is normal also to analyse ground water. Concrete foundations may be attacked by acidic or sulphate-bearing waters. The relative acidity of ground water is indicated by the pH value; the lower the pH then the higher the acidity (the pH value for distilled or neutral water is 7). Sulphate concentration may be expressed in grams of SO_4 per litre in a 2:1 water:soil extract. Where there is more than 1.2 grams per litre refer to the Table in BRE Digest 363 (July 1991) *Sulphate and acid resistance of concrete in the ground.*

Table 3.2 Minimum width of strip foundations

Type of subsoil	Condition of subsoil	Field test applicable	*Total load of load-bearing walling not more than [kN/linear metre]* 20	30	40	50	60	70
			Minimum width of strip foundation (mm)					
I rock	not inferior to sandstone, limestone or firm chalk	requires at least a pneumatic or other mechanically operated pick for excavation	in each case equal to the width of wall					
II gravel sand	 compact compact	requires pick for excavation. Wooden peg 50 mm square in cross section hard to drive beyond 150 mm	250	300	400	500	600	650
III clay sandy clay	 stiff stiff	cannot be moulded with the fingers and requires a pick or pneumatically operated spade for its removal	250	300	400	500	600	650
IV clay sandy clay	 firm firm	can be moulded by substantial pressure with the fingers and can be excavated with graft or spade	300	350	450	600	750	850

V					
sand silty sand clayey sand	loose loose loose	can be excavated with a spade. Wooden peg 50 mm square in cross section can be easily driven	400	600	
VI silt clay sandy clay silty clay	 soft soft soft soft	fairly easily moulded in the fingers and readily excavated	450	650	*Note* In relation to types V, VI and VII foundations do not fall within the provisions of this section if the total load exceeds 30 kN/m
VII silt clay sandy clay silty clay	 very soft very soft very soft very soft	natural sample in winter conditions exudes between fingers when squeezed in fist	600	850	

Chapter 4

Direct distance measurement

4.1 Introduction

Horizontal distances may be determined by *direct* or *indirec*t measurement. *Direct measurement* of a line means to find its length by comparing it with something of known length, such as a wooden rod, a rope, a steel chain or today a glassfibre or steel measuring tape. *Indirect measurement* means deducing the length of a line from the measurement of *other* quantities – see optical distance measurement and electromagnetic distance measurement later.

Before the seventeenth century lines were actually measured by wooden rods or knotted ropes, then the *Gunter's Chain*, 66 feet long, was introduced by Aaron Rathbone and became the standard measuring device until steel tapes were developed last century. As the chain was the standard instrument (various forms, 66 feet, then 100 feet, then 20 metres or 30 metres) for so long, it was customary to term surveys based on linear measurement only as *chain surveys*.

The chain being obsolete, surveys based on direct distance measurement are now termed *linear surveys*, or sometimes *tape and offset surveys*, although many of the terms used in chaining work are still used. Thus the distance to a point along a measured survey line is often called the *chainage* of the point, even though one speaks about *taping a line* rather than chaining a line.

4.2 Equipment

The following sections describe the equipment used for the direct linear measurement of survey lines. Linear measuring equipment should generally conform with British Standard Specification 4484:1969, which was produced to assist the metrication of construction site survey work.

4.2.1 The land chain

The current typical *metric chain* is 20 m in length and consists of 100 steel wire links, each of which is joined to its neighbours by three oval rings (*Figure 4.1*). Swivelling brass handles are fitted to each end of the chain and its total length is

Figure 4.1

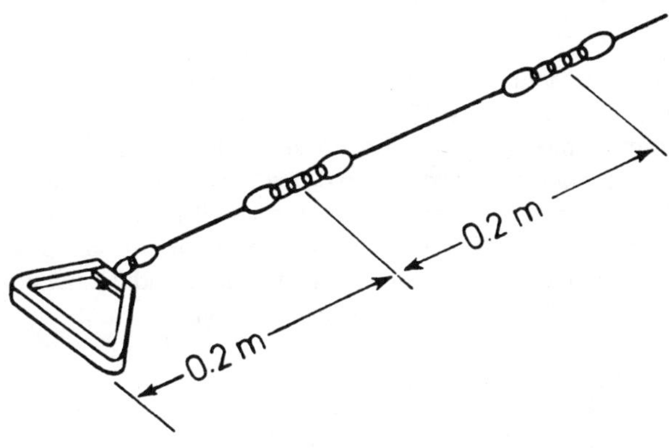

Figure 4.2

measured from the outside of one handle to the outside of the other. *Figure 4.2* shows a part of such a chain.

Since the chain bears no graduation marks, red numbered tallies are attached at every 5 metres, and plain plastic tallies are fitted at single metre intervals. The chain may be used to measure in either direction. It is very robust and probably its most useful role today is in laying grids for levelling spot heights.

A 'chainage' figure may be read by estimation to 0.05 m, but in practice chain measurements are often correct only to 1:500 or less, thus it is evident that the chain is not a very accurate measuring instrument.

4.2.2 Steel tapes and bands

Carbon steel tapes and *bands* are both formed of a continuous ribbon of steel, and typically graduated at 5 mm intervals, the first metre at each end being further subdivided to single millimetres.

The *steel tape* is fixed into and carried in a steel, leather or plastic case, one end of the tape being fixed into the case. Lengths are typically from 10 m to 30 m, the most suitable for general use being 30 m probably. Tapes are normally made with the measurement commencing from the *outside* of the ring attached to the end of the tape as in *Figure 4.3*, but some have the zero set in 100 or 150 mm from the end of the ring – this should be checked carefully. Cased tapes are usually coated in white or yellow, the measurements figured on the coating, and the whole protected with a transparent coating.

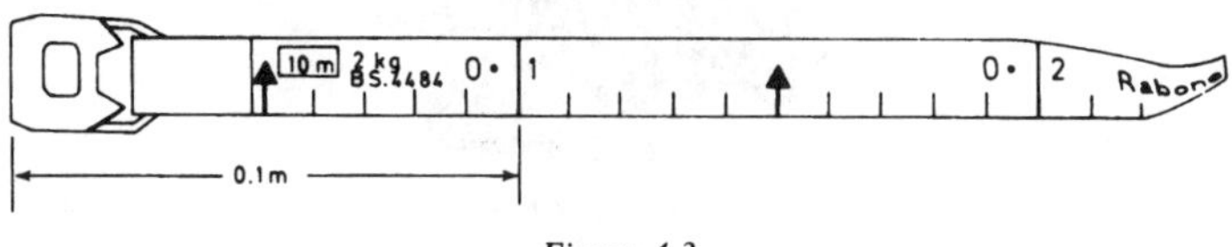

Figure 4.3

The *steel band* is held on an open cross-frame and usually detached from the frame when in use. Bands are fitted with a small steel ring at each end, and a variety of fittings may be attached to these, including clip-on handles, leather loop grips, etc. Lengths may be up to 100 m, but again the most convenient is probably 30 m and the actual graduations start about 150 mm in from each end of the band. *Figure 4.4* shows a steel band and winding frame. Bands are sometimes made of plain uncoated steel with the figures and graduations etched into the surface.

Figure 4.4

Steel tapes and bands should carry markings showing the overall length and the standard tension to be applied. The standard temperature for all steel tapes and bands is 20°C, and a common tension is 50 N. High accuracy bands can be obtained complete with a certificate as to the true lengths at various points along the band. In practice, both tapes and bands are referred to as tapes, and measuring operations as taping, and these conventions will be followed in the remainder of this text.

As steel tapes rust easily they should be wiped dry and clean after use, even the coated types. They also kink easily and this frequently leads to their breaking. Even a bicycle passing over an untensioned steel tape lying flat may break the tape.

Invar tapes made from a nickel steel alloy were traditionally used for measurements of the highest accuracy, since they have an extremely low coefficient of thermal expansion and are almost unaffected by temperature changes. Such measurements are typically made by electromagnetic distance measurement (EDM) methods today.

4.2.3 The synthetic tape

Similar in shape and size to the steel tape, the synthetic tape is manufactured from strands of glass fibre coated in PVC. It is often used to take offsets in detail survey, the most suitable length for this being 10 m, but longer lengths are made. Synthetic tapes are susceptible to length distortion, excessive tension causes them to stretch, and strong winds may blow them off line. Synthetic tapes should be used in place of the steel tape where it is necessary to take measurements in the vicinity of electric fences and railway lines, but they are not as accurate. In such circumstances, the surveyor should ensure that the tape is dry. Graduation is as in *Figure 4.3*.

4.2.4 Arrows

The *marking arrow* is a steel wire pin, roughly 0.35 m in length, as shown in *Figure 4.5*, used to mark the end of a tape length laid down. Arrows are also used to record the number of tape lengths laid down when measuring a line. One of the commonest mistakes when measuring long lines is to forget exactly how many tape lengths have been laid down. This may be overcome by always using a standard number of arrows, say five or ten. To aid visibility arrows traditionally had a piece of red bunting tied through the top ring, but today they are normally painted fluorescent red.

Figure 4.5

4.2.5 Ranging rods or poles

These are one- or two-piece poles of wood or aluminium, pointed at one end and made in various lengths, but typically 2 m (*Figure 4.6*). They are painted in bands of 0.2 or 0.5 m width, alternately red and/or black and white. They are used for ranging lines (setting out a straight line between two end points where the endpoints are more than a tape length apart), marking points on lines and marking the endpoints of lines to be measured. The preferred form is the 2 m rod with 0.2 m banding, since in addition to its main function this type is useful for measuring short offsets between a survey line and detail. The banding is an aid to visibility at long sights, but it may be better to attach a flag on very long sights. On hard surfaces, a tripod-form ranging rod support must be used.

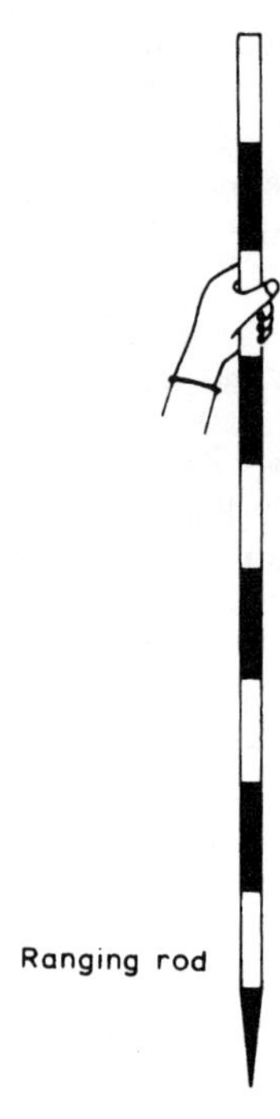

Figure 4.6

4.2.6 The optical square

Several types of optical square are made, the best form having two pentagonal prisms mounted one above the other in a metal or plastic housing (*Figures 4.7* and *4.8*). The instrument is used for establishing and checking alignments and for 'raising' offsets.

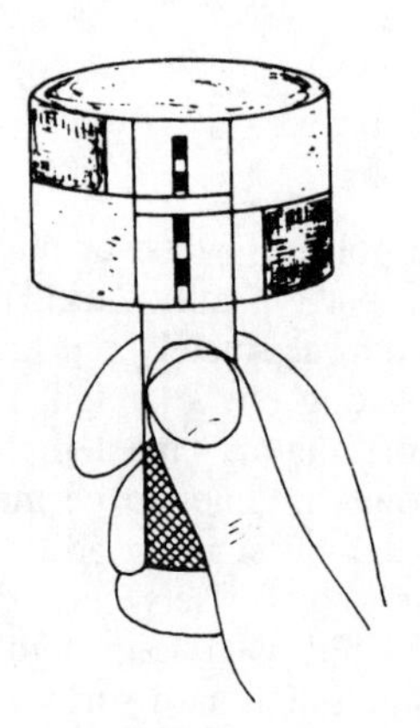

Figure 4.7

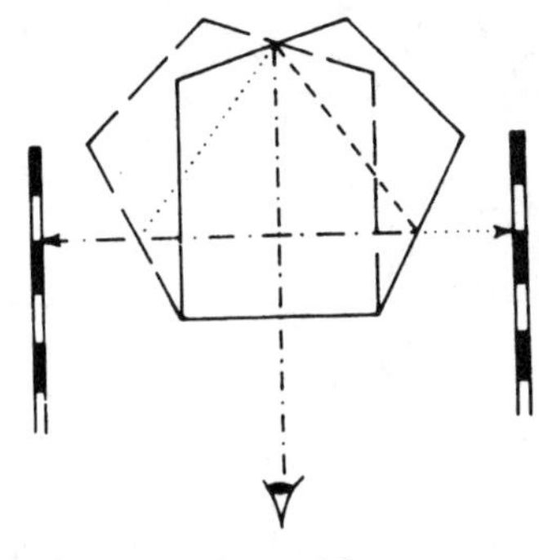

Figure 4.8

To use the instrument as a *line ranger*, i.e. to fix a point on a survey line without sighting from one end of the line, the surveyor holds the instrument in front of his eye and walks across the line and at right angles to it. If a ranging rod has been placed at each end of the line, then when the instrument is exactly on the line each prism will show an image of one of the rods, and the images will be in line vertically.

To raise an offset, i.e. to drop a perpendicular from a detail point to a chain line, one of the prisms only is used. The surveyor places himself on the chain line at the approximate position of the perpendicular and sights a rod at the far end of the chain line while simultaneously looking into the prism. Since the prism provides a view at right angles, the surveyor may move forward or backward until the image of the detail point appears in the prism and lines up with the directly viewed rod at the far end of the chain line.

4.2.7 The Abney level or clinometer

The *clinometer* is a hand-held instrument, used to observe the angle of slope of the ground along which a straight line is to be measured. The *Abney level* (*Figure 4.9*), the most popular type of clinometer, comprises a rectangular sighting tube, a graduated semi-circular vertical arc with vernier scale, a bubble tube, and a mirror. The mirror mounted in the tube enables the bubble to be observed in coincidence with the object being viewed (*Figure 4.10*).

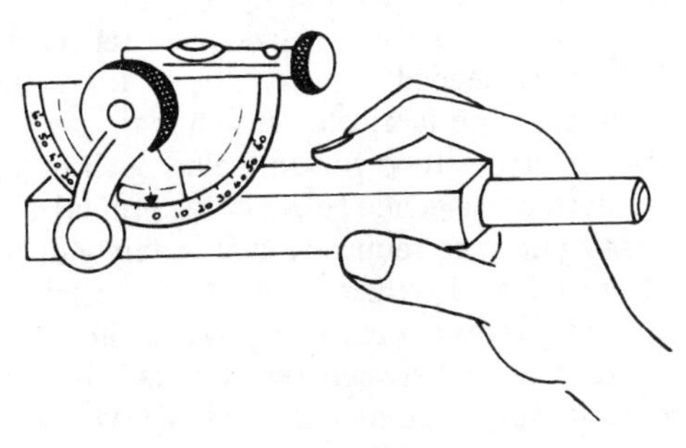

Figure 4.9

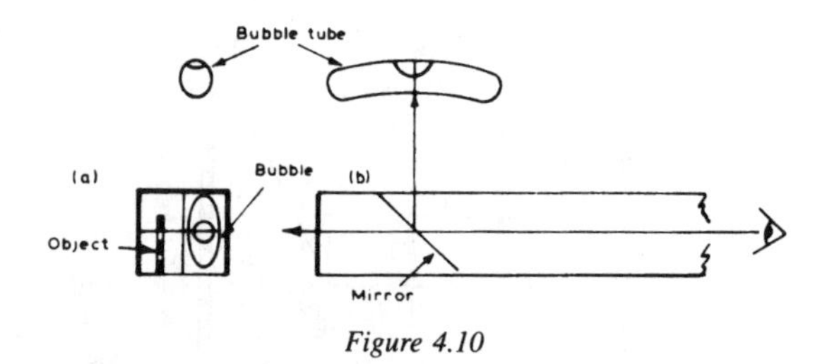

Figure 4.10

The semi-circular arc is graduated in degrees, being read to 5 or 10 minutes of arc by estimation, using the vernier scale and a magnifying glass. The instrument is not suitable for use where slopes are excessively steep, or where high accuracy is demanded. Many Abney levels also have a scale of gradients on the graduated arc.

4.2.8 Ancillary equipment

This may include a *taping thermometer*, used to determine the air (and hence the tape) temperature while measuring a line, a *spring balance* to ensure the correct tension is applied to the tape, and a *tape grip*. The tape grip is a device to allow a spring balance to be attached to the tape at any point in its length, e.g. when tension must be applied to a short length of tape.

4.2.9 Additional materials, tools

Other items required may include ground station marking materials, i.e. hammer, chisel, nails, wooden pegs, steel pegs, grease (wax) crayon, chalk, paint, plumb-bob and possibly cement and sand for semi-permanent station marks.

4.3 Direct linear measurement fieldwork

4.3.1 Surface taping

In the direct measurement of lines today, the tape is normally laid on the ground and fully supported by the ground while the measurement is made, this method being known as *surface taping*.

Surface taping can achieve accuracies of up to 1 in 10 000 with relative ease on the majority of survey sites, and this is adequate for most tasks. Accuracies greater than 1 in 10 000 are needed occasionally, as for example in some base lines and in setting out some prefabricated structures.

In the past, the customary method for obtaining such higher accuracies was *catenary taping*, with the tape suspended clear of the ground, since this eliminated the effects of surface irregularities, reduced the uncertainty in the determination of the tape temperature, and allowed greater flexibility in the positioning of the line to be measured. (The *catenary* is the curve formed by a chain, tape or rope of uniform section and density suspended between two points.) Today, higher accuracy measurement is more usually carried out using EDM methods. However, if necessary, surface taping can achieve accuracies greater than 1 in 10 000, provided the conditions are suitable, i.e. long, even ground slopes, smooth ground surface, shady conditions, a standardized tape, etc.

4.3.2 Line clearing and ranging

A line to be measured is normally marked on the ground by placing a ranging rod at each end, and it should preferably be possible to sight from one end to the other. Lines are occasionally obstructed by vegetation which must be cleared to allow vision and accurate taping and it may be necessary to cut bushes and very long grass. A bush-knife (machete or panga) is most useful for this purpose, but a billhook will do. Care must be exercised with standing crops to avoid flattening wheat, maize, etc., or cutting cultivated trees and shrubs. On occasion it may be worth considering the use of EDM.

If it is not easy to see directly from one end to the other, additional rods may need to be placed on the line, and this is termed *ranging the line*. If a survey line is very long higher accuracy will be achieved if the line is ranged by theodolite.

Where the ends of the line are not intervisible due to intervening high ground it is necessary to *lift the line* over the hill. This procedure is illustrated in *Figure 4.11*. A and B are the rods at the line ends, C and D rods held at intermediate points by assistants, such that A is visible from D and B is visible from C. The assistant at C directs the rod-holder at D into line between C and B, then the assistant at D directs the rod-holder at C into line with A and D, and so on until no further movement is possible.

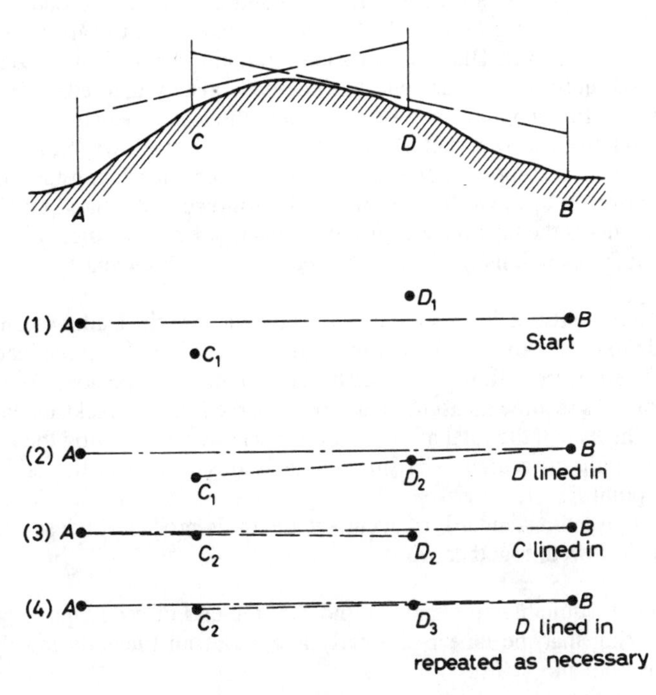

Figure 4.11

4.3.3 Line measurement

A minimum team of two is required for line measurement, the *leader*, who pulls the tape along and marks or reads the actual measurement, and the *follower*, who directs the leader on line and holds the zero of the tape at the rear point. Assuming a horizontal line is to be measured between two points A and B, the basic measurement procedure is as follows:

(i) The line between two points A and B having been ranged (if necessary) and rods placed at A and B, the tape is unwound from its case or frame and dragged forward along the line towards B while the follower holds the zero point of the tape graduations at A. When the tape zero is exactly at the mark at A it may then be more convenient for the follower to stand on the handle or leather grip to keep the tape from moving.
(ii) Holding the forward handle together with ranging rod C in one hand and the set of arrows in the other hand, the leader drags the tape forward. When the leader still has about two metres to go to reach a tape length, the follower shouts 'check' and the leader turns to face the follower and inserts rod C vertically in the ground.
(iii) Sighting over the mark at point A towards the rod at B, the follower directs the leader to move the rod C left or right until it is on the correct alignment, keeping the sight line as close to the ground as possible.
(iv) The leader then crouches, applies appropriate tension to the tape and lays it against the side of rod C so that the tape is tensioned and also lying correctly on line.
(v) The leader inserts an arrow in the ground exactly at the end graduation mark, or on hard ground a chalk T mark is made and the arrow laid on the ground close to it. The follower should remain standing on the zero end of the tape until the forward end has been adequately marked.
(vi) With the first length of any tape line laid, the leader selects a 'back' object beyond the follower before either moves. This allows him to position himself on the approximate alignment for each succeeding tape length.
(vii) When ready to move forward again the follower calls 'next tape' and the leader drags the tape forward using the back object as a guide, repeating the previous operations for laying the second and subsequent tape lengths.

Note that a rod must be left at A to provide guidance for the leader when sighting back, and possibly for measurement in the other direction. When the leader drags the tape forward, the follower releases his end so that the tape does not loop onto obstructions. As the line measurement proceeds, the follower picks up each arrow placed by the leader, the total number of arrows collected showing the number of full tape lengths measured. The line length is this total plus the odd end bay length to point B.

For the minimum standards of accuracy, the basic procedures described above should be followed, together with:

(i) Making a suitably fine mark to indicate the end of each tape length.
(ii) Checking that the tape is correctly aligned at all times during the actual measurement.
(iii) Making the required corrections when measuring on sloping ground.

For higher accuracy work, the following additional procedures may be employed:

(i) Applying the correct measured tension to the tape when marking each tape length.
(ii) Applying corrections to the measured length to allow for the effect of temperature changes causing expansion or contraction of the tape.
(iii) Standardizing the tape before starting the work.
(iv) Measuring the line length twice, once in each direction.

4.3.4 Obstacles to measurement

The methods described below have been applied by the authors and are considered to be the best. There are many alternatives but with some there is difficulty in finding a site large enough to set out the required geometric shapes, and in others there will be problems in adhering to the principles of surveying. Note that it is always best to use the simplest and most direct method – if it is possible to 'throw a tape across a river' then there is no point in using one of these methods, and where possible it is always better to measure straight through a building rather than around it.

The methods outlined below are typically used where a straight line XABY can be ranged visually, but it is not possible to measure the section AB directly.

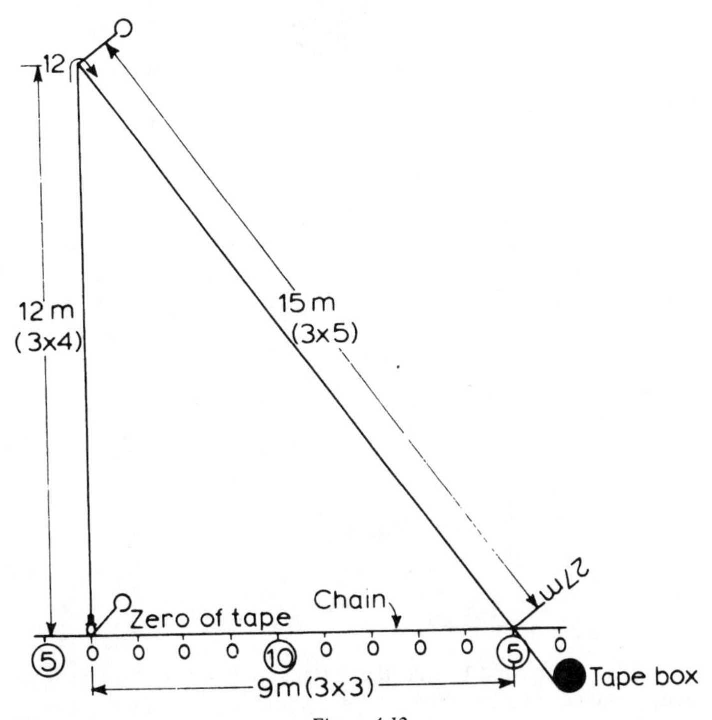

Figure 4.12

(A and B may be located on either side of a deep river or pond, a heavily trafficked road, or some similar obstacle to direct measurement.)

In some of these methods it is necessary to lay out one line at right angles to another (*raise a perpendicular* or *raise an offset*). If low accuracy is adequate (say, correct to 50 mm) then the methods of raising an offset as used in detail survey may be appropriate, as described in Section 10.2. If higher accuracy is demanded then the right angle should be set out by theodolite or by constructing a triangle with sides in the ratio 3:4:5 using one or two tapes. The latter technique is illustrated in *Figure 4.12*.

4.3.4.1 Single 'A' method

In *Figure 4.13* it is required to find the distance AB along the line XY. Construct a well-conditioned triangle ABC, such that the line DE from the mid-point of the line AC to the mid-point of the line BC is clear of obstacles. Measure DE.

Now AB = 2 × DE, since the triangles ABC and DEC are similar.

Note: DE should be measured to twice the accuracy required for the measurement of the line XY.

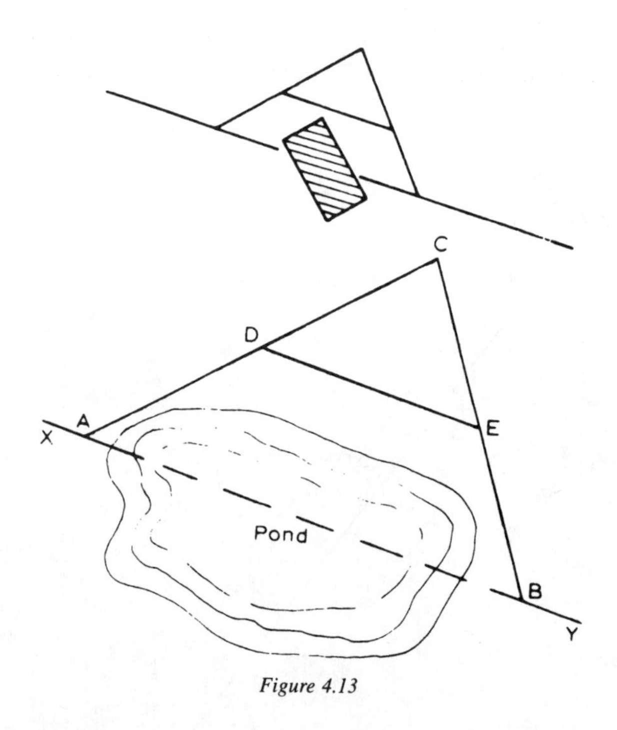

Figure 4.13

4.3.4.2 Inaccessible point method

In *Figure 4.14* it is required to find the distance AB along the line XY. Erect a perpendicular, AC, on the line XY. At the point C, construct a right angle BCD, such that D lies on the line XY.

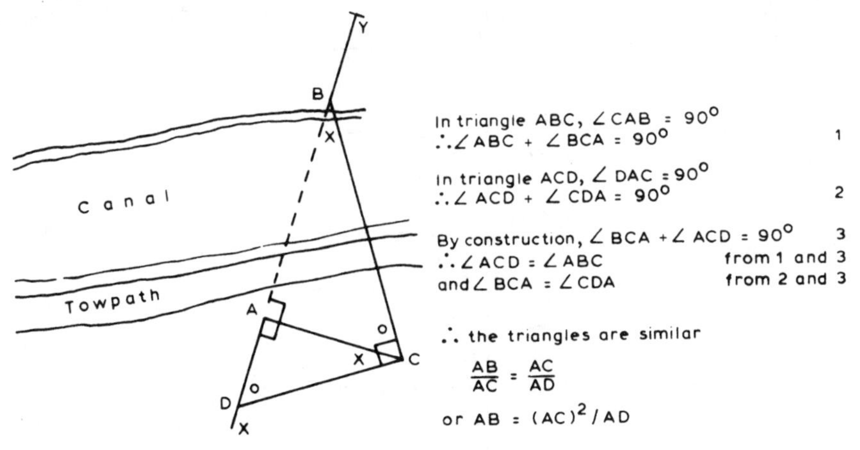

Figure 4.14

Now AB = $(AC)^2/AD$, since the triangles ABC and ACD are similar.

The distance AC should be as long as possible, but it need be no longer than the unknown length AB. The construction of the right angles and the measurement of distances, for which the steel tape is used, should be to a fairly high standard of accuracy. Consider the following numerical examples:

if AC = 8.00 m, and AD = 4.00 m, then
AB = 16.00 m,

but

if AC = 8.01 m, and AD = 3.99 m, then
AB = 16.08 m.

Thus a 10 mm error in each measurement, or in the raising of the right angle, can cause an error of 80 mm.

4.3.4.3 14 02 10 method

This simple method (*Figure 4.15*) requires the use of a theodolite, but the reader who is not yet familiar with the instrument may nevertheless appreciate the method and put it into practice when he has had the opportunity to handle a theodolite. The name of the method comes from the fact that a theodolite must be used to set out an angle of 14° 02' 10".

It is required to find the distance AB along the line XY. At B, construct a right angle using a tape or the optical square. At A lay off (set out) an angle of 14° 02' 10", such that it intersects the right angle raised from B at the point C, and measure BC.

Now AB = 4 × BC.

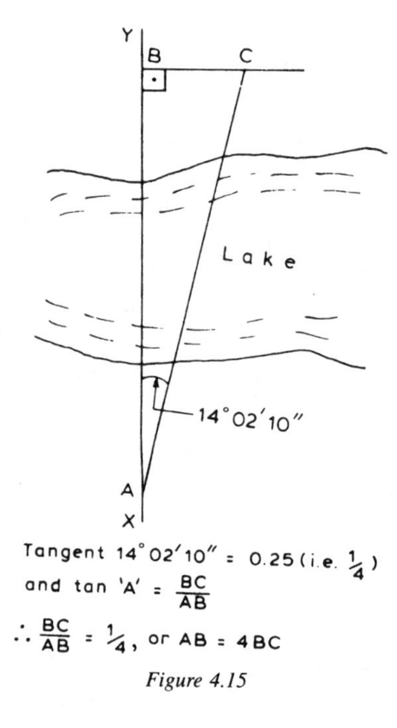

Figure 4.15

In practice, angles other than 14° 02' 10" could be used, but the advantage of using this particular angle is that the perpendicular BC is relatively short as compared with the unknown distance, therefore errors in the setting out of this right angle will usually be minimal. The multiplication factor of 4 is also an easy number to multiply by, and to maintain the accuracy of the line XY it follows that the perpendicular BC must be measured to four times the accuracy required of the line itself, and again this is usually attainable.

4.4 Errors in measurement and corrections

To compensate for the inevitable errors in measurement, it is generally necessary to apply *corrections* to measurements made by direct linear methods, the particular corrections to be made depending upon the standards of accuracy demanded of the work. An accuracy of 1:2000 can be achieved simply by using a steel tape, provided a reasonable tension is applied by hand, corrections for slope are applied, the tape is in good order and the standard procedures are followed with care.

An accuracy of 1:5000 requires as a minimum a standardized steel tape in good order, with corrections for slope applied. An accuracy of 1:10 000 or more would, in addition, require that the correct tension (specified by the maker) be applied, and corrections be applied for variation of temperature from standard. To achieve better than 1:10 000 the line positions must be such that the ground slopes are long, even and smooth, the work is not carried out in direct sun, and sag corrections are applied where appropriate.

The following sections identify sources of error and corrections which may be applied. It should be noted that where a *correction factor* is obtained the measured line length is to be multiplied by the factor, but where a *correction* is calculated, the correction is to be added to or subtracted from the measured line length. As a general rule it is better practice to calculate corrections rather than modify the line length by multiplying it by a factor.

4.4.1 Slope

Measurements made on slope are always longer than the required horizontal distance, thus measurements on slope must be *reduced to the horizontal*. Surface taping measurements being made on the surface of the ground, the angle of slope of the ground may be read by use of an Abney level or a theodolite by observing a line parallel to the line measured with the tape as in *Figure 4.16*.

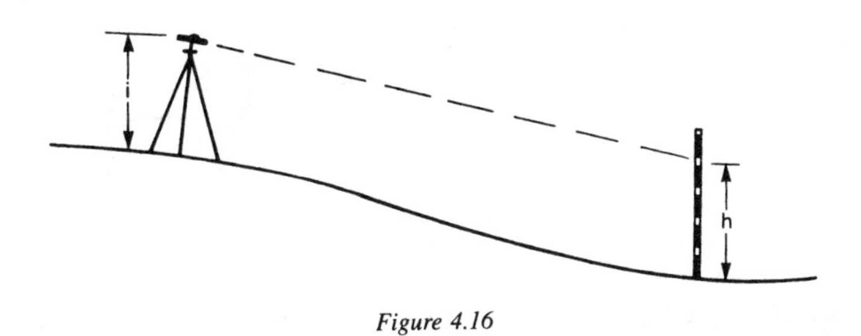

Figure 4.16

Horizontal distance = slope distance × cosine of the angle of slope, thus *slope correction factor* = cosine of the vertical angle, and *correction for slope* = $-L(1 - \cos\alpha)$, where the measured length on the slope is L and vertical angle is α. This correction is always negative.

Although slope correction by the cosine of the slope angle is easily applied, the accuracy of the slope correction itself depends upon the accuracy with which the slope angle is measured and the steeper the slope the more accurately the angle must be measured.

Where a measured length is required to be accurate to 1:5000 or 1:10 000, then the slope correction must be accurate to within about 2 mm/30 m for each measure of the line. To achieve this, the vertical angle (α degrees) must be correct to within approximately $600/\alpha$ seconds of arc, thus on a slope of about 2° the angle must be measured to within 300" or 5' of arc. Similarly, a slope of about 10° must be measured to within 60" or 1' of arc.

In *Figure 4.16*, the distances h and i should be equal to within about $300/\alpha$ mm for every 100 m of line length. Thus for a 2° slope, h and i may differ by up to 150 mm/100 m, but on a 10° slope not more than 30 mm/100 m of line length.

A common problem is that of identifying where changes of slope actually occur and thus defining the separate slopes. No serious guidance can be given in this respect, however, only practice will develop some skill.

On occasion the difference in height of the two ends of the line, d, is obtained, then:

$$\textit{correction for slope} = -(d^2/2L + d^4/8L^3)$$

where L is the measured slope length. If the slope is less than about 10% the second term of the expression may be ignored.

As an alternative to calculating corrections as above, a sloping line may be step taped, as shown in *Figure 4.17*, i.e. measured as a series of short horizontal steps or bays. The line should be measured in a downhill direction, the suspended bays must be kept short, say 10 m or less, and the plumb point at each bay length should be found using a plumb line or a weighted *drop arrow*. The drop should not exceed about 1.5 m.

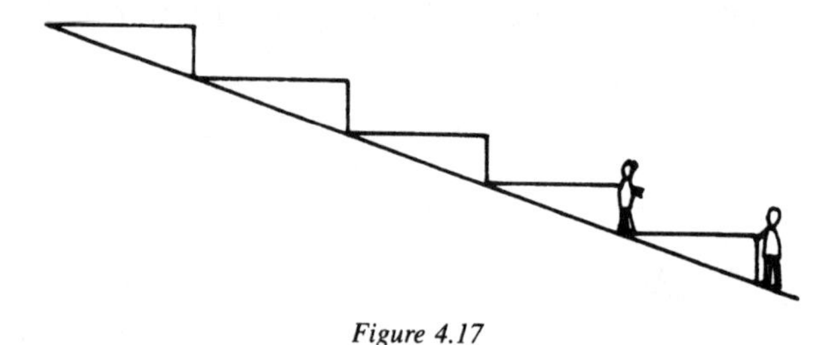

Figure 4.17

4.4.2 Standardization

If a steel tape is longer or shorter than its *nominal length* (the length marked on it) a cumulative error will result when it is used to measure lines. Tapes should be checked against some standard from time to time, then appropriate corrections can be applied to the measured distances. The *standard* may be a steel tape reserved solely for this purpose, never used in the field, and the determination of the correction is termed *standardization*. Standardization may be used to find the true length of the tape, or to find a *standardization factor* such that

$$\text{true line length} = \text{measured line length} \times \textit{standardization correction factor}$$

The standardization correction factor is obtained by setting out a horizontal bay, slightly shorter than the length of the tape to be standardized, on smooth, firm level ground, then measuring the bay length with the standard tape and the field (work) tape, in each case applying the correct tension for the particular tape and allowing for temperature differences. The bay ends should have fine markers, preferably in concrete blocks. Then

$$\textit{standardization correction factor} = \frac{\text{length recorded on standard tape at } x°\text{C}}{\text{length recorded on field tape at } x°\text{C}}$$

An answer to five decimal places is usually adequate, and the result cannot be expected to be more accurate than that attained with the standard tape. A new tape is often used as a standard and manufacturers state generally that this may be accurate only to ±2.5 mm or so per 30 m length. As stated earlier, some manufacturers will issue a certificate for a tape if a higher accuracy is required.

Alternatively, if the standardized length of the field tape has been found to be l' by measuring an accurate base, while its nominal length is l, then for a line length measured as L,

$$\textit{standardization correction} = L(l' - l)/l$$

If the tape is too long (stretched) the correction is positive, if short the correction is negative.

Unless a tape is badly damaged, or very old, standardization is only necessary when accuracies of 1:5000 or better are demanded.

4.4.3 Incorrect tension

Field tests have shown that failure to apply the correct tension with a spring balance may result in readings which are in error by as much as 15 mm/30 m, i.e. 1:2000. This suggests that where an accuracy of greater than 1:2000 is required a spring balance should be used to apply the correct tension to the tape. The balance is attached to the forward end of the tape for tensioning, using a tape grip if necessary. The tension to be used is stated on the tape, e.g. 50 N or 5 kgf, or 70 N or 7 kgf are typical values.

Errors which arise from a failure to use the standard tension may be corrected by re-standardizing the field tape at the incorrect tension. Alternatively,

$$\textit{correction to measured length} = \pm\, l(T_f - T_s)/AE$$

where l = measured length, T_f = tension used in the field, T_s = standard tension, A = cross-sectional area of the tape, and E = Young's modulus of elasticity. The units must be compatible, thus l and the correction in metres, T in newtons, A in mm^2 and E in newtons per square millimetre.

If an accuracy of 1:10 000 is required, then the tension should be correct within ±1.5 kgf or 15 N. Spring balances should be checked against one another occasionally.

4.4.4 Temperature variation

Temperature variation results in expansion and contraction of the steel tape and consequent measurement errors. Temperature may be observed by placing a taping thermometer on the ground some time before measurement, to allow it to adapt to the air temperature, or again for every line measured, subject to the accuracy required. The typical taping thermometer is encased to prevent it coming in direct contact with the ground.

For an accuracy of 1:5000, the tape temperature should be known to within ±5°C, and for 1:10 000 it should be known to within ±3°C.

Note that when a tape lies on the surface of the ground in the sun, then the tape temperature may be 5°C or more greater than that of the ground surface. When measuring over concrete in these conditions it is advisable to support the tape on small wooden blocks to insulate it from the concrete temperature.

Manufacturers state that a tape reads correctly at a particular temperature under a particular tension, this particular (*standard*) temperature being 20°C. The coefficient of linear expansion for steel tapes is generally 0.000 011 2 per degree Celsius, and the correction for temperature variation is *line temperature correction* = temperature variation from standard × coefficient of linear expansion × measured length of line or

$$\begin{aligned} c &= \Delta t \times 0.000\,011\,2 \times l_m \\ &= 0.000\,011\,2\,(t_f - t_s)\,l_m \end{aligned}$$

in which t_f and t_s are the *field* and *standard* temperatures, respectively, and l_m is the line length as measured.

If the corrected length = l_c, then

$$\begin{aligned} l_c &= l_m + 0.000\,011\,2(t_f - t_s)\,l_m \\ &= l_m\,\{1 + 0.000\,011\,2\,(t_f - t_s)\} \end{aligned}$$

If the temperature is higher than 20°C the distance as measured will be less than the true distance, and if it is lower the measured distance will be greater than the true distance. The sign of the correction is indicated by the relative magnitudes of the temperatures.

As a rough indication, a 100 m tape will be 22 mm shorter when the temperature is 0°C than when the tape is at the standard temperature of 20°C. Occasionally surveyors combine the above two corrections to produce the temperature at which the field tape reads its true length.

4.4.5 Sag

If the whole tape length is suspended in catenary, perhaps due to obstacles at ground level, a *sag correction* must be applied. If the tape ends are level the correction is $-W^2l/24T^2$, where W = tape weight, l = tape length and T = applied tension. W and T must both be in the same units and T should be the standard tension specified for the tape. If the tape ends are not level, the above correction must be multiplied by $\cos^2 \alpha$, where α is the angle of slope.

When taping over broken ground, it is permissible to occasionally suspend a short length of the tape in catenary, typically up to 15 m or so in length. This depends upon the tape weight per unit length, however, and the greater the weight the less must be the length of the suspended span. For the highest accuracy a sag correction would have to be calculated.

4.4.6 Bad alignment

Where tape lengths are aligned by eye, a common practice over short distances, lines may be misaligned by as much as 100 mm/100 m, a bow of 300 mm in a 300 m line. This has the effect of increasing the apparent line length by about 0.4 mm/100 m, an amount which may be ignored in all except the most accurate work.

Poor alignment or bowing may also occur along each tape length, that is to say individual tape lengths may not be laid straight. Causes may be a strong wind blowing across the line, the line passing through a fence post or blocked by a car wheel or vegetation, these causing the line to be bent round the obstruction. All these forms of misalignment should be guarded against. Note, however, that the odd bowing of a tape may be ignored on occasion, a bow of 50 mm in a 30 m tape length along a 100 m line producing an error of less than 0.2 mm in the total line length.

Vertical misalignment occurs when the line of sight of the angle measuring instrument is not parallel to the ground, as in *Figure 4.16*, if h is not equal to i. Surface irregularities, undulations, cause effects similar to horizontal misalignment, and all result in cumulative error.

4.4.7 Marking tape ends, reading the tape

Each time a tape is laid down, errors may arise in marking the end of the tape and in reading the graduations. These errors may tend to compensate, but it is good practice to measure each line twice, particularly if an accuracy of 1:10 000 is required, and demand that the two measures agree within some standard such as 2 mm/30 m tape or part tape length.

4.4.8 Projection and scale

When tying linear measurements to a higher order of control, such as that of the Ordnance Survey, a *projection* or *scale correction* is necessary, as referred to in Section 2.3. Every reduced horizontal distance in the survey must be multiplied by the Ordnance Survey *local scale factor* to give its equivalent National Grid distance. This factor F_p may be calculated using the formula:

$$F_p = F_o \{1 + (E_p - 400\,000)^2/2R^2\}$$

where E_p is the NG eastings in metres and $F_o = 0.999\,601\,3$, and R is taken as 6 381 000 metres.

4.4.9 Altitude (height above sea level)

In higher accuracy work, measured lengths may need to be reduced to the equivalent distance at mean sea level, as in *Figure 4.18*.

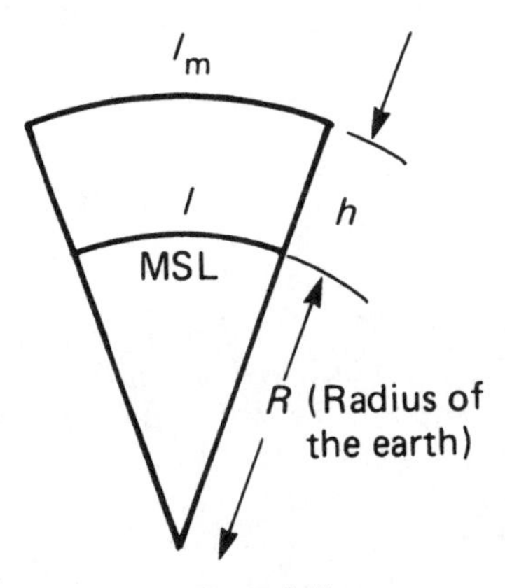

Figure 4.18

If l_m = length measured and h = mean altitude of the measured line above mean sea level, then the correction to reduce to mean sea level is $-l_m h/R$.
R may be taken as 6 381 000, then correction = $-l_m h/(6381 \times 10^3)$ metres.

4.4.10 Combining corrections

It is normal to measure the line length l, then separately calculate the various corrections to be applied, then add them to the observed line length l in the following sequence:

final line length = l – slope correction ± standardization correction ± tension correction ± temperature correction – sag correction

The other corrections for scale and altitude above sea level may then be applied if required.

Chapter 5

Height measurement

5.1 Introduction

The preferred method of measuring height in construction and similar work is by *levelling* using a *surveyor's level.* Other methods of determining heights are outlined in the subsections immediately below.

5.1.1 Trigonometric heighting

This method is not generally as accurate as levelling, but it may occasionally be useful when the ground is undulating or very steep. As shown in *Figure 5.1*, instead of a surveyor's level, a *theodolite* is used to read angles of elevation or depression (i.e. angles above or below the horizontal) to a target; then provided the horizontal or slope distance is known, the difference in height can be calculated by elementary trigonometry.

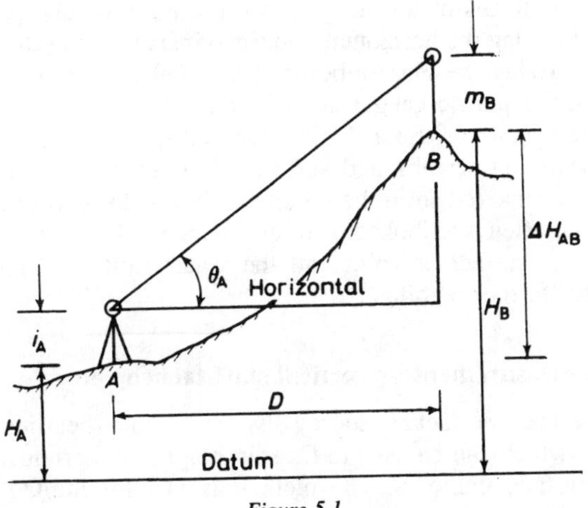

Figure 5.1

Given that the horizontal distance D in the figure is known, then the required height difference between the instrument station at A and the ground point at B is

$$\Delta H_{AB} = H_B - H_A$$

but

$$H_A + i_A + D \tan \theta_A = H_B + m_B$$

thus

$$\Delta H_{AB} = D \tan \theta_A + i_A - m_B$$

This result is not very accurate, as the vertical angle should be corrected for curvature and refraction, and the correction cannot be reliably calculated. Accordingly, the recommended method is to make *simultaneous reciprocal observations* from each end of the line. If these are truly simultaneous, then the mean of the two results is free from error due to refraction.

5.1.2 Barometric heighting

Atmospheric pressure varies with height, thus a *barometer* which records atmospheric pressure may be used to obtain height measurements. Instead of being calibrated in inches, millimetres of mercury or millibars, it can be calibrated in metres of height above sea level, then the instrument is known as an *altimeter.* Anyone who has studied a barometer or a weather map will appreciate that atmospheric pressure changes continually without any change in height, hence the method is often not of sufficient accuracy. However, the method has been used in many parts of the world for heighting points for small-scale maps.

5.1.3 Hydrostatic (water) levelling

One of the oldest forms of levelling is to use a glass U-tube partly filled with water, then sight along the horizontal line (imaginary) joining the tops of the two water surfaces. Today the glass tube has been replaced by a flexible one with glass or transparent plastic gauges at each end.

Building site operatives use a flexible tube of approximately 30 m in length, typically for levelling screeds and surfaces. It is essential to note that all air bubbles must be removed from the tube, and this is difficult over long lengths. Another method, often overlooked, is to use the natural level of a body of water, provided that the surface is calm and the water flow is negligible. See the bibliography for further reading.

5.1.4 Stadia measurement or vertical staff tacheometry

This is a method of detail survey by radiation (bearing and distance measurements) which can be used to fix spot heights and produce a survey plan showing heights and contours. The method is fast for limited areas, but the heights obtained are not of high accuracy, see Section 7.2.

5.1.5 EDM or electronic tacheometry

Detail survey by radiation from an instrument station can be carried out using an electromagnetic distance measuring instrument, providing both distance and height measurement. The heights obtained are considerably more accurate than those obtained with stadia measurement. See Section 7.3.

5.1.6 Global positioning systems

GPS surveying is described in Chapter 9 on Control Surveys. While primarily intended to locate a point in plan, GPS surveying actually gives all three co-ordinates, but on the world GPS datum WGS84, and heights obtained will need to be converted to the local system.

5.1.7 Steel tape measurement

On occasion it may be necessary to make a vertical measurement which exceeds the length of the normal levelling staff, e.g. the height to the underside of a high overhead bridge. In this case, heights may be measured by suspending a steel tape vertically from the high point, applying appropriate tension to the bottom of the tape, then reading the tape in the same way as a levelling staff.

5.2 Levelling definitions

Levelling is the name given to the method of determining heights, or rather *differences in height*, by the use of the instrument known as the *surveyor's level.*

Level line: one which is of constant or uniform height relative to mean sea level and is therefore a curved line concentric with the mean surface of the earth. More formally, a level line is a line which lies on one level surface and is normal (at right angles) to the direction of gravity at all points in its length.

Horizontal line: a line through a point, tangential to the level line passing through the same point and normal to the direction of gravity at that point. The difference between a horizontal and a level line passing through the same point must be appreciated. The greater the distance from the common point, the greater the discrepancy. In ordinary levelling, with sights less than 60 m or so, the difference is negligible for practical purposes and may be ignored, except for the most precise work, see *Figure 5.2*.

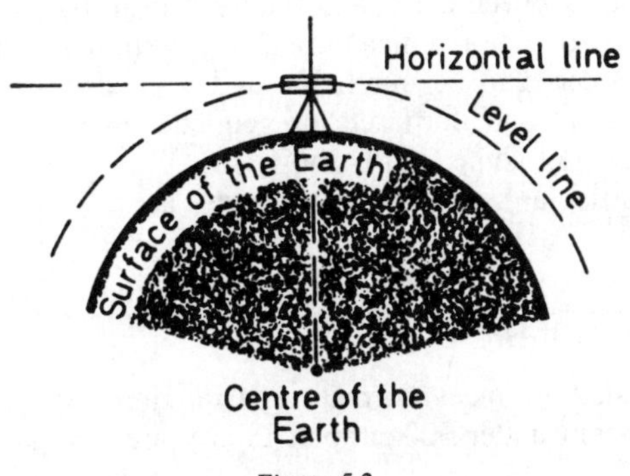

Figure 5.2

Datum surface or *datum line*: a level surface or line from which heights are measured, or to which heights may be referred.

Height: the vertical distance of a point above or below a datum surface.

Reduced level (RL): the calculated height of a point above or below a datum as deduced from the surveyor's field observations.

Mean sea level (MSL): the mean level of the sea as determined at some selected place from observations over a period of time, used as a datum surface for levelling work. The concept is used by many national mapping organizations.

Ordnance Datum (OD): the current datum for heights used by the OS of Great Britain. It is based on the mean level of the sea at Newlyn in Cornwall, calculated from hourly tide gauge readings recorded between 1915 and 1921.

It should be noted that heights are not always measured from MSL – as an example, in Hong Kong the reference datum for all heights and levels on land is the 'Hong Kong Principal Datum', which is actually 1.23 m below the mean sea level.

AOD: seen occasionally on a plan or map after a height value, means *height above Ordnance Datum.*

Bench marks (BMs): fixed points whose heights relative to a datum surface have been determined using a surveyor's level.

Ordnance bench mark (OBM): a bench mark established by the OS, the height of the bench mark relative to Ordnance Datum being known accurately. The OS have established OBMs all over the mainland and inshore islands of Great Britain, and levelling operations may be referred to these known points.

Temporary bench mark (TBM): a bench mark set up by a surveyor for his own use on a particular job. The TBM height may be established from an OBM, then levels on site may be referred back to the TBM without checking back to the OBM every time. TBMs should be stable, semi-permanent marks, such as concreted pegs or features on a permanent building.

5.3 Bench marks

The OS having established bench marks all over the country, any levelling operations may be referred to an OBM of known height above mean sea level. The OS levelling consisted first of lines of primary geodetic levelling, then secondary levelling between these, and finally 'fill-in' by tertiary levelling. The work was originally based on a mean sea level determined at Liverpool in 1844, but this was considered to be unreliable and was superseded by the Newlyn datum for the second and third sets of geodetic levelling. Slight differences appear in the heights of some OBMs shown on OS map sheets, depending upon the date of the particular sheet. Post-1956 levels are based on the third geodetic levelling.

5.3.1 Density of OBMs

OBMs are provided to meet normal user requirements and the density of provision varies from under 300 m apart in city areas to over 1000 m apart in rural areas.

5.3.2 Types of OBM

Six different types of BM have been set up by the OS, similar marks being found in other parts of the world.

Cut bench marks, the commonest form, consist of a horizontal line incised in a vertical surface such as a brick or stone wall. The traditional government 'broad arrow' is cut below the centre of the horizontal line, point upwards, see *Figure 5.3(a)*.

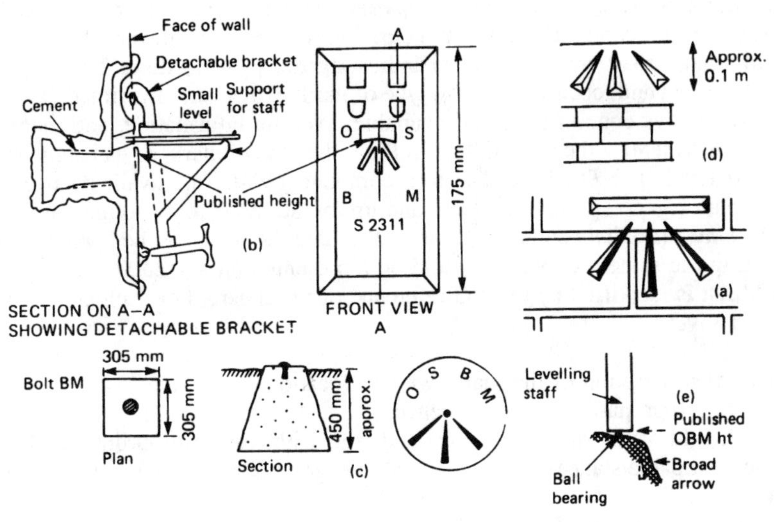

Figure 5.3

Fundamental bench marks (FBMs) are marks placed on solid rock, at points roughly 40 km apart, and they provide control for the whole of the OS levelling network. The mark consists of three reference points, two of which are in a buried chamber not accessible to the public. The third mark is a brass or gunmetal bolt set on top of a low granite or concrete pillar, available for public use.

Bronze flush bracket OBMs, of lower order than FBMs, are levelling control points set into the side of large public buildings (churches, etc.) and OS triangulation pillars. The published height is to the small horizontal platform at the point of the broad arrow marked on the face of the bench mark (*Figure 5.3(b)*). A special staff support (*hanging bracket*) should be used with this type of bench mark. *Brass* or *gunmetal bolts* are used as an alternative to the flush bracket OBMs, where the structure provides no suitable site for the flush bracket. These are 50 or 60 mm diameter mushroom-headed bolts set in solid rock or concrete (*Figure 5.3(c)*).

Brass rivets are an occasional alternative to the standard cut OBM, used where the bench mark must be located on a horizontal surface. If possible, a broad arrow is cut alongside (*Figure 5.3(d)*).

Pivot bench marks are used on horizontal surfaces such as soft sandstone, where the insertion of a rivet would break away the stone. Such an OBM actually consists of a small hole or depression cut to take a pivot, a steel ball bearing of ⅝ in (approximately 16 mm) diameter. In use, the pivot is placed in the depression and the staff held on top of the pivot. The published height refers to the top of the ball bearing (*Figure 5.3(e)*).

In Hong Kong and many other areas, the common form of government BM is a large metal staple cemented into walls.

5.3.3 Sources of OBM information

OS large-scale 1:1250 and 1:2500 maps and many large-scale plans produced by private survey companies and municipal authorities indicate bench marks by means of a broad arrow, the head indicating the plan position. The height is usually given and, occasionally, the type of mark. However, the printed OS maps did not always contain the latest available levelling information, and maps did not always identify the date of the levelling, the type of mark, or its height above ground level. The OS have therefore compiled *OBM lists*, available from OS Superplan Agents, giving complete and up-to-date levelling information.

A *bench mark list* is published for each square kilometre of the National Grid, covering the same area which would be represented on a 1:2500 OS map. The OBM list is identified by the NG reference of the square. For each bench mark, the list gives:

a brief description of the mark, with its location,
the full 10 m national grid reference,
the height of the mark, in metres and feet, with respect to Ordnance Datum,
the vertical distance of the mark above ground level, in metres and feet, and
the date of the levelling.

5.4 Types of levelling

There are two basic types of levelling, *precise* (or *geodetic*) and *ordinary* (or *simple*) *levelling*. The latter is often simply termed *levelling*, and both forms are known as *spirit levelling*, dating from the historic use of the spirit level.

5.4.1 Precise or geodetic levelling

This is the highest order of levelling work, with readings generally observed and recorded to decimals of a millimetre. This form is used for the basic levelling framework of a country, such as the establishment of fundamental bench marks. It is outside the scope of this book, but some information has been provided since it may be necessary to use the techniques in work such as, for example, checking on the deflection of beams, certain engineering works and irrigation schemes over very flat areas.

5.4.2 Ordinary or simple levelling

This term covers all levelling work which is not regarded as being precise levelling, with readings taken, at best, to 1 mm. Ordinary levelling may be categorized by its purpose or use, e.g. section levelling, area levelling, construction levelling, etc.

5.5 The principle of levelling

In all forms of levelling, the typical problem is that the height of one point above datum is known and it is required to find the reduced levels of other points with respect to the same datum. It will be evident that if a level surface or line is established and the vertical distances from all the points to the line are measured, a little simple arithmetic will enable the desired heights or reduced levels to be calculated.

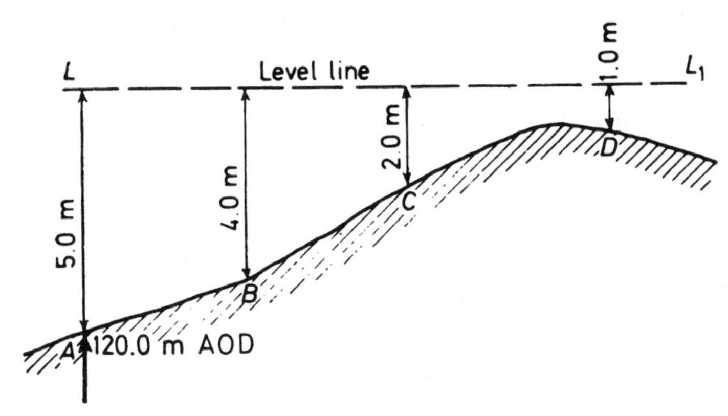

Figure 5.4

In *Figure 5.4*, L–L_1 is a level line, A is a point at 120 m AOD, and the heights of B, C and D above Ordnance Datum are required. The vertical distances from A, B, C and D to the level line are measured using a suitable graduated rod. The distances are, respectively, 5.0 m, 4.0 m, 2.0 m and 1.0 m. The reduced level of B is

120.0 + 5.0 – 4.0 = 121 m

the reduced level of C is

120.0 + 5.0 – 2.0 = 123.00 m or 121.0 + 4.0 – 2.0 = 123.0 m

and the reduced level of D is

120.0 + 5.0 – 1.0 = 124.0 m or 123.0 + 2.0 – 1.0 = 124.0 m

Note that the simple calculation may be done in either of two ways – the significance of these two procedures will be considered later.

In practice, it is not possible to establish a level line, but it is practicable to set up a horizontal line or plane through a point. Since at normal range (under 60 m or so) the horizontal and level lines through a point are almost indistinguishable, a horizontal plane or line is set up and the vertical distances are measured from the ground points to the horizontal line using a graduated rod.

5.5.1 Methods of obtaining a horizontal plane or line

The force of gravity may be used in several simple mechanisms to define a horizontal line.

5.5.1.1 The plumb line and cross-piece

A weighted pendulum, freely suspended, defines the direction of gravity. If a cross-piece is fixed accurately at right angles to the pendulum, a horizontal line is defined by sighting along the cross-piece, but the device is clumsy and inaccurate.

5.5.1.2 The water level

Variations on this system, mentioned in Section 5.1.3, were used by the Romans and into the seventeenth and eighteenth centuries, but the disadvantages for survey applications are evident.

5.5.1.3 The spirit level vial

The spirit level vial, first developed in 1666, is a simple, yet most effective, device for defining a horizontal line. In the early versions, a glass tube was bent to a slight curve, the tube then part-filled with fluid and re-sealed. The bubble formed in the tube seeks the highest part of the tube and when the bubble is centred in the length of the tube the longitudinal axis of the bubble is horizontal. This may be seen in any mason's level on a building site. In modern versions the tube is actually ground to a barrel shape rather than being a simple bent tube.

The spirit level vial may be attached to a straight-edge, allowing levels to be transferred over a metre or so. The distance can be increased by fixing sights to the straightedge, but the naked-eye range is very limited, particularly for reading graduations on a staff.

5.5.2 The surveyor's level

The obvious way to increase the sighting range of the mason's level was to attach the level vial to a telescope rather than to a straightedge, and to provide a diaphragm with a reticule pattern within the telescope to act as a 'sight' for aiming the line of sight or *collimation line*. The level vial axis (more properly the 'principal tangent to the level vial') must be arranged parallel to the collimation line, of course.

5.5.2.1 The surveying telescope

The telescope was invented in Holland in 1608, the form adopted for surveying instruments being that suggested by Johannes Kepler (1571–1630), known as the

Keplerian or *astronomical telescope*. The original telescope had an objective lens at one end and an eyepiece system at the other end, and the distance between them had to be varied according to the range to the object being viewed.

Figure 5.5 is a diagram of the essentials of a modern tilting level with an *internal focus telescope*. The telescope is a tube of fixed length, with an *objective lens* (1) at the end nearest the object being viewed and an *eyepiece lens system* (4) at the other. Within the tube is fitted an *internal focus lens system* (2), capable of being moved within the tube so as to permit focusing of the distant object. This movable lens is adjusted in position by the movement of a control wheel at the side in levels, and in theodolites often by movement of a sleeve fitted around the telescope barrel. Traditional telescopes provided an inverted image, but it is common practice today to include extra lenses so as to give an erect image, although with some consequent slight loss in light transmission. Telescope lenses are normally coated to reduce reflections and hence improve light transmission. For surveying purposes, the magnification and resolving power of a telescope are important considerations, magnification typically varying from 10× to 40× depending upon the class of instrument.

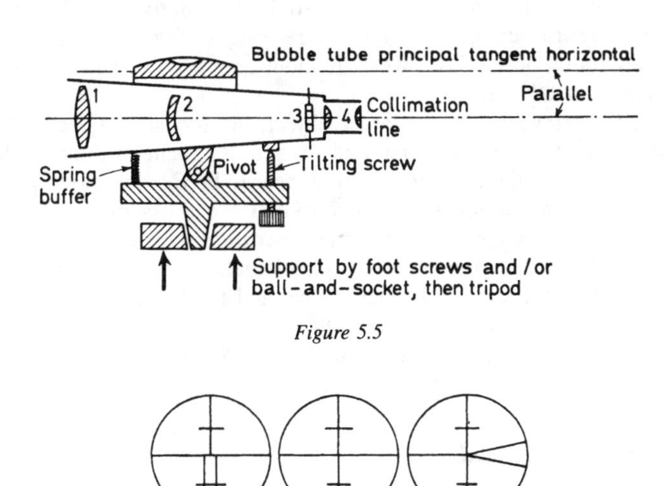

Figure 5.5

Figure 5.6

The sight line through the telescope is defined by a glass circle with engraved cross lines, inserted between the focusing and eyepiece lens systems. This *diaphragm* (3) bears a graticule or reticule of fine engraved lines, commonly called *cross-hairs*, since the original diaphragm consisted of a brass ring with spiderwebs stretched across it. Some common arrangements are shown in *Figure 5.6*. The short horizontal lines are *stadia hairs*, used for optical distance measurement, as in Chapters 7, 8 and 10. The intersection of the long horizontal and vertical lines defines the collimation line (sight line) of the telescope.

The telescope may be fitted with a pair of external sights, used for pointing the telescope initially towards the target to be sighted. This is important due to the narrow field of view of the surveying telescope, often only about 1.5° of arc or a field width of about 2.6 m at a distance of 100 m. If the telescope is pointed by looking towards the target (a levelling staff in this case) along the telescope body using the sights and then clamped, then when the telescope is focused the target staff should be visible within the field of view.

5.5.2.2 Telescope focusing

For accurate work when observing a target, both the cross-hairs of the reticule and the distant target must be in sharp focus and achieving this entails two separate operations. These are (a) to focus the eyepiece on the cross-hairs, and (b) to focus on the target object, i.e. bring the image of the target object into exactly the same plane as the cross-hairs.

To focus the *eyepiece* on the cross-hairs, the telescope is pointed towards a light background with the telescope racked out of focus, then the eyepiece is screwed fully in clockwise, then screwed out again anticlockwise until the cross-hairs appear sharp and black. To make quite certain, this point should be overrun slightly and then the eyepiece screwed in slightly again to reach the sharp black position. This adjustment depends upon the eyesight of the user and it should be set at the beginning of the day's work. It should not need re-adjustment unless a different observer takes over.

To focus the *telescope* on the target object, the focusing wheel or sleeve is rotated until the object appears clear and sharp in the plane of the reticule. When this is done correctly, the cross-hairs will appear to be fixed to the target and will not move across the target when the observer's head is moved vertically or laterally. If there is any apparent movement, *parallax error* exists and if it cannot be eliminated by use of the focusing wheel then the cross-hairs must be re-focused with the eyepiece as detailed above and the target focused again. The telescope must be re-focused each time the telescope is pointed to a new target, using the focusing wheel, since the position of the internal focusing lens must be varied as the distance to the target varies.

5.5.2.3 The levelling head

As described, the surveyor's level consists of a telescope with a spirit level vial attached to it in such a manner that the collimation (sight) line and the bubble tube axis are parallel. In use, the instrument is mounted on a tripod so that the telescope eyepiece is at a convenient height for the user.

In order to allow the instrument to be tilted until the spirit bubble is central in the length of its tube, thus making the bubble axis horizontal and therefore the collimation line horizontal, the base of the instrument incorporates a *levelling head*. The principal forms of levelling head used are the *three-foot-screw tribrach* and the *ball-and-socket mounting*. The choice as to whether to use an instrument with one or the other type of mounting is a matter of personal preference for the surveyor.

In use, the tripod is set up with its top plate approximately level, the instrument bolted to the tripod head, then the instrument manipulated so that the sight line is horizontal. The markings on a graduated staff are then observed through the telescope and give the vertical distance from the ground point to the horizontal collimation line.

An instrument with three-foot-screw mounting is shown in *Figure 5.7*, the third screw being hidden by the body of the instrument. The vertical axis of the instrument can be tilted in any vertical plane by appropriate rotation of one or more of the footscrews.

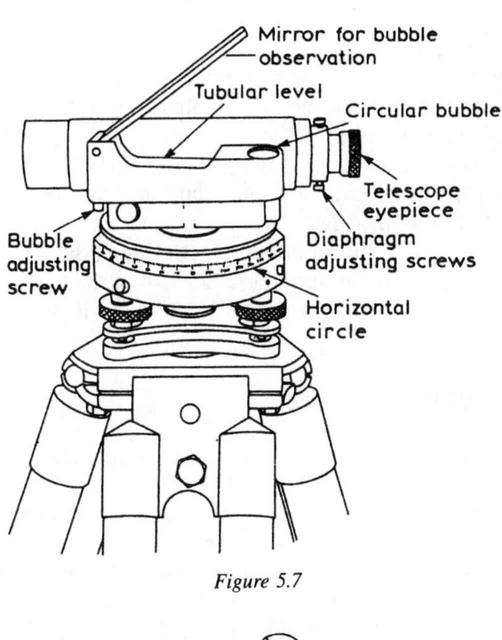

Figure 5.7

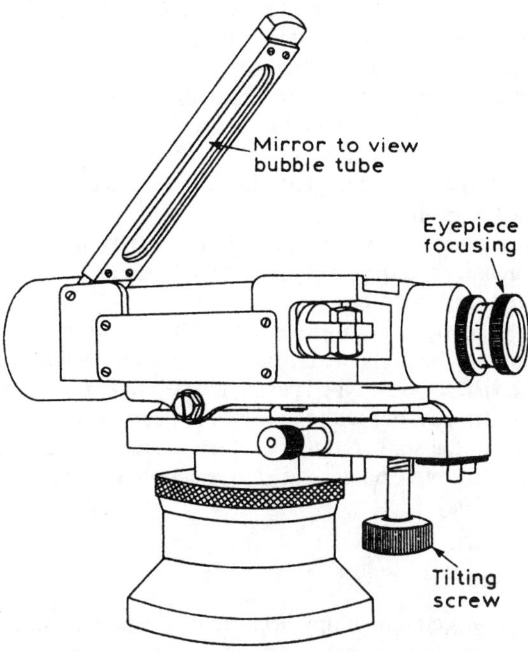

Figure 5.8

A ball-and-socket mounting is shown in *Figure 5.8*. This is similar to the mountings used on the head of camera tripods, thus the instrument can be moved in any direction in a vertical plane and clamped in the desired position with a locking ring.

5.5.2.4 The tripod

The modern tripod typically has three framed (crutch-like) timber legs, which may be rigid (fixed length) or telescopic, but metal tripods are also available. The rigid form is more stable in use, but the telescopic pattern is more adaptable, e.g. if the surveyor is very tall or short, or the ground slopes or undulates. The tripod generally has a captive bolt in the top plate, and this screws into the base of the levelling head to hold the instrument securely to the tripod. The foot of each leg is fitted with a pointed steel shoe, and a projecting lug which is used to press the shoe firmly into the ground. On very hard, smooth surfaces it may be necessary to use a triangular tripod support frame to prevent the legs sliding outwards.

5.6 Modern surveyor's levels

Two principal types of surveyor's level are in general use today, the *tilting level* and the *automatic level*. Occasionally a *dumpy level* may be encountered, but these are generally considered to be obsolete now and will not be described here. Most manufacturers have ceased making dumpies, and a number have even stopped making tilting levels, both of these being replaced by the automatic level.

The construction of an automatic level is basically similar to that of the traditional spirit levels described above, but it does not actually use a spirit level vial to determine the horizontal. Instead an automatic compensator arrangement of reflecting prisms is fitted within the telescope barrel, the prisms being arranged in the line of sight from the surveyor's eye to the distant staff. If the telescope is tilted from the horizontal by a small amount, the prism arrangement adjusts its position automatically to compensate for the deviation from the horizontal. The design of the compensating unit differs considerably between the various manufacturers' models.

The latest automatic levels are *digital levels*, where the staff reading can be scanned electronically and the surveyor need only set up the instrument and point it on the required staff positions.

Laser levels generate a beam of light which may be rotated in a horizontal plane. The beam may be invisible infrared, or a visible helium neon beam of red light. Where the visible beam type is used, the height of the beam can be read off a graduated staff directly, but with the invisible beam type a photoelectric sensor must be attached to the staff. Laser levels are mainly used in setting out works, and will be considered in Chapter 12. (*Laser = light amplification by stimulated emission of radiation.*)

5.6.1 The tilting level

This instrument replaced the dumpy because it is generally quicker in use due to the shorter time required to set it up and prepare it for use. The principal constructional features are shown diagrammatically in *Figure 5.5*.

The telescope, bubble tube and diaphragm of the tilting level are all as described earlier, and the levelling head may be of either the ball-and-socket or the three-foot-screw type. A tilting level with ball-and-socket mounting is sketched in *Figure 5.8* and an instrument with three-foot-screw mounting in *Figure 5.9*.

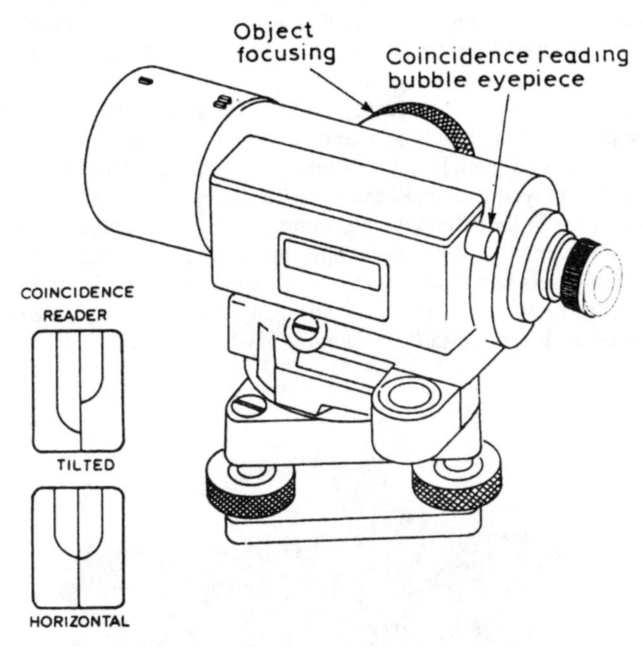

Figure 5.9

The telescope is not rigidly fixed to the levelling head, but instead is supported on a pivot which allows the telescope to be tilted at an angle to the levelling head – hence the name *tilting*. The bubble axis and the collimation line should always be parallel. In use, when the instrument is first set up at a point it is levelled-up roughly (i.e. the vertical axis made approximately vertical) by the footscrews or the ball-and-socket, as judged by a small circular spirit level attached to the lower part of the instrument, and then it is levelled exactly by the tilting screw *immediately before making each observation on the distant staff.*

The rough levelling by reference to the circular bubble takes very little time and is done each time the tripod is set up, but it is essential that the long bubble be centred by the tilting screw immediately before every observation. The tilting of the telescope at every sighting should have no effect on the height of the collimation line and the centring of the bubble takes only a second or two.

It will be evident that one condition is critical for work of any accuracy, and this is that the collimation line and the bubble tube axis must be parallel. This condition can be difficult to maintain and it must be checked and corrected at intervals. Such a correction is called a *permanent adjustment*, and when once made should remain correct for months or longer, depending on how the instrument is used. The actual adjustment, in practice, is generally made by moving the bubble tube in its mounting, using the adjusting screw(s) located at one end of the tube (the other end has a hinge mount).

Some levels have a clamp and slow-motion screw for controlling the rotation of the telescope about the instrument's vertical axis, while others simply use a 'friction-braked' vertical axis and a slow-motion screw. A simple horizontal circle of degrees may be fitted, useful when setting out the occasional horizontal angle. The tilting screw may be graduated in such a way that gradients can be set out and it is then termed a *gradienter screw.*

Tilting levels for higher accuracy work are generally equipped with a coincidence prism bubble reading system as shown in *Figure 5.9*. In this system, the bubble tube is totally enclosed, and images of half of each of the two ends of the bubble are displayed in a small eyepiece beside the telescope eyepiece. When the two images are in line, the bubble is central. This allows much more accurate levelling of the sight line than the simple open bubble tube. The very highest accuracy tilting levels used for geodetic levelling, such as the Wild N3 precision tilting level illustrated in *Figure 5.10*, are also equipped with a *plane parallel plate micrometer*, described later in Section 5.8.1.

Figure 5.10

5.6.2 The automatic level

The essential features of this instrument are shown in *Figure 5.11* and a good example is the Wild NAK2 shown in *Figure 5.12*.

The telescope is rigidly fixed to the levelling head, with its horizontal axis at right angles to the vertical axis of rotation of the instrument, and the levelling

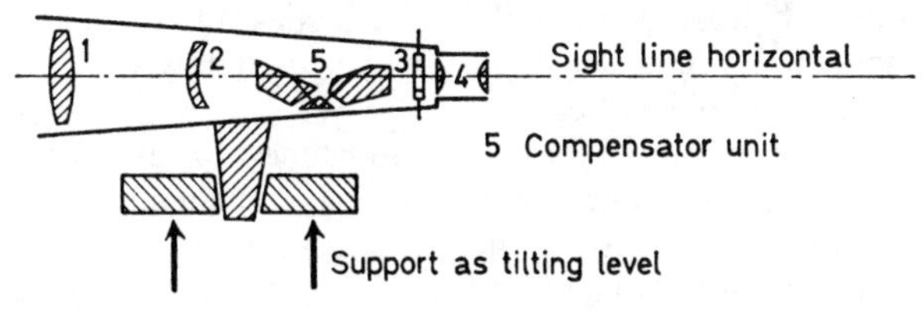

Figure 5.11

Figure 5.12

head may be either a ball-and-socket arrangement or the three-foot-screw type. Like the tilting level, a small circular bubble is fitted for levelling the instrument up approximately after it has been attached to the tripod. Once the instrument has been approximately levelled-up, the compensator unit operates automatically, its movement being slowed down by damping mechanisms.

When the telescope is aimed in the required direction, a horizontal ray of light entering the centre of the objective lens (1) is passed through the internal focus lens (2) then through a system of fixed and suspended prisms (5) and is directed by these to the centre of the cross-hairs in the diaphragm (3), where it is observed through the usual eyepiece (4). Individual manufacturers' arrangement of the mechanisms vary, but in general, provided the telescope is levelled up initially to within ±10 minutes of arc of the horizontal (as can be achieved using the circular bubble, since these typically have an accuracy of ±8 or 10 minutes of arc per 2 mm run), then in whatever direction the telescope is turned, the cross-hairs will sweep out a horizontal plane of constant height.

Despite the damping device, vibrations due to blustery weather conditions or perhaps traffic or site plant, may make sighting difficult. This vibration may be restricted by laying a hand lightly on the tripod, but it must be remembered that this should never be done with a tilting level, since it would disturb the level bubble setting.

Some early instruments had problems with excessive friction at the pivots of the suspended prisms, but these appear to have been overcome now. Some makes of automatic level are fitted with a push button – if this is pressed immediately before taking the reading, the compensator is given a small push to check that it is moving freely. If there is a possibility of a sticking compensator with other makes of automatic level, the tripod may be tapped lightly with a finger to check that the compensator swings freely.

There is usually only one permanent adjustment to an automatic level, to ensure that the collimation line is horizontal when the instrument is set up. The mode of adjustment depends on the manufacturer. In some cases there are diaphragm adjusting screws and it is simply a matter of moving the diaphragm up or down with these, thus changing the position of the collimation line within the telescope. In other instruments an adjustment must be made to the actual compensator unit itself. In all cases, the instrument handbook should be consulted.

Fieldwork may be carried out about twice as fast with the automatic level as compared with the tilting level and fewer mistakes occur because there is no bubble to be continually checked and adjusted. Automatic levels are generally more expensive than tilting, but their advantages are such that they may be expected to replace the others completely and indeed, some manufacturers produce only automatic levels now.

5.6.3 The digital level

The digital electronic level, illustrated in *Figure 5.13*, first introduced by Wild in 1990, is essentially an automatic level, and may be used as an ordinary optical automatic level with the usual type of levelling staff. However, the telescope incorporates a beam splitter which separates the light entering the telescope into its infrared and visible components. The infra-red component is deflected to a detector and if a special bar-coded staff is used then an electronic image-processing system will compare the scanned staff image with the image of the staff pattern stored in the instrument and finally give a digital display of the staff reading and the horizontal distance. The time taken for one observation may vary from 4 ms to 2 s, depending upon the illumination available.

Provided the surveyor has levelled up the instrument, pointed it on the staff, and focused the telescope, then the reading of the staff, the recording of the measured data and the computation of the levelling results can all be carried out automatically, eliminating errors in reading and recording the data, and also

Figure 5.13

speeding up the levelling fieldwork. The instrument incorporates software to carry out all these operations, and the data may either be displayed for recording by the surveyor, or stored in a plug-in data recorder module for later transfer to a computer, or transmitted directly to a computer by an RS232 interface. Apart from routines for normal line levelling with backsights, intersights and foresights, the software includes routines for setting-out and also a check module for use in the *two-peg test* described later.

Heat shimmer reduces the contrast of the staff image, and in such an event the Wild instrument's *repeat measurement program* can be used to take multiple readings and improve the accuracy of the reading. The instrument then displays the mean observed value, the standard deviation of the set of readings, and the number of measurements taken. If part of the staff is obscured, as with overhanging branches, it may still be possible to level, since the instrument takes an image of a discrete length of the staff, not merely the division at one point.

5.6.4 Classes of level

In addition to levels being known as dumpy, tilting, automatic or digital, they may be classified according to a theoretically obtainable accuracy in a double run of levels over a specified distance, typically one kilometre. The three main classes are Class I, Class II, and Class III, otherwise described as precise or geodetic levels, general purpose or engineer's levels and construction or builder's levels. (Hence descriptions such as 'Builder's Automatic Level', 'Engineer's Tilting Level', etc.)

Table 5.1 Data on Leica Wild levels

Level model code	*Standard deviation per 1 km double run levelling*	*Magnification*	*Clear objective aperture*	*Setting accuracy – compensator or bubble tube*	*Description*
NK01	10 mm	19×	25 mm	± 20 seconds	Builder's Dumpy
NAK2	0.7 mm	32× or 40×	45 mm	± 0.3 seconds	Universal automatic
With pppm	0.3 mm				
N3	0.2 mm	11× to 40×	52 mm	± 0.2 seconds	Precision tilting
NA2002		28×	36 mm	± 0.8 seconds	Digital level
Electronic levelling	Using Invar staff 0.9 mm Using glassfibre staff 1.5 mm				
Optical levelling	Any staff 2.0 mm				

Notes: (1) Pppm = Plane parallel plate micrometer.
(2) The N3 has a panfocal telescope, hence magnification varies with the focusing distance.
(3) The letter K in the model code indicates that the instrument is fitted with a horizontal circle of degrees.

Where a manufacturer quotes a standard deviation figure of less than 1 millimetre per double run of levels over a kilometre, then it may be considered to be a precise or Class I level. Between 1 and 5 mm would indicate an engineer's level, and 5 mm or more a builder's level. It should be noted, however, that on site a good set of unadjusted field readings may have a misclosure of up to 2.5 or 3 times the manufacturer's quoted figures.

Table 5.1 shows the significant data of a range of Leica Wild levels.

5.7 The levelling staff

As explained earlier, the vertical distance from the ground point to the collimation line is measured with a graduated rod. This rod is termed a *levelling staff* or *levelling rod*.

5.7.1 Ordinary levelling staves

These are made of wood, aluminium alloy, or glassfibre, usually between 2 and 5 m in length, the construction being telescopic, or rigid one-piece, or hinged, or even multiple socket-jointed sections. Since timber swells when wet, telescopic timber staves should not be placed in water as the sections will tend to stick and can then be neither closed nor extended. Telescopic staves are fitted with a brass spring catch to keep the extended section in position and this catch must be checked to ensure that it 'clicks home' and locks the sections properly. Aluminium staves are not unduly heavy, they are resistant to water and are strong and durable.

A variety of graduation patterns are available, the commonest being the 'E' pattern specified by BS 4484: Part 1: 1969 and the 'E-and-checkerboard' patterns popular in Europe, both shown in *Figure 5.14*. These types of staff have 10 mm graduations, read by estimation to 1, 2, or 5 mm, generally figured in black on white, although sometimes alternate metres are in red on white. Some users prefer staves with a yellow background, as this can provide a better contrast than white. Reflective facings are available, for night work with appropriate illumination.

These relatively coarse graduations are used because of the wide range of distances over which they can be read in ordinary levelling.

5.7.2 Precise staves

These are made of a strip of invar steel supported by a wooden or metal frame. In precise levelling, the sighting distances are short and it is usual to graduate staves in the same way as scales, as shown in *Figure 5.15*. Note that the precise staff has two sets of graduations, offset from one another, to enable a check observation to be made. Precise staves with a single set of graduations are also obtainable.

5.7.3 Digital levelling staves

Leica supply a glassfibre and plastic combination staff for use with the NA2002 digital level. This is made in three socketed 1.35 m lengths which may be used

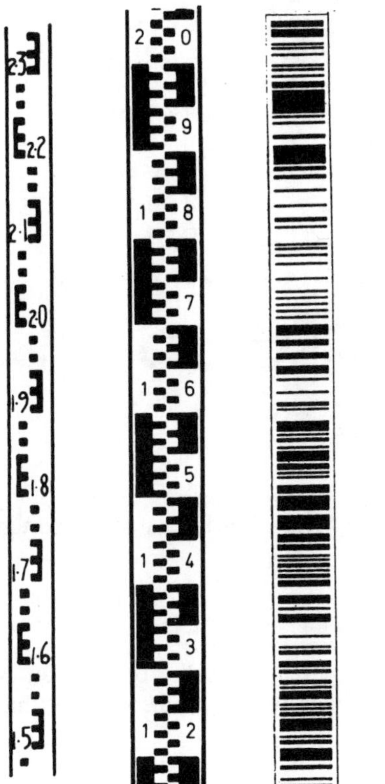

Figure 5.14

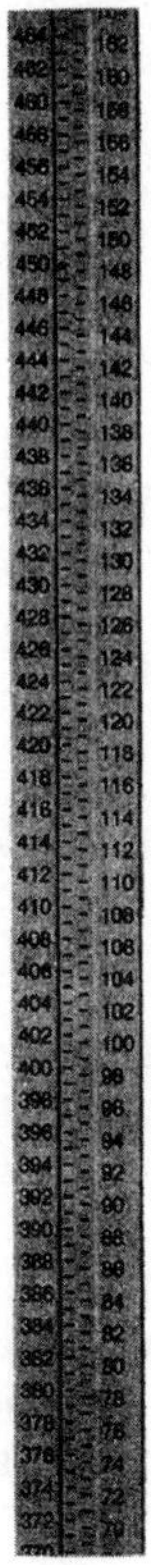

Figure 5.15

to give a staff of 1.35, 2.70 or 4.05 m length as required. One face of the staff is binary (bar) coded for electronic levelling, the other face has the conventional metric staff graduation for optical levelling. Binary-coded invar staves are available for precision levelling. A binary-coded staff pattern is shown in *Figure 5.14*. Similar staves are made by other manufacturers.

5.7.4 Other staff types

A variety of other types of staff have been made, the most notable probably being the Philadelphia staff, fitted with a target and vernier, used for long sights as in river crossings.

5.7.5 Levelling staff accessories

5.7.5.1 The staff support or change plate

This is a triangular steel plate with a raised centre and the three corners turned down, used to support the staff on soft ground and prevent it sinking. Also known as a shoe or a crow's foot, it is generally fitted with a length of chain for carrying

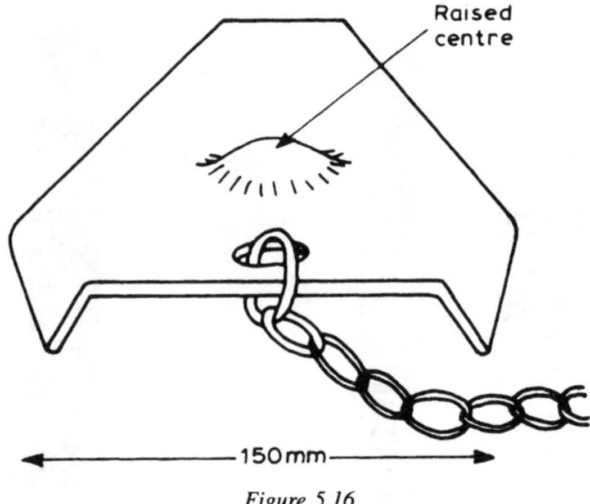

Figure 5.16

and for pulling it free from the ground (see *Figure 5.16*). A large, round-topped stone provides a suitable alternative.

5.7.5.2 The staff bubble

This is a circular spirit level, used to check the verticality of the staff when making observations. It is attached to the back or side of the staff, or may be hand-held against the staff.

5.7.5.3 The hanging bracket

This is used with an OS bronze flush bracket BM, as shown in *Figure 5.3(b)*.

5.7.5.4 Handles and steadying rods

These are usually supplied with precise staves. Handles are sometimes permanently attached, or may be clamped on to all types of staff. Steadying rods are normally used only in precise work, the staff holder holding a handle and a steadying rod with each hand so that a tripod-like structure is formed, with the staff vertical. In ordinary work, in windy conditions, it may be helpful to use ranging rods as steadying rods.

5.8 Level accessories

5.8.1 The plane parallel plate micrometer

Even with a good telescope and a staff marked in fine divisions, staff readings cannot be made finely enough by simple telescope for precision levelling demanding accuracies such as 0.5 mm per km or so. The *plane parallel plate micrometer* is an attachment to a level telescope which typically permits the determination of level staff readings to 0.1 mm directly and by estimation to 0.01 mm (0.000 01 m).

The device is simply a piece of glass with parallel plane faces, placed in front of the telescope objective and supported on horizontal pivots with the plane faces at right angles to the collimation line. Since glass refracts a ray of light entering it, rotation of the parallel plate causes the collimation line to be raised or lowered while still remaining parallel to its original path. The physical constants of the glass being known, the vertical displacement of the collimation line can be calculated for a known tilt of the plate. The plate is tilted by a micrometer screw which registers the displacement of the collimation line rather than the amount of tilt.

The simplest version, often used as an attachment to a level, has a displacement scale engraved on the edge of the micrometer screw operating the plate. When the device is permanently 'built-in' to the level, the plate is generally linked up to an optical scale viewed in an eyepiece alongside the telescope eyepiece.

Figure 5.17 shows the system used on a precise level. The total vertical displacement possible is 10 mm, and the eyepiece scale is graduated 0, 1, 2, ... 10, each number representing 1 mm of vertical displacement.

Each division is further subdivided into ten parts of 0.1 mm, and these may be subdivided by eye to 0.01 mm. Operation is extremely simple – after carefully

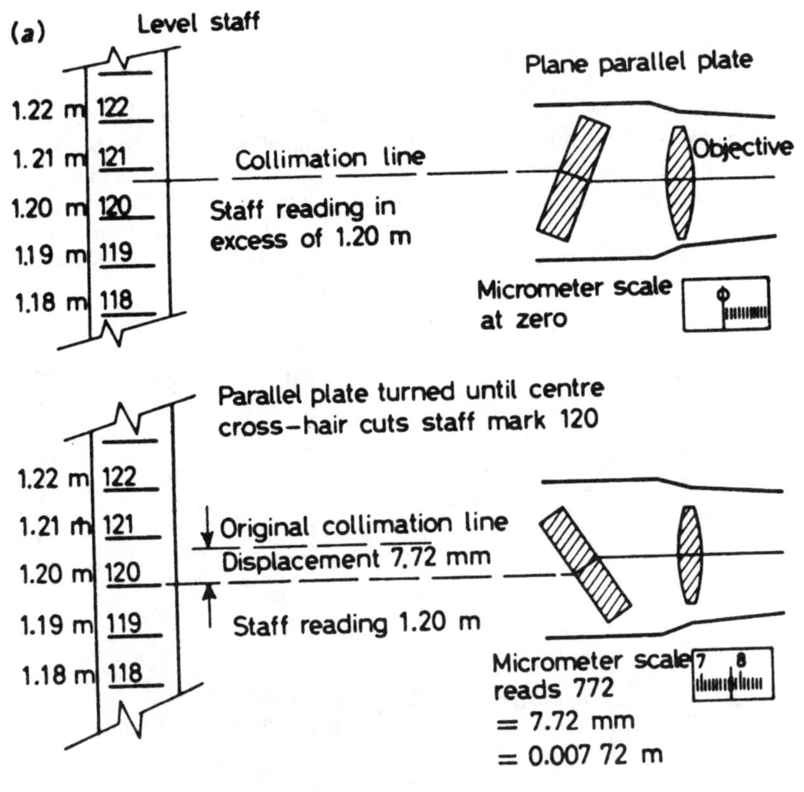

Figure 5.17

focusing and levelling the instrument, turn the micrometer screw until the central horizontal cross-hair cuts a 10 mm mark on the observed staff, note that reading and add on the reading from the micrometer scale. On drum instruments, take the reading from the edge of the micrometer drum.

5.8.2 Eyepiece accessories

A variety of eyepieces are available for level telescopes, allowing the magnification of the telescope to be varied according to the conditions. A *laser eyepiece* may be attached to some levels, giving a visible collimation line on targets and staves.

5.8.3 Staff illumination

The Wild GEB89 Staff Illuminator allows measurements to be made on survey staves regardless of the lighting conditions, indoors, in tunnels, etc. It is battery powered and can be attached to the staff to light up 1.4 m of the staff length, allowing readings at distances of up to 40 m with digital levels and 60 m with optical levels. Some manufacturers provide a form of torch which can be clipped to the telescope barrel to direct a beam of light along the collimation line.

5.9 Levelling fieldwork

All levelling operations consist of observing and recording height readings at two or more staff positions from each instrument station. The work may involve only one instrument station, or there may be several instrument stations involved. Each instrument station with its associated set of staff readings may be completely independent of any other set of readings from another instrument station, but more often separate sets are linked by observations on

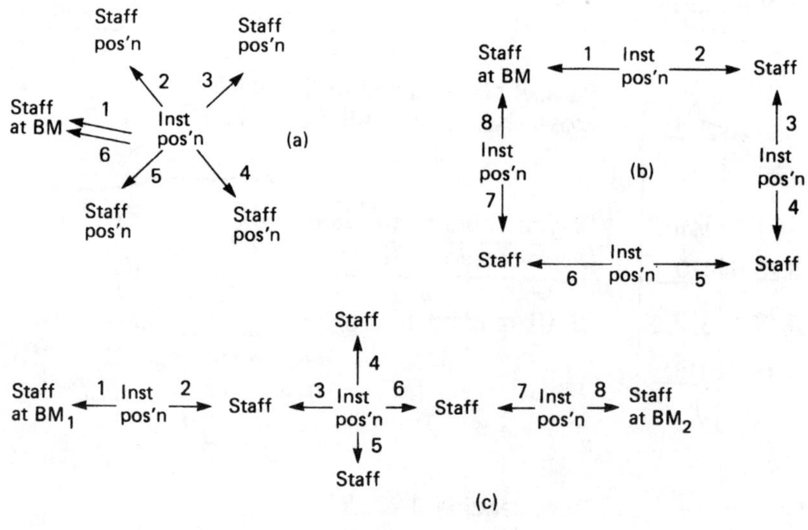

Figure 5.18

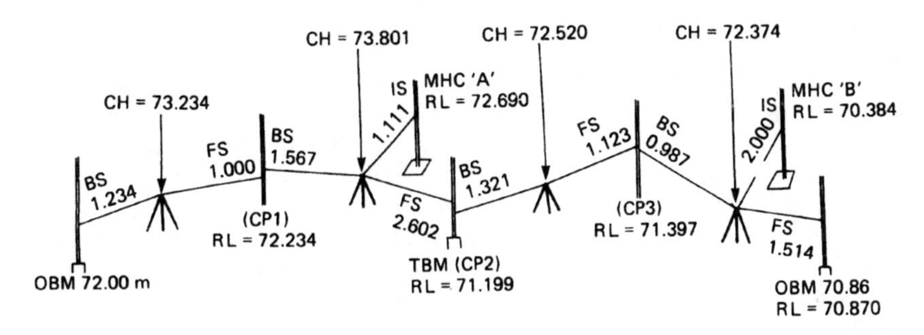

Figure 5.19

a staff position common to both sets. Examples of these situations are illustrated in *Figures 5.18* and *5.19*. *Figure 5.18(a)* illustrates what is sometimes known as *one set-up levelling* and the other figures as *series levelling* or *lines of levels*. A line of levels run simply to check the accuracy of the work is also known as *flying levels*.

Levelling operations may include establishing TBMs, finding heights along the line of a section, fixing height pegs on a construction site, checking a particular gradient, etc. While tasks may differ in the actual detail of distances to BMs, size of job, labour and equipment available, and so on, in each case the principles of the operations are essentially the same.

5.9.1 Terms used in levelling

Some levelling terms were defined in preceding sections of this chapter. The following should now be noted.

Backsight: the first reading taken from an instrument station.

Foresight: the last reading taken from an instrument position.

Intermediate sights: readings which are neither the first nor the last to be taken from an instrument position.

Changepoint (CP): a staff position on which first a foresight reading from one instrument position and then a backsight reading from another instrument position are taken.

Collimation height: the calculated height of the line of collimation above or below the datum surface.

Rise and fall: the vertical distance between two consecutive staff positions is either a rise or a fall, a *rise* being a positive difference (the second point being higher than the first) and a *fall* being a negative difference (the second point lower than the first).

These last two terms are also used to identify the two methods commonly used for calculating reduced levels, as will be seen in Section 5.9.7.

5.9.2 The location of staff and instrument positions

The location of staff and instrument positions will vary with the task to be carried out, the equipment in use and the climatic and environmental conditions. However, the instrument and staff must always be set up on firm ground if this is at all possible and the task should be carried out with as few instrument stations as practicable, within the limitations on observing imposed by the length of the levelling staff and the horizontal distance. Distances between instrument and staff should be kept uniform, as far as possible, especially when levelling over long distances, or levelling up or down steep gradients, or when establishing temporary bench marks.

The purpose of equalizing these distances is to minimize instrument errors and also to reduce the effect of the Earth's *curvature* and the *refraction* ('bending') of light by the Earth's atmosphere. In most surveying tasks the effects of curvature and refraction are generally so small that they may be ignored for practical purposes. The combined error due to refraction and curvature in levelling is approximately 24 mm at a distance of 600 m and 0.25 mm at 60 m distance.

The ideal length of sight between instrument and staff is from 45 to 60 m, longer sights tending to lead to inaccuracies in reading and shorter sights implying more instrument stations and a more costly job in terms of time and money. The length of the levelling staff will impose restrictions on the length of sight on steep slopes, especially if equal sight lengths are to be maintained. Further, it is best to avoid reading the lower 0.2 m of the staff since refraction has its greatest effect near the ground, particularly in precise work. *Grazing rays* (sight line skimming the ground) should be avoided when sighting over the crest of a hill. To maintain equal sights, it may be necessary to level up or down hill in a zig-zag pattern in plan, or alternatively to select the high spots, if any, as instrument stations.

5.9.3 Setting up the level and tripod

The operations involved in setting up the level and tripod are termed the *temporary* or *station adjustments*.

5.9.3.1 Setting up

The tripod must be set up with the feet of the legs about one metre apart, roughly forming an equilateral triangle in plan, and so that it is stable and the eyepiece of the instrument is at a convenient height for the observer. The tripod feet should be well pressed into the ground and on slopes, one leg should point uphill for stability. The legs should be oriented in such a way as to avoid the surveyor having to straddle a leg while observing, and the top must be approximately horizontal. In very hot climates it should preferably be set up in a shady position, or a survey umbrella used for protection of the instrument. Normally no attempt is made to set the tripod exactly over a particular ground mark since this is not required in levelling. However, if the level has a horizontal circle it can be used to set out or measure the occasional horizontal angle, and in this case it will have to be centred over a point. In this case, refer to Chapter 6 on the theodolite.

The level is usually attached to the tripod head by some form of captive bolt in the tripod head, screwed up into the underside of the level. On older levels, the top of the tripod may be threaded and the whole instrument screwed onto the tripod head.

5.9.3.2 Levelling-up

To level up tilting and automatic levels after they have been attached to the tripod, with the top plate of the tripod approximately horizontal, it is only necessary to get the small circular bubble approximately central.

If the levelling head is of the three-foot-screw type, then the footscrews should be used as needed to centre the bubble. The simple way to do this is to try to visualize the imaginary line passing through the centre of the circular bubble and also through the centre of the bubble's casing, then manipulate the footscrew nearest to this line to move the bubble towards the centre, repeating the operation as necessary.

Where a ball-and-socket levelling head is fitted, the clamping ring or fastening screw, as appropriate, should be eased off with one hand while the other hand holds the telescope and tilts it as necessary until the circular bubble is central. Finally, the ball-and-socket should be clamped without disturbing the centring of the bubble.

In the case of an automatic level, if the instrument is in adjustment the compensator unit should now ensure that the line of collimation is horizontal. With the tilting level, the tilting screw must be used to centre the main bubble immediately before taking each reading.

When using an automatic level for precise work, it may be sensible to assume that the compensator unit has some residual error causing the line of sight to be inclined to the horizontal. This can then be considered to be a systematic error, eliminated by an appropriate observing technique, provided that the small circular bubble is always levelled in the same manner; for example, the final adjustment of the circular bubble always being made towards No. 1 staff. (In precise work, the use of two levelling staves is recommended.)

5.9.3.3 Focusing and the elimination of parallax

These are as described earlier in Section 5.5.2.2.

5.9.4 Observing the staff

The procedure to be used depends upon the type of instrument in use, as shown in the following sections.

5.9.4.1 Tilting level

Point the telescope carefully at the staff by aiming over and along the telescope barrel. Look through the telescope, then focus carefully on the staff as described earlier. Turn the telescope in the horizontal plane until the vertical cross-hair bisects the staff. Centre the longitudinal bubble using the tilting screw. Levels without a coincidence prism bubble reading system generally have a mirror (at 45° above the bubble tube) in which the bubble may be viewed, as shown in *Figure 5.8*. The mirror should be used for all bubble viewing, since its actual position and its apparent position as viewed in the mirror are not identical.

Read and record the staff graduation at the centre horizontal cross-hair, then check the reading again.

(If the instrument is fitted with a horizontal clamp and slow motion screw, it will be necessary to apply the clamp after pointing the staff, then bisect the staff using the slow motion screw.)

5.9.4.2 Automatic level

Aim, focus, bisect the staff as above. If the instrument is fitted with a push button, press the button immediately before taking the reading, to check that the compensator is free. If there is no push button, the tripod may be tapped lightly – the image in the telescope should move vertically slightly but stabilize quickly. Read and record the staff graduation.

5.9.4.3 Digital level

A digital level may be used in the same way as a normal optical automatic level as described above. When used for electronic levelling, the operations required will vary with the different manufacturers and models, and the maker's handbook should be consulted. Even in electronic levelling, however, the telescope must be pointed and focused by the user before the measuring cycle can start.

5.9.4.4 Telescope reticule and stadia measurement

The upper and lower horizontal *stadia hairs* visible in the reticule may be used to determine the distance between the instrument and the staff, correct to within 0.2 m over the recommended maximum sighting distance of 60 m, an accuracy more than adequate for maintaining equal sight lengths. The instrument makers have positioned the horizontal lines of the reticule so that they are equally spaced and in such a manner that if the staff is read against all three hairs, then the horizontal distance from instrument to staff is normally (but not invariably, check the handbook) equal to 100 times the staff intercept between the upper and lower hair readings (*Figure 5.20*).

It should be noted that, except in precise levelling, the sighting distances are not checked in this way at every station, rather it is customary to rely on the staff holder to ensure equal distances by pacing. The stadia hairs may also be used, particularly for an uncertain beginner, to provide a check on the centre hair reading, since the average of the upper and lower hair readings should be equal to the centre hair reading.

In precise work it is usual to observe as follows:

No. 1 staff, stadia hairs, centre hair
No. 2 staff, centre hair, stadia hairs
No. 2 staff, centre hair
No. 1 staff, centre hair.

In all cases in precise work, of course, both the staff and the parallel plate are read and recorded. This technique gives a check on both the distances and the centre hair readings. It will be noted that the centre hair readings on both occasions are read consecutively to minimize errors resulting from the staves or the instrument sinking or rising.

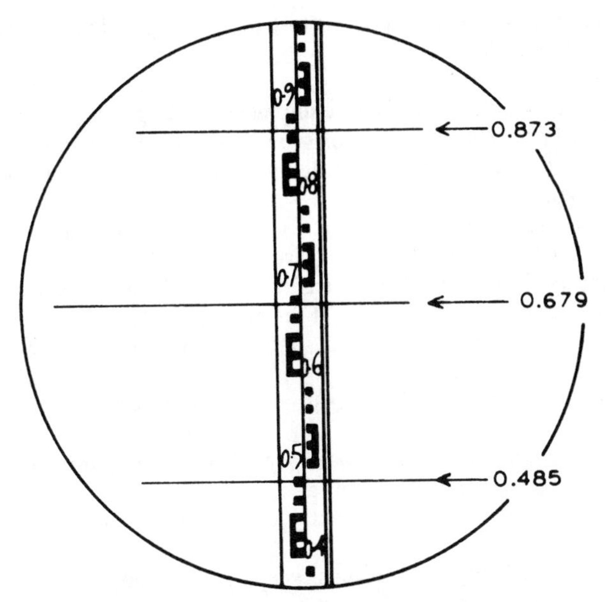

1 Has the centre hair been read correctly ?

$$\begin{array}{r} 0.873 \\ + \underline{0.485} \\ 1.358 \end{array} \div 2 = 0.679$$

= centre hair

2 What is the horizontal distance ?

100 x (0.873 − 0.485)

= 100 x 0.388

= 38.8m

Figure 5.20

5.9.5 Moving the level to a new position

When an instrument is to be moved to a new position, any hinged mirror should be closed down, any footscrews centred in their run, and the bolt holding the instrument to the tripod head checked for tightness. Thereafter the level and tripod may be carried together to the new position. Makers recommend that an instrument is carried by inserting a shoulder between the tripod legs and gripping one leg, so that the other legs hang freely and the weight is taken by the shoulder.

5.9.6 Duties of the staff holder

A good staff holder is essential if the surveyor is to carry out the levelling accurately and quickly and such an assistant will often anticipate the surveyor's requirements. Where possible, the surveyor should advise the staff holder, before the commencement of the levelling, as to where staff positions will be required. The staff holder should:

ensure that the staff and ancillary equipment are in good working order;
lay the staff on a clean, dry area of ground when it is not in use;
carry the staff between stations, holding it vertically and retracted if it is a telescopic type;
check before setting up the staff that its zero end and the raised centre of the staff support (if used) are free from dirt;
select a staff position which may be easily re-located later, if requested to do so;
use a staff support, or a large stone, if the ground is not as firm as the surveyor would wish for a staff position (a staff support should always be used in precise work);
extend or clamp the staff as necessary, ensuring that this has been done correctly;
understand the simple system of hand signals used for communication in the field (similar to tape and offset survey);
place the zero end of the staff gently on to the selected staff position, as required;
hold the staff vertically for readings, keeping the hands off its face so as to avoid dirtying it or obstructing vision;
check that the staff remains vertical during readings by occasionally glancing at the staff bubble;
move the staff or its support only when instructed to do so;
inform the surveyor if the staff (or its support) is displaced or tending to sink (possibly under its own weight, or through the staff holder inadvertently leaning his weight on it);
maintain equal distances (within about a metre) between an instrument and its foresight and backsight staff positions by careful pacing; and
wipe clean and dry off the staff and ancillary equipment upon completion of the day's levelling.

5.9.7 Level booking, recording and reducing field observations

For the majority of levelling tasks readings are recorded in a ruled level book, the form of ruling depending on the surveyor's preference. The standard patterns are *Rise and Fall*, and *Collimation Height*. In precise work a modification of the Rise and Fall form is commonly used so as to enable the stadia hair and parallel plate micrometer readings to be booked. Because the recording and reduction of levelling observations is a relatively simple task, it is possible to program programmable pocket calculators, or hand-held computers, to store and compute the data. A good program should prompt the user as to the data to be entered and, at completion, determine the misclosure and display or print out the reduced levels. There are a variety of data recorders available from the different manufacturers, together with appropriate software.

5.9.7.1 The Rise and Fall method

The line of levels shown diagrammatically in *Figure 5.19* is used to illustrate this method in *Figure 5.21*. The following points should be noted in *Figure 5.21*.

Date __________ Levels taken for __________

From __________ To __________

BACK SIGHT	INTER-MEDIATE	FORE SIGHT	RISE	FALL	REDUCED LEVEL	REMARKS
1.234					72.000	OBM St Johns Church
1.567		1.000	0.234		72.234	CP1
	1.111		0.456		72.690	MHC 'A'
1.321		2.602		1.491	71.199	TBM (CP2)
0.987		1.123	0.198		71.397	CP3
	2.000			1.013	70.384	MHC 'B'
		1.514	0.486		70.870	OBM (70.86) 'The Ring Of Bells' 'PH'
5.109		6.239	1.374	2.504	72.000	
6.239			2.504		−1.130 ✓	
−1.130 ✓			−1.130 ✓			

Figure 5.21

Each line of the book represents a staff position, and that position is identified in the 'Remarks' column. The first reduced level is entered from the given data, the remainder being calculated as follows:

Obtain the rises or falls, thus

1.234 – 1.000 = + 0.234 = rise
1.567 – 1.111 = + 0.456 = rise
1.111 – 2.602 = – 1.491 = fall
1.321 – 1.123 = + 0.198 = rise
0.987 – 2.000 = – 1.013 = fall
2.000 – 1.514 = + 0.486 = rise

Note carefully where these calculated values have been entered on the booking sheet, also note that while each backsight or foresight is used only once, each intermediate sight value is used twice.

Using the calculated rises and falls, calculate the reduced levels in succession from:

Reduced level = reduced level on previous line + rise between the two staff positions, *or* – fall between the two staff positions.

e.g. 72.000 + 0.234 = 72.234
72.234 + 0.456
= 72.690
72.690 – 1.491
= 71.199
71.199 + 0.198 = 71.397, etc.

The calculations may be summarized as:

(BS1 or IS1) – (FS2 or IS2) is a rise if positive *or* a fall if negative
RL1 + rise2 *or* – fall2 = RL2.

The commonest mistakes which occur in level booking and reduction are arithmetical, hence every arithmetic operation must be checked. If the calculations above are correct, then:

> The sum of the backsights – the sum of the foresights = the sum of the rises – the sum of the falls = the last reduced level – the first reduced level.

Note that in practice, the correct routine is to calculate the rises and falls, then compare the difference between the backsight and foresight sums with the difference between the sum of the rises and the sum of the falls, to ensure that the rises and falls have been accurately computed. Only when this arithmetic has been checked should the reduced levels be calculated. Finally, the checking should be completed by comparing these differences with the difference between the first and last reduced levels. The Rise and Fall method provides a complete check on the arithmetic of the reductions, but it must be appreciated that it does not check the accuracy of the actual observations. These may be checked or 'proved' only by levelling back to the opening bench mark or completing the line of levels on to another point of known height and comparing the calculated and known heights.

It will be noted that in the above example there is a *misclosure* of 10 mm, since the calculated reduced level at the end of the line of levels is 70.870 m, while the given height is 70.86 m. The method of dealing with this is described in Sections 5.9.7.5 and 6.

5.9.7.2 The Collimation Height or Height of Instrument method

In this method, the level bookings are exactly the same as for the Rise and Fall method, the difference lying in the method of reduction. Again, the line of levels in *Figure 5.19* is used in *Figure 5.22* to illustrate booking and reduction by this method.

As before, each book line represents a staff position, that position being identified in the 'Remarks' column. The first reduced level is again entered from the given data, but the remainder are calculated as follows:

> Reduced level + backsight = *collimation height*, and collimation height – foresight (*or* intermediate sight, as relevant) = reduced level.

Thus
72.000 + 1.234 = 73.234 (inst ht)
73.234 – 1.000 = 72.234 (RL of CP 1)
72.234 + 1.567 = 73.801 (inst ht)
73.801 – 1.111 = 72.690 (MHC 'A')
73.801 – 2.602 = 71.199 (RL of TBM)
71.199 + 1.321 = 72.520 (inst ht)
72.520 – 1.123 = 71.397 (RL of CP 3)
71.397 + 0.987 = 72.384 (inst ht)
72.384 – 2.000 = 70.384 (MHC 'B')
72.384 – 1.514 = 70.870 (RL of OBM)

BACK	INTER-MEDIATE	FORE	COLLIM-ATION	REDUCED LEVEL	REMARKS
1.234			73.234	72.000	OBM St Johns Church
1.567		1.000	73.801	72.234	CP1
	1.111			72.690	MHC 'A'
1.321		2.602	72.520	71.199	TBM (CP2)
0.987		1.123	72.384	71.397	CP3
	2.000			70.384	MHC 'B'
		1.514		70.870	OBM (70.86) 'The Ring of bells'
5.109		6.239		72.000	
6.239				-1.130	✓
-1.130					
	Check on intermediate sights				
	3.111	6.239	438.124	428.774	
				3.111	
				6.239	
				438.124	

Figure 5.22

Again, the position where these values have been entered on the booking sheet should be noted carefully. The calculations could be summarized as

RL + BS = CH
CH1 − FS2 = RL2, or
CH1 − IS3 = RL3

As always, the arithmetic should be checked. The difference between the sum of the foresights and the sum of the backsights should equal the difference between the first and last reduced levels. If this is so, it checks the calculation of the changepoint reduced levels, but it does not check the calculation of the reduced levels of the intermediates.

A method is available for checking the intermediate reductions, the rule being 'The sum of each collimation height multiplied by the number of reduced levels obtained from it is equal to the sum of all the intermediate sights, foresights and reduced levels excluding the first reduced level' (see *Figure 5.22*).

This check is so tedious that it is doubtful if it is used at all in low accuracy work and precise work is reduced by the Rise and Fall method. The Collimation Height method is widely used in ordinary building works and since this is often levelling carried out from a single instrument station then collimation height reduction is fast and easy. The commonest errors in levelling, however, are arithmetical, particularly with individuals who seldom use a level. Such persons and beginners, would be best to use the Rise and Fall method to eliminate such mistakes.

5.9.7.3 *Checking level entries extending over more than one page*

Where level book entries extend over more than one page, each page should be separately checked when reducing. If each page is checked (and this is strongly recommended), then three or four lines must be left clear at the bottom of each page or booking sheet to allow for totals and differences. Where a page is to be checked separately, then the entries on the page must commence with a backsight and end with a foresight. It will be evident that the last reading on a page will generally be either a reading to a changepoint or a reading to an intermediate sight.

If the last entry is for a changepoint, then the foresight reading to that point will be the last reading on the page and the backsight reading to the same point must be entered as the first reading on the next page. If the page ends at an intermediate sight, then it must be booked as a foresight and repeated again as a backsight as the first reading on the next page. Note that a page can only be checked by the methods above if there are the same number of backsights and foresights on the page.

5.9.7.4 *Using the staff in the inverted position*

On occasion it is convenient to use the staff upside down where a point to be levelled is above the collimation line, e.g. bridge soffit levels, or a line of levels run across a high wall. In this case, the inverted staff readings should be booked as negative values and due account of the negative sign taken in the reduction calculation. A sketch may help to clarify the required arithmetic.

5.9.7.5 *Permissible error in levelling*

The example reductions in *Figures 5.19*, *5.21* and *5.22* show a misclosure of 10 mm or 0.010 m. Since the arithmetic has been checked, the 0.010 m must be an error in the levelling and not in the calculations. Since there are always errors in survey, a limit has to be set for the *permissible* (i.e. acceptable) *error* in any levelling job. The actual error permissible depends upon the type of job.

British Standard 5606 : 1990 *Code of Practice for Accuracy in Building*, requires that

(i) the accuracy of construction site TBMs with respect to Ordnance Survey bench marks should be within ±10 mm,
(ii) the accuracy of spot levels with respect to TBMs should also be ±10 mm, and on hard surfaces 90% of values should be within ±5 mm, and
(iii) the accuracy of a line of levels run with an engineer's level should be within $\pm 10\sqrt{K}$ mm, where K is the distance levelled in kilometres.

These standards can be obtained with equal backsight and foresight distances and careful estimation of the third decimal place (i.e. reading carefully to 1 mm or 0.001 m).

In general line levelling an accuracy of $\pm 12\sqrt{K}$ mm is often considered acceptable, and up to $\pm 24\sqrt{K}$ where the work is carried out on broken ground or steep slopes. By contrast, the typical allowance in first-order precise levelling might be $\pm 4\sqrt{K}$ mm or better. In survey work on small sites, where distances are short, closing errors not exceeding $\pm 5\sqrt{s}$ mm may be reasonable, where s = the number of instrument set-ups used.

Commonsense must be applied in deciding whether a misclosure is acceptable or otherwise. When the permissible error for a task has been exceeded and the error cannot be located in one part of the levelling, it is necessary to repeat the whole of the levelling. For this reason, it is best to carry out a check over the same changepoints as the original levelling – the error may be located in one section.

5.9.7.6 *Adjustment of the level book*

In ordinary levelling, the misclosure is likely to be a combination of both random errors, e.g. small errors in reading the staff, and cumulative errors, e.g. the instrument and/or staff sinking during the intervals between taking readings.

Although random errors may occur in all readings, such errors in intermediate readings do not affect the overall misclosure, since they are not carried forward and do not influence the reduced levels of the changepoints. However, if a staff sinks while being held at a changepoint, this will affect the backsight reading to the staff and the intermediates and foresight following and this error will be carried forward as well as any random errors in the changepoint readings.

In practice, it is the *reduced levels* which are adjusted rather than the staff readings. Since the individual errors are never known, generally the same adjustment is applied at each changepoint and at the closing BM, the sum of these adjustments being equal to the magnitude of the misclosure but of opposite sign. In ordinary work, adjustments are made to the nearest millimetre only.

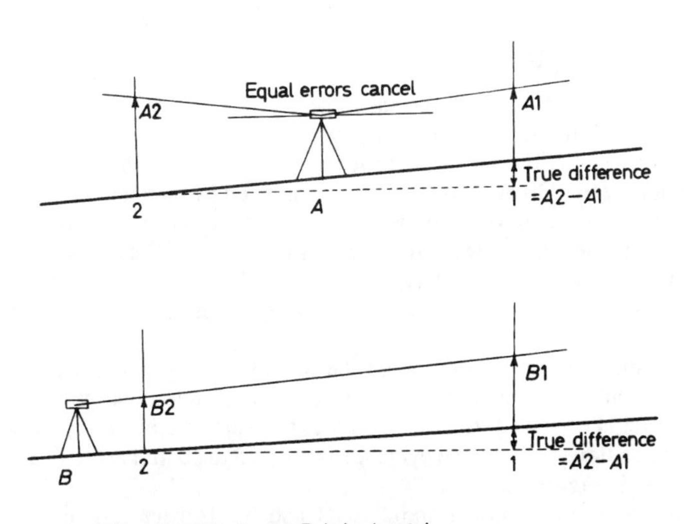

Figure 5.23

The result of applying the above rules to the line of levels in *Figures 5.19, 5.21* and *5.22* with an error of +0.010 m could be:

Opening BM	*Adjustment*	*Running total*
RL at CP1	–0.002	–0.002
RL at CP2	–0.003	–0.005
RL at CP3	–0.002	–0.007
Closing BM	–0.003	–0.010

There are four points at which the adjustment is incremented (three CPs and the closing BM), and +10 mm misclosure, giving –2.5 mm at each, rounded here to –2, –3, –2 and –3. As the adjustment at a CP is applied, all reduced levels following that point are adjusted by the same amount, the process being repeated through all the reduced levels. Some surveyors may adjust intermediates differentially, on the assumption of cumulative error, but no rule can be laid down for this. In precise levelling there should be no intermediate sights and cumulative error should be kept to a minimum, then this latter problem does not arise.

5.10 Permanent adjustments to the level

As previously stated, there are certain adjustments which must be maintained in good order if an instrument is to provide accurate work. These are the *permanent adjustments* of the level and the checking and correcting of these varies with the type of instrument. Levels should be checked regularly, but it should be noted as regards the following sections that for ordinary levelling collimation adjustment is not made unless the error exceeds about 1 mm per 15 m of sight distance.

5.10.1 The tilting level

This instrument has one requirement only, that the bubble tube axis and collimation line should be parallel (*collimation adjustment*).

To check, on reasonably level ground set out two stable marks, generally wooden pegs, about 30 m apart (or as recommended by the manufacturer). Set up and level the instrument midway between the pegs, at point A as in *Figure 5.23*. Observe the readings on a level staff held in turn on peg 1 and peg 2, note these as readings A1 and A2. The difference between the readings A1 and A2 will be the true difference in level between the pegs, regardless of any collimation error in the level.

Now set the level up at point B, on the line of the pegs but outside them and as close to point 2 as the short focus distance of the telescope will allow when the staff on 2 is observed. (Alternatively, the level could be set up between the marks but positioned so that the eyepiece is only 25 to 50 mm away from the staff held on peg 2, *Figure 5.24*.)

Note the staff readings on 1 and 2 as B1 and B2. If there is no error, then the difference between B1 and B2 will be the same as that between A1 and A2. If they are not the same there is *collimation error*.

In the alternative method, the reading on to the staff at peg 2 should be taken through the *objective end* of the telescope. In reading the staff this way, the

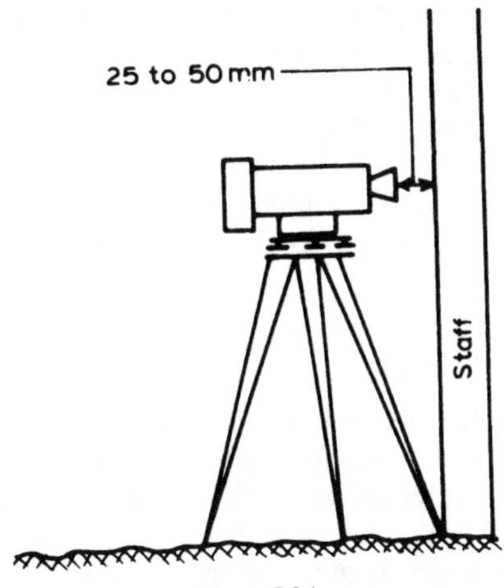

Figure 5.24

surveyor will not be able to see the cross-hairs, but this is not important since the field of view will be very small and it is easy to judge the centre of the visible circle. A pencil point laid against the staff face, *Figure 5.25*, will aid in determining the reading, since the pencil will be visible through the telescope even if no graduation figures can be seen.

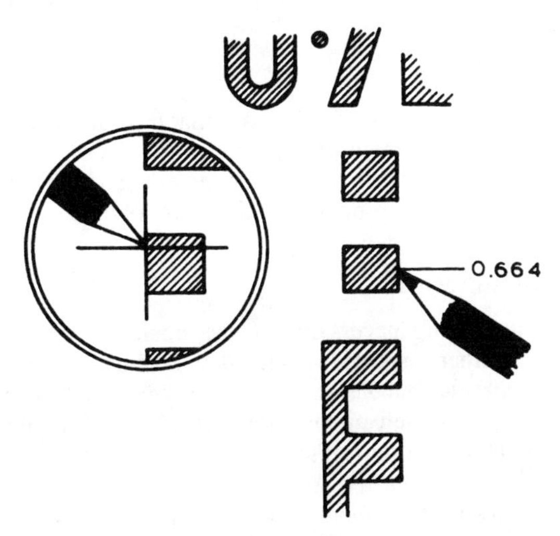

Figure 5.25

The pencil is moved until its point is central then the graduation at the pencil point is observed directly by eye. (For obvious reasons this method of test is termed the *Two-peg Test.*)

To adjust the instrument, calculate what the staff reading on B1 should be, assuming there is no error in reading B2. (There may be an error in B2, depending on the method used, but for most practical purposes it can be ignored.) Using the tilting screw, bring the central horizontal cross-hair to this calculated reading. This will result in the bubble moving off centre. Centre the bubble carefully again by means of its own adjusting screws and the operation is complete. Repeat the check and adjustment again as necessary.

Makers' instruction manuals should be consulted for detailed guidance on adjustment of any gradienter screw and for the adjustment of precise levels with bubble scales.

5.10.2 The automatic level

This instrument also has one requirement only, that the collimation line be horizontal when the circular bubble has been centred, a slight variation on the collimation adjustment described above. To check, carry out the Two-peg Test as before.

To adjust, calculate the required correct reading on staff 1. Move the collimation line until the correct reading is at the central horizontal cross-hair. The method of moving the line of collimation to get the cross-hairs on to the reading, however, varies between different instruments and reference should be made to the maker's handbook. Some instruments are adjusted by moving the diaphragm, while in others it is the compensator unit which must be adjusted. In some, both must be adjusted. It should be noted that in some automatic levels the eyepiece and the objective lens do not lie in the same horizontal plane, hence the method of looking at the staff through the objective lens is not feasible.

5.10.3 The digital level

The requirement for the digital level is again that the collimation line should be horizontal when the circular bubble has been centred. These levels generally have a program for the Two-peg Test built into the instrument – the maker's handbook should be consulted for the particular details to follow.

5.10.4 Other adjustments

The principal adjustments necessary for the operation of levels have been described above. Other aspects which may require attention include the adjustment of the circular bubble on tilting and automatic levels, the take-up of wear on footscrews, etc. Where an instrument is to be used for any length of time it is advisable to obtain a copy of the maker's instruction manual.

5.11 Sources of error in levelling

As with linear and angular measurement work, levelling is never free from error. Some errors are due to carelessness, some have cumulative or constant or systematic effects on the results of the levelling. *Table 5.2* lists the common sources of error in levelling, together with the precautions to be taken to minimize their effects.

Table 5.2

Source	*Precaution*
Errors attributable to the surveyor:	
Mistakes in reading the staff	Read all three hairs
Mistakes in booking readings	Book stadia readings in 'remarks' column
Disturbing level or tripod or both	Check position of bubble; do not lean on or kick tripod
Failure to level the bubble	Check before reading
Incorrect focusing	Eliminate parallax
Mistakes in reducing levels	Carry out mathematical checks
Errors attributable to the staff holder: (*see Section 5.9.6 – Duties of the staff holder*)	
Not holding the staff upright	Check staff bubble
At change points, not ensuring that the staff is held in exactly the same position for both back- and foresights at a point	Use staff support, always used in precise work
Unequal back- and foresights	Surveyor check occasionally using stadia hairs, always in precise work
Staff not properly extended	Surveyor check as needed by viewing connecting portion(s) of staff through telescope
Errors attributable to the ground or climatic conditions:	
Sinking/rising of the instrument and/or staff	Set up on firm (not frozen) ground, use staff support
Strong winds, staff and instrument vibrating	Find sheltered spot for instrument and staff, set tripod up low with its legs spread and tread its feet well in, if automatic level, hold tripod lightly; brace staff with two steadying rods; if wind becomes too strong, cease work
Heat shimmer, staff graduations unsteady (appear to bounce)	Reduce length of sights; try to keep line of sight well above ground level
Direct heat of the sun causing differential expansion of instrument parts, bubble tube in particular:	Hold field book to shade bubble tube when levelling up; survey umbrellas can be bought
Curvature and refraction (but no visible disturbance of image)	These errors generally negligible, but reduce or eliminate by using equal sight lengths; keep sights short (max. 60 m); avoid grazing rays (sights near ground), avoid continually reading zero end of staff, e.g. when going uphill. Steep hills should be avoided in precise levelling. In precise work each line is measured twice, at different times on different days Sun near horizon may make sighting impossible. Raindrops on objective may make reading difficult. Avoid sighting near the sun, use telescope rayshade. Use rayshade and umbrella
Errors attributable to the level:	
Faulty permanent adjustments	Check adjustments from time to time (see Section 5.10); keep backsight and foresight lengths equal
Errors attributable to the tripod:	
Play in the joints	Check occasionally; tighten as necessary
Errors attributable to the staff:	
Longitudinal warping of the staff	Errors usually insignificant, but they are cumulative; if serious, discard staff
Graduation errors	An error in the 'zero point' of the staff has no effect, but errors at other points on staff may be cumulative; occasionally check graduations against steel tape
Staff bubble out of adjustment	Check occasionally; error minimal but cumulative

5.12 Computer applications

Level data can be stored in a data recorder for later reduction by computer, a variety of software being available from instrument makers and others. The modern digital level has levelling programs built into the instrument, which will compute normal lines of levels, etc.

Chapter 6

Angular measurement

6.1 Introduction

Angular measurement covers the measurement of angles in both the horizontal and vertical planes. The traditional instrument for angle measurement is the *theodolite*.

The theodolite is designed specifically for the measurement of horizontal and vertical angles in surveying and in construction works. Although the instrument's basic principles are simple, it is the most versatile of survey instruments, capable of performing a wide range of tasks. In addition to the measurement of angles it may be used in setting out lines and angles, levelling, optical distance measurement, electromagnetic distance measurement, plumbing tall buildings and deep shafts, and geographical position fixing from observations of the sun or stars, etc.

There are two types of theodolite in use today, *optical theodolites* in which the observer must read the angle values against graduated scales and *electronic theodolites*. With the latter, the user levels up and points the theodolite manually but the theodolite determines the angle value and displays the reading in digital form like a calculator.

It can be an extremely accurate piece of equipment, although specifications vary considerably, some instruments being capable of being read to the nearest 5 minutes of arc, while others can be read to 0.1 of a second of arc. In UK practice theodolite circles are graduated in the sexagesimal system, 0 to 360 degrees in one revolution, but in some parts of Europe they use gons, 400^g to one revolution.

Optical theodolites are typically classed according to their reading accuracy, and each class has a generally accepted range of characteristics.

Precision theodolites may be read directly to better than one second of arc and they are typically used by national mapping and land survey organizations for geodetic surveying.

Universal theodolites, also known as 'single second' instruments, allow readings directly to 1" of arc. They are occasionally used on site survey and engineering work, especially when extreme angular accuracy is needed, such as possibly on very long sights.

General purpose theodolites or *engineer's theodolites* generally read directly to about 20", and they are fast and easy to use, ideal for general survey work.
Builder's theodolites are of a comparatively low order of accuracy, reading direct to 1, 5 or 10' of arc and by estimation to 30" or 1'. They are usually rugged, simple to operate, and relatively cheap.

Electronic theodolites may similarly be classed according to the accuracy attainable.

6.2 The basic construction of the theodolite

Figure 6.1 shows the basic construction of early instruments, first made about 400 years ago, and modern instruments are merely refined versions of this original concept.

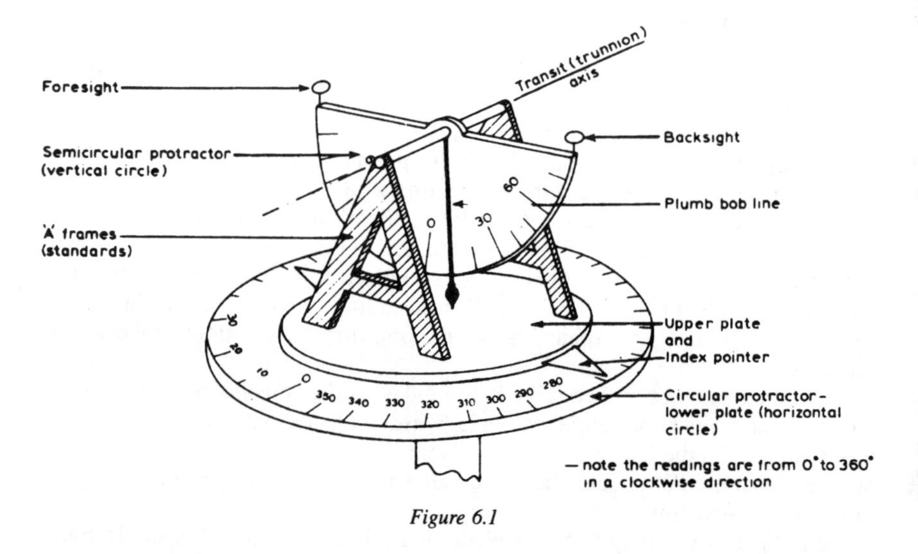

Figure 6.1

A circular protractor, today called the *horizontal circle* or *lower plate*, was supported horizontally with its engraved surface lying uppermost. Above this was an *upper plate* with an index pointer, the plate being rotatable to allow the pointer to be laid against any graduation on the lower plate. The upper plate carried two vertical A-frames (now known as *standards*) supporting between them a horizontal axis, known as the *transit* or *trunnion axis*, at the centre of which a semi-circular protractor was rigidly fixed at right angles to it. A weighted plumbline was also carried by the transit axis. (The modern protractor is a full circle and is termed the *vertical circle*.) On its diameter the semi-circle carried two sights, as on a rifle, which defined the line of sight or *line of collimation*. The upper part of the theodolite, rotatable in the horizontal plane, is termed the *alidade*, originally an Arabic word meaning a sighting rule.

To read at the instrument station a horizontal angle between two distant targets, the sights were aimed alternately at the left-hand and right-hand targets and the horizontal circle readings to them were noted, the difference between these readings giving the horizontal angle between the directions to the targets as measured at the instrument station. To read a vertical angle, the sights were aimed at a target and the angle of elevation or depression read off the vertical circle against the weighted plumbline.

The *telescope* was added to improve the line of sight in distance and clarity of definition, and at an early stage the horizontal and vertical circles were fitted with metal graduated scales with attached vernier scales for more accurate reading of the angles. These instruments were termed *vernier theodolites*, and, if the telescope could be rotated completely about the horizontal axis (i.e. *transited*), they were *transit theodolites* – simply called *transits* in American practice. In European practice, all theodolites are assumed to be able to transit.

In the early 1920s Heinrich Wild introduced glass circles instead of metal, together with microscope reading systems, the whole being totally enclosed to keep out dust and moisture. These circles could be graduated more finely and light could be reflected through the circles to show the markings with greater clarity. These instruments are termed *optical* or *glass arc theodolites*, and *Figure 6.2* shows the essential components of such an instrument.

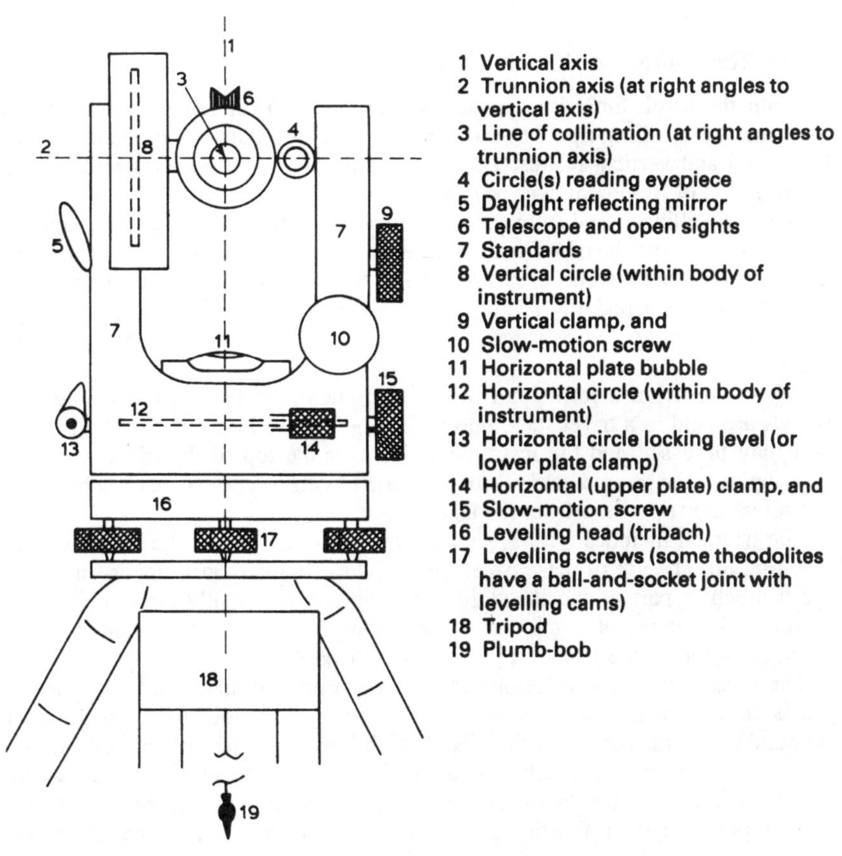

Figure 6.2

The latest generation of theodolites use *electronic digital reading systems*, and these are superseding the glass arc instruments still in common use today. As the cost of optical components has risen, and that of electronic parts has fallen, some electronic theodolites are now cheaper than equivalent optical theodolites, and some manufacturers have ceased production of glass arc instruments.

6.2.1 The tripod

The *tripod*, described earlier in Chapter 5, is used to support the instrument at about eye height over survey marks. Some manufacturers use the same tripod for both their levels and their theodolite, but some theodolites require special tripods.

The tripod should be set up in the same way as for the level, except that since it is used to measure angles at a point, it must normally be set up with the top plate centred over a ground survey mark. It is important to ensure that there are no loose tripod fittings which might allow the tripod head to twist or turn in use.

In some jobs, it is necessary to place the theodolite on top of a wall or pillar and in these circumstances a special *base-plate* (or *wall plate* or *trivet*) is used instead of a tripod.

6.2.2 The tribrach or levelling head

As with the level, the prime purpose of the *tribrach* (or levelling head) is to facilitate the levelling-up of the instrument, so that readings are taken in truly horizontal and vertical planes. The *levelling-up* operation involves making the vertical axis of the instrument properly vertical.

When travelling to and from survey sites, the theodolite is clamped securely in a carrying case and the tripod is carried separately by its carrying handle or strap. The tribrach usually forms the lower part of the theodolite, the instrument being clamped to the tripod by a captive bolt screwed into the tribrach when the instrument is set up, but on some theodolites the tribrach is part of the top of the tripod.

Where the tribrach forms part of the instrument, the bottom end of the foot screws are held in a trivet stage into which the captive bolt screws. The captive bolt may be eased and the instrument slid over the top of the tripod to assist centring over the ground mark. With this form of construction, care must be taken to protect the tripod top from damage.

The tribrach includes *levelling screws* or *cams*, used to tilt the instrument in levelling up. The use of these is described below. In many instruments in which the tribrach is part of the theodolite, the tribrach is actually detachable. This permits other items of equipment (traverse targets, etc.) to be mounted in the tribrach on top of the tripod in place of the theodolite.

The theodolite in use must be centred over a ground mark, and traditionally this is done by reference to a *plumb-bob* and *string* suspended from the centre of the underside of the tribrach. The plumb-bob is generally carried in the instrument case or in a pouch attached to the tripod. Instead of the plumb-bob, many modern instruments are supplied with a telescopic *centring-rod*, (like a non-supporting central fourth leg) which has an attached circular bubble and is used for centring the instrument over the ground mark.

Centring the instrument is achieved by appropriate movement (laterally, or by extending or shortening) of one or more of the tripod legs, until the plumb-bob is exactly over the ground mark, with the final fine adjustment being made by unclamping the tribrach and sliding it across the top of the tripod head. Throughout the centring operation, of course, the top plate of the tripod must be kept horizontal.

If the centring rod is used a similar process is followed but the point of the rod is placed on the ground mark and the tripod legs adjusted to make the circular bubble of the rod central. On some few instruments, the upper part of the instrument may be slid over the top of the tribrach, termed 'centring above the foot screws'.

6.2.3 The horizontal plate(s)

Directly above the tribrach is that part of the instrument known as the *horizontal plate* or *plates*, which may include items 11 to 15 shown in *Figure 6.2*. If there are two plates, they are known as the *upper* and *lower* or *top* and *bottom* plates, respectively.

The existence of two-plate construction is indicated by the presence of two horizontal clamps and two slow-motion screws for the control of rotation about the vertical axis, such an arrangement being termed *double centre*. If an instrument has only one clamp and slow-motion screw, then it is of single-plate construction. This type will have an additional control for horizontal movement, either a *circle locking lever* (*Figure 6.9*) or a *circle-orienting drive* protected by a hinged cover (*Figure 6.10*). These control arrangements affect the methods to be used in setting or reading horizontal angles, shown later. The circle locking lever is sometimes termed a *repetition clamp*.

6.2.3.1 The horizontal plate bubble

The upper plate carries a bubble tube (the *plate bubble*) which is used for levelling up the instrument. The traditional procedure, illustrated in *Figure 6.3*, is as follows:

(i) Set up the instrument, lay the plate bubble parallel (in plan) to the line joining any two foot screws.
(ii) Centre the bubble by turning these two foot screws slowly in opposite directions, at the same speed. Note that the bubble will move in the same direction as the user's left thumb.
(iii) Turn the plate through 90° in plan, centre the bubble again, but using the *third footscrew only*.
(iv) Repeat (ii) and (iii) in the same quadrant in plan until the bubble remains centred.
(v) Turn the plate through 180° – if the bubble stays central, the instrument is levelled up, check in any other position.

If the bubble does *not* stay central, note the position of the bubble with respect to the graduations marked on the tube, see *Figure 6.3*, then:

(i) Move the bubble half-way back to centre using the two foot screws; note its new position against the tube division as the *adjustment position*.

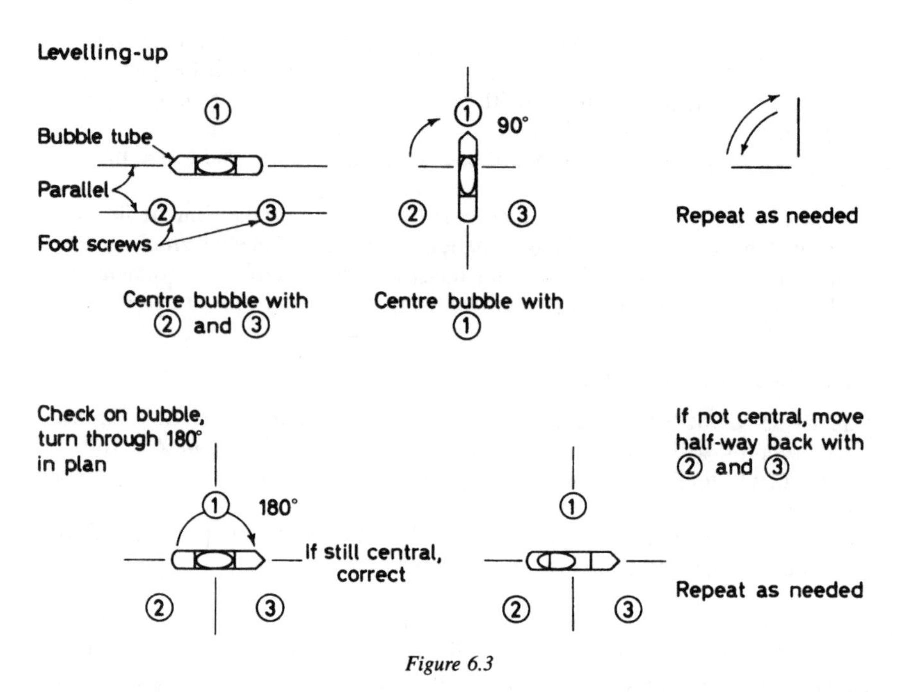

Figure 6.3

(ii) Turn the plate through 90° in plan, bring the bubble to the adjustment position using the third footscrew.

If the bubble is always brought to the 'adjustment position' on levelling-up, instead of to the centre of its run, then the vertical axis will be properly vertical. A small circular bubble is often attached to the tribrach for rough levelling-up, and this reduces the time taken for the accurate levelling-up.

In practice, the instrument must be set up and centred over the ground mark as described earlier, then levelled up by the foot screws, then the centring checked again. It may require some repeats before the instrument is both centred and levelled correctly.

6.2.3.2 The optical plummet

Most modern theodolites are fitted with an *optical plummet* within the lower part of the instrument, as in *Figure 6.4*. This is a small telescope which provides a line of sight down the vertical axis of the instrument, the horizontal view being deflected through 90° by a prism. It is an alternative to the plumb-bob for centring the instrument, giving greater accuracy, particularly in windy conditions, but it is only effective if the instrument has been levelled-up so that the vertical axis is truly vertical. Optical plummets may be fitted in tribrachs, or they may be separate, specialized pieces of equipment which can sit in a detachable tribrach.

If the plumb-bob is detached from the instrument, then the optical plummet and the foot screws may be used both to centre the theodolite and to level it up. The instrument should be set on the tripod with the tribrach roughly horizontal

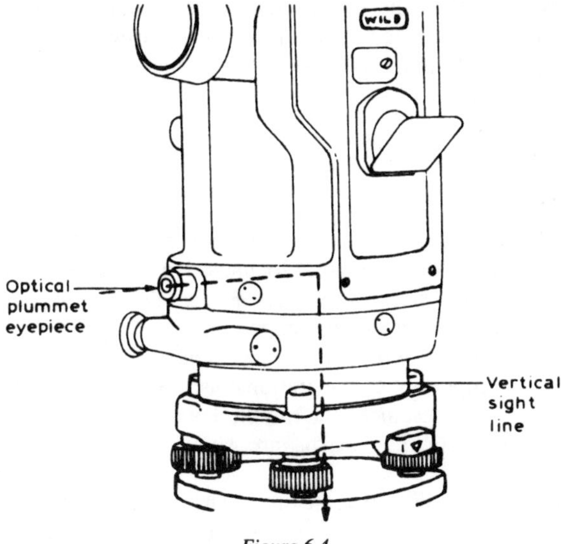

Figure 6.4

and approximately centred over the ground mark, the optical plummet telescope focused on the ground mark, then, looking through the optical plummet, the foot screws should be adjusted to bring the ground mark central in the field of view. The procedure is then:

(i) Turn the instrument in plan until the bubble tube lies parallel to the line joining the feet of any two legs of the tripod.
(ii) Roughly centre the bubble by extending or shortening one of these two tripod legs.
(iii) Turn the plate through 90°, again roughly centre the bubble by extending or shortening the third leg.
(iv) Level the instrument using the foot screws as described in Section 6.2.3.1.

It will be found that if the tripod legs are firmly positioned in the ground then when one leg is extended or shortened, the ground mark will often appear to remain stationary when viewed through the optical plummet. Due to this feature, when the above operations have been completed the ground mark will still lie close to the centre of the optical plummet field, and then:

(v) Ease off the clamping bolt and, without rotating, slide the instrument over the tripod head gently to bring the optical plummet cross-hair exactly on to the centre of the ground mark.
(vi) Re-level the instrument as necessary with the foot screws, check the centring, repeat operations (v) and (vi) as needed.

Note that the optical plummet should be checked by turning the plate through 180° and observing whether the cross-hair indicates the same ground point as before. If it does not, then the true plumb point is mid-way between the two indicated positions. As with the bubble tube, this 'adjustment position' should be used until the optical plummet has been adjusted.

6.2.4 The telescope and vertical circle

6.2.4.1 The telescope

Modern theodolite telescopes are essentially the same as those in surveyor's levels, but the movement of the internal focusing lens is more commonly carried out by the movement of a sleeve fitted around the telescope barrel rather than a focusing wheel on the side of the telescope.

For the theodolite the *magnification* and *resolving power* and *field of view* of a telescope are very important considerations, as the theodolite is typically used on much longer sighting distances than is the level.

Typical diaphragm reticule patterns are shown in *Figure 5.6*, as for the level.

The telescope should be fitted with a pair of external sights, or more commonly today a *finder-collimator* device, used for pointing the telescope initially towards the target to be sighted, as described for the level.

6.2.4.2 Focusing

The procedure for focusing the eyepiece on the reticule at the start of work, and focusing on individual targets thereafter, should be as described in Section 5.5.2.2. Parallax error should be checked for by moving the head to observe whether the target and the cross-hairs appear to be 'glued together'.

6.2.4.3 Face left and face right

The transit or trunnion axis carrying the telescope is at right angles to the telescope collimation line, and it is supported at each end by the standards. (In older instruments the telescope, transit axis and vertical circle could be removed from the standards, but this is not feasible in modern instruments.) Generally one standard is bulkier than the other, since it encases the vertical circle mounted on the transit axis. The telescope and vertical circle together rotate in a vertical plane about the transit axis, the movement being controlled by a clamp and slow-motion screw fitted to one of the standards.

The normal observing position when looking through the telescope is to have the vertical circle located at the observer's left-hand side, and this is described as *observing with face left*, simply annotated as *FL* in the field book. If the telescope is now rotated about the transit axis through an angle of 180°, termed *transiting the telescope* and the plate turned until the eyepiece is at the observer's eye again, then the vertical circle will lie to the observer's right, and this is *observing with face right*, annotated in the field book as *FR*. Movement in the horizontal plane is controlled by the clamp(s) and slow-motion screw(s) on the horizontal plate.

In the normal FL position the controls will be found to be in suitable locations for a right-handed observer. It should be noted, however, that angular observations are generally taken on both faces, the results of the two observations being meaned and this process will eliminate or minimize most instrumental errors such as circle graduation error, circle eccentricity, collimation line not at right angles to the transit axis, or dislevelment of the transit axis, as well as providing a check on the surveyor's readings.

6.2.4.4 *The centring thorn*

For centring an instrument underneath a mark in the roof of a tunnel, etc., some telescopes are fitted with a *centring thorn*. This is merely a projecting stud on the top of the telescope, such that when the telescope is horizontal the thorn defines the centre of the instrument in plan. The thorn is sometimes connected to a small mirror inside the telescope and turning the thorn adjusts the mirror so as to reflect light on to the reticule and illuminate the cross-hairs in night work or when artificial illumination is being used.

6.2.4.5 *The altitude bubble*

In addition to the plate bubble, some older instruments are fitted with another bubble tube at the top of the standard supporting the vertical circle. This is the altitude bubble, and it is linked to the vertical circle and to an *altitude bubble adjusting screw* on the standard, similar to a slow-motion screw.

When the altitude bubble is central it indicates that the vertical circle is properly zeroed and vertical angles read off the vertical circle will give the correct values. The altitude bubble must be centred by its adjusting screw immediately before taking a vertical angle reading.

The altitude bubble was sometimes read through a *coincidence prism reading system*, instead of being a simple open bubble tube. Like the similar systems used on some levels, these give an image of the two ends of the bubble tube, as in *Figure 6.5*.

Most modern instruments are fitted with self-zeroing vertical circles (*automatic vertical indexing*) which make use of gravity-operated liquid compensators and these avoid the need to centre an altitude bubble before reading a vertical angle. However, the instrument must be levelled up with care to ensure that the working range of the compensator is not exceeded.

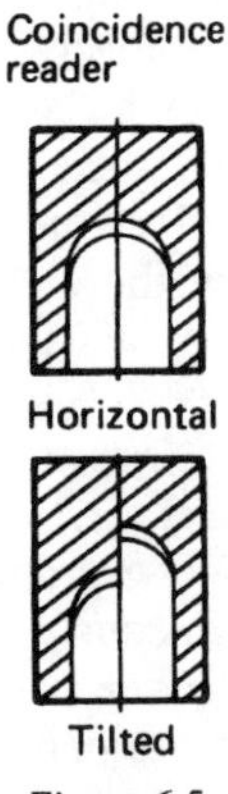

Figure 6.5

6.2.4.6 Vertical angles

The vertical angle from the instrument centre to a distant point may be measured in two ways. If the angle is measured from the horizontal it is termed an *angle of elevation*, or an *angle of depression*, according to whether the distant point is above or below the horizontal plane through the instrument's centre. In *Figure 6.6* these angles are denoted as + and –. If the vertical angle is measured downwards from the zenith, it is termed a *zenith distance* or *zenith angle*, denoted here by z. In this connection it must be noted that the direction of the line through the instrument centre to the centre of the earth (downwards) is termed the *nadir*, and the opposite direction vertically upwards from the instrument centre is the *zenith*.

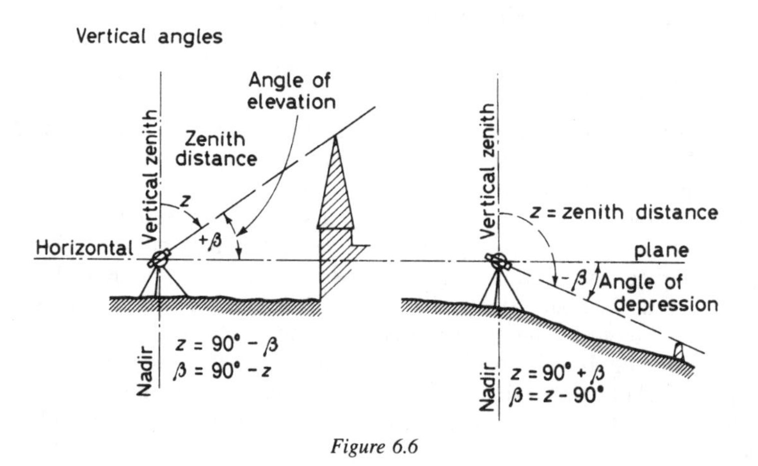

Figure 6.6

6.3 Reading systems – optical theodolites

Most of the instruments still in use today have *optical reading systems*, where the graduated circles are of glass and are totally enclosed so that circle readings must be made directly by the surveyor, through one or more small telescope-style microscope eyepieces located either alongside the telescope, or on a standard, or on the horizontal plate. Small mirrors direct light into the instrument, then prism systems deflect the light to illuminate the graduations on the glass circles. The images of the circle graduations are then directed to the field of the reading microscope, and the surveyor observes and books the readings.

Some optical instruments have one microscope eyepiece to view the horizontal circle and another to read the vertical circle. In others, a single eyepiece viewer is fitted, either alongside the telescope proper or on one standard. These latter types may show both circles simultaneously or else alternate circle views may be obtained by operation of a changeover knob or switch. A variety of reading systems are used on different instruments, according to the accuracy and performance required of the instrument.

6.3.1 Circle microscope system

This is the most basic optical reading system and has a simple fixed hair-line in the microscope, against which the circle graduations are read directly. It is widely used on builder's theodolites.

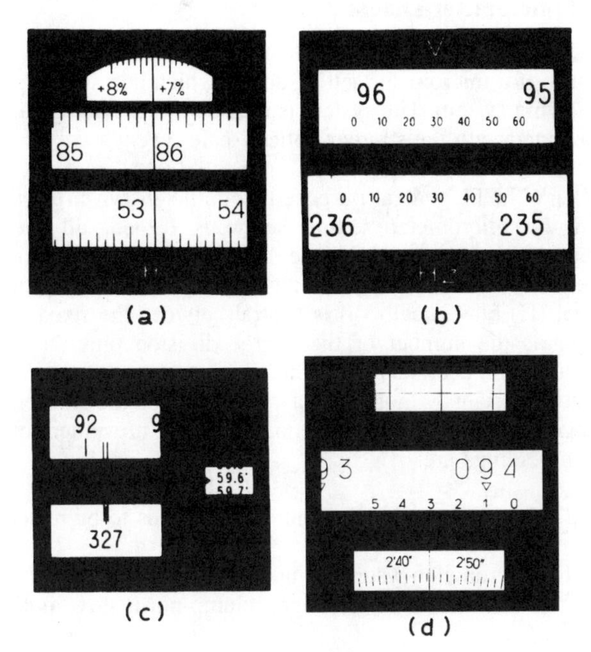

Figure 6.7

Figure 6.7(a) illustrates a typical set of readings viewed in the microscope of a construction theodolite with circles graduated to five minutes of arc, allowing readings to be estimated to the nearest minute of arc. This instrument has three scales – the lowest being the horizontal angle, then the vertical and finally a vertical gradient scale showing percentage gradients. (This last is not common on theodolites.) The readings are:

horizontal 53° 12'; vertical 85° 48'; gradient + 7.35%

6.3.2 Optical scale system

A development from the preceding type, this has a diaphragm fixed in the microscope, with a scale of divisions on it rather than a fixed hair-line.

Example: *Figure 6.7(b)* shows a typical optical scale reading system, with the glass circle divided into single degrees only and the optical scale extending over one degree but divided to single minutes of arc. The number of the degree division is noted, then the number of minutes at the degree division and finally the tens of seconds of arc are estimated (only one degree division can cut the optical scale unless the reading is an exact number of degrees). The readings are:

horizontal 235° 56' 20"; vertical 96° 06' 30"

6.3.3 Optical micrometer system

This combines the optical scale principle with lateral movement of the scale under the control of a micrometer setting screw which adjusts the position of a prism in the viewing system. The system is intended to provide greater accuracy than can be obtained with the simpler optical scale type.

Example: The Wild T1 general purpose theodolite shown in *Figure 6.7(c)* is typical. When the micrometer setting screw is turned, all the numbered graduations appear to move while the central single/double line remains stationary. The micrometer screw is turned until a degree division of the required scale (horizontal (H) or vertical (V)) is central between the fixed pair of lines, then the reading is the number of the degree division plus the minutes and seconds (or minutes and decimals of a minute in this example) shown in the third 'window'. It is important to note that a movement of the micrometer screw moves the images of the circles, it does not cause any movement of the circles themselves. The readings are:

horizontal 327° 59.6'; vertical – the micrometer needs to be re-set

The presence of a micrometer system is indicated by the extra control knob, the micrometer setting screw, which is neither a clamp nor a slow-motion screw.

6.3.4 Coincidence (or double-reading) optical micrometer system

This system is used on precision and universal theodolites where a high accuracy of reading is required, since it provides the mean of two readings taken at diametrically opposite points on the circle, thus minimizing errors of eccentricity. The optical reading eyepiece presents graduations from both sides of the circle and these must be placed equally on either side of a fixed hair-line or, alternatively, brought into coincidence using the micrometer setting screw.

Example: *Figure 6.7(d)* shows the reading system of the Wild T2 universal theodolite, the upper window showing the graduations brought into coincidence with one another, while the central window shows the degrees and tens of minutes and the lowest window shows minutes and seconds which can be read direct to one second.

The reading is: 94° 12' 44.4"

6.4 Electronic theodolites

An electronic theodolite is much like an optical theodolite in general appearance, but instead of having reading microscopes for the circles it is equipped with a liquid crystal display screen and a small keyboard rather like an electronic calculator. An encoded glass circle is used and the light passing through this is detected by *photodiodes* and converted to electrical signals. Measurement systems, either *incremental* or *absolute*, measure these signals and a microprocessor converts them into a digital angular output display.

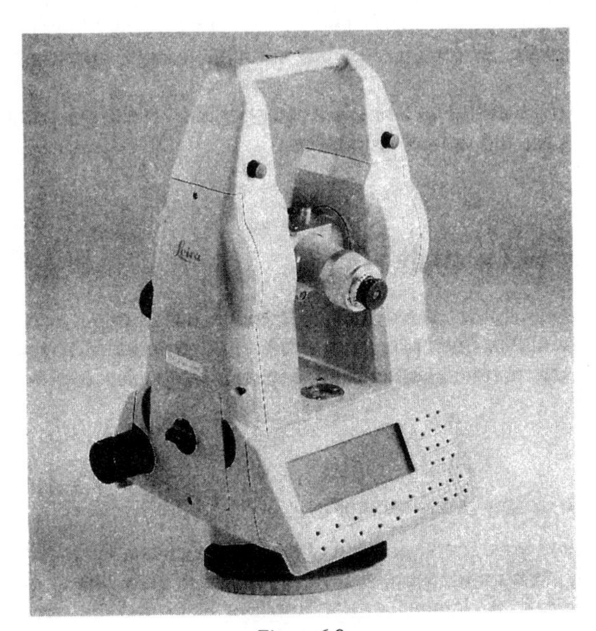

Figure 6.8

A typical example is the Leica T1800 theodolite shown in *Figure 6.8*. The features of this instrument include:

Measurement of horizontal and vertical angles continuously, with absolute encoders, with a standard deviation of 1".
Automatic corrections are made for all axial errors and vertical index error.
The electronic level has a least count of 2".
Display contains eight lines of 30 characters.
Keyboard is fitted on both faces.
Clockwise or anticlockwise readings as required.
2 Mb PCMCIA memory card for data storage.
1 Mb internal memory for programs and data.
RS232C interface for data transfer.
Battery module allows 12 hours continuous use.
Telescope 30×, objective 42 mm.
EDM may be attached, for survey and setting out.
Motorized version (TM1800) available, aligns itself in the required direction.

All electronic theodolites have illuminated displays and cross-hairs. Keys are provided to reset the horizontal circle to 00° 00' 00", to set a required reading on the circle and to lock the reading on. Data can be read from the display and booked manually, or alternatively stored in a data recorder or a computer connected to the instrument data port.

Electronic theodolites vary in their specifications and constructional details, most manufacturers making several different models. As an example, in contrast to the T1800, the Leica T100 electronic construction theodolite has an incremental encoder, an accuracy of 10", no data storage, and uses AA dry cell batteries.

The setting up, levelling up and telescope operation of the electronic theodolite are the same as for the optical theodolite.

6.5 Setting on an angle

It is often necessary to point the instrument on a target with a specific reading already set on the horizontal circle. The process is termed *setting on an angle* and the method to be used depends upon the construction of the instrument.

6.5.1 'Double centre' instruments

These instruments have upper and lower horizontal plates, each controlled by its own clamp and slow-motion screw. If only the lower plate clamp is applied then the horizontal circle is clamped to the levelling head and the upper part of the instrument may be rotated about its vertical axis. If the instrument is turned, then the readings on the horizontal circle viewed through the optical reading eyepiece will alter. However, if only the upper plate is clamped and the lower plate released, then the horizontal circle is clamped to the upper plate and rotates with the telescope, and in this case the readings will not alter as the instrument is turned. When both plates are clamped it should not be possible to turn the instrument (except by applying excessive force).

Accordingly, to set on a particular angle, first apply the lower clamp, then turn the alidade until the horizontal circle reading approximates to the required reading and then apply the upper clamp.

If the instrument is a simple circle microscope or optical scale theodolite, use the upper plate slow-motion screw to set the desired reading exactly then unclamp the lower plate. The alidade may now be turned and the required reading will remain set in the field of view of the optical reading eyepiece; the operation is complete.

If the instrument is an optical micrometer type, then use the micrometer setting screw to set the minutes and seconds and turn the upper plate slow-motion screw until the required degree value is exactly central between the pair of lines (*Figure 6.7(c)*). The required horizontal angle value is set and the operation is complete when the lower plate is unclamped. It may be noted that some theodolites have a milled rim fitted to the circle, this saving the operator from having to walk around the instrument looking for the desired angle reading.

6.5.2 Single plate instruments with a circle locking lever

These instruments have a single horizontal plate clamp and slow-motion screw and a circle locking lever. The circle locking lever replaces the lower plate clamp and slow-motion screw (*Figure 6.9*) thus the method is basically similar to that in Section 6.5.1 above. Various designs of locking lever are used, but in all cases they may be used either to lock the circle to the levelling head or to lock the circle to the horizontal plate.

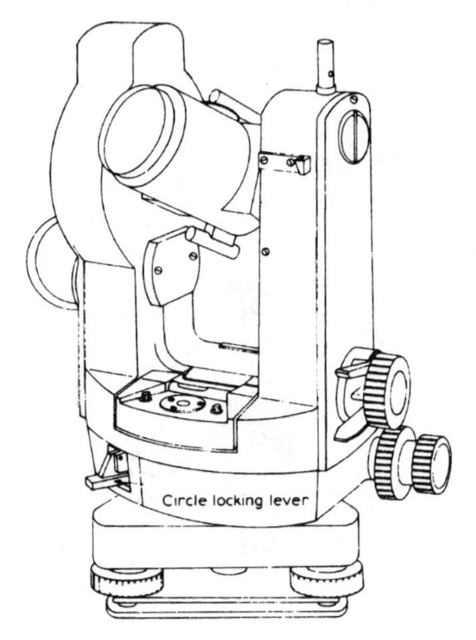

Figure 6.9

To set on a particular angle, lock the circle to the levelling head using the circle locking lever, then release the horizontal plate clamp and turn the instrument until the horizontal circle reading approximates to the required reading, then apply the horizontal plate clamp. Use the horizontal slow-motion screw to set the desired reading exactly, then use the locking lever to lock the circle to the horizontal plate and release the horizontal clamp. The theodolite may now be turned and the required reading will remain set in the field of view of the optical reading eyepiece.

6.5.3 Instruments with a circle orienting drive

These are normally universal theodolites with coincidence optical micrometer systems. Since the horizontal circle of instruments of this type can only be moved by the circle orienting drive and the circle cannot be swung freely as in the other types, it is not possible to set on an angle before pointing the target.

Accordingly, before setting on the angle, the target must be bisected, as explained below.

To set on a particular angle, having pointed the target, use the micrometer setting screw to set the required minutes and seconds, then unclamp (or lift the protective cover from) the circle orienting drive and use it to set the required degree value (*Figure 6.10*). The drive should then be re-locked (or the cover replaced) to prevent accidental movement of the circle. Note that in this case, if the horizontal clamp is released and the telescope is swung, the readings will not remain the same, but if the target is re-pointed the angle originally set should re-appear in the reading eyepiece.

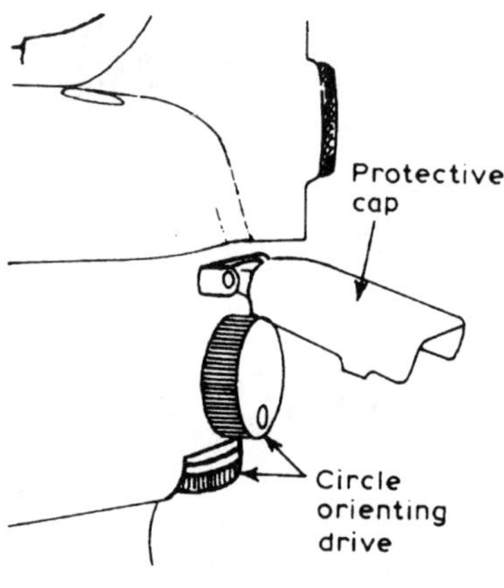

Figure 6.10

6.5.4 Electronic theodolites

These instruments may be set to any desired horizontal angle reading, then pressing a 'Hold' key keeps that reading in memory as the telescope is pointed in the required direction. Pressing a 'Reset' key sets the horizontal circle to zero.

6.6 Measuring angles

For most survey work an observing technique is required which is easy to use, provides some form of check on reading and instrumental errors and will allow an increase in accuracy to be achieved readily when necessary.

6.6.1 Sighting targets

A sighting target for angle measurement must provide a fine mark which can be accurately bisected by the central cross-hair of the reticule. For horizontal angle measurement a target should be placed vertically above the centre of the station marker it identifies and the telescope should be aimed at the base of the target. Non-vertical targets lead to centring/bisection errors, and these are most significant on short lines.

Probably the best target in construction work is a small nail protruding about 10 mm from the top of a wooden peg placed exactly on the station point, or a plumb-string suspended over the marker from an instrument tripod. Manufacturers supply special targets for fitting into tribrachs for traverse and similar work.

A ranging rod is suitable only on long lines, preferably held in a rod support, and the point of the rod should be bisected with the central vertical hair.

6.6.2 Terminology

The following additional terms are used in theodolite work and it will be necessary to define them before the observation routines can be fully explained.

Reference Object (RO), back object and *back station* are all terms used to denote the first target the theodolite is pointed at when measuring the horizontal angle subtended at a point or station by two distant target points or stations.

The *zero* of a measurement is the reading on the horizontal circle when the instrument is pointed on the RO.

Forward object and *forward station* are terms used to denote the second target point observed when measuring a single horizontal angle.

Swinging, or *turning*, means rotating the alidade about its vertical axis. Accordingly, swinging (turning) right means rotating the instrument about its vertical axis in a clockwise direction in plan, while swinging (turning) left indicates that the instrument is being rotated in an anti-clockwise direction in plan.

6.6.3 Single horizontal angle

A *single horizontal angle measurement* is made by pointing on the RO, booking the horizontal circle reading, then turning the instrument clockwise to point the forward object and again booking the circle reading, all on FL. The difference between the two circle readings is the required horizontal angle and the reading when pointed on the reference object is the *zero* of the measurement. (Remember that theodolite horizontal circles are normally graduated clockwise from 0° to 360°.) If, as an example, the circle read 5° when pointed on the RO and 65° when pointed on the forward object, then the horizontal angle is 60°, measured from a *zero of 5°*. In practice, the *zero* is not normally set to the actual value of 0°, but instead to slightly more than 0° or to some other convenient value, as will be seen later.

6.6.4 Single horizontal angle by simple reversal

Simple reversal means to measure a single horizontal angle as above, then transit the telescope on to FR, point the forward object again, read and book the circle, then point the RO and read and book the circle. This procedure gives two measures of the angle, the first being the difference between the two FL readings and the second the difference between the two FR readings, the mean of the two values being accepted as the value of the angle if there is no gross error. Note that the pairs of readings (FL and FR) for each target direction provide checks on reading and instrumental errors. A value of an angle obtained in this way, with the graduated horizontal circle kept stationary in one position throughout the operation, is a value obtained by *simple reversal on one zero.*

The accuracy of this first value may be improved by moving the horizontal circle to a new zero setting, repeating the whole operation to obtain a second value, then taking the mean of the two values. The final value of the angle may be said to have been taken by *simple reversal on two zeros* which, barring gross errors, is more accurate than using only a single zero. As a general rule then, angles should be measured by simple reversal and for higher accuracy, two or more zeros should be used and the set of results meaned.

The recommended procedure for observing a horizontal angle is as follows:

(i) Centre and level-up the instrument over the station, focus the eyepiece on the cross-hairs.
(ii) With the instrument on face left, set on a small angle (the first zero) and sight on to the RO. *If the instrument has circle orienting drive, however, sight on to the RO and then set on the small angle.*
(iii) Read and record the horizontal angle reading to the RO.
(iv) Unclamp the telescope, release the horizontal or upper plate clamp, swing right and sight on to the forward object through the open sights or the finder-collimator.
(v) Clamp the telescope, apply the horizontal or upper plate clamp, focus accurately on the distant target.
(vi) Bisect the target with the vertical cross-hair, using the horizontal or upper plate slow-motion screws.
(vii) Read and record the horizontal angle reading to the forward object.
(viii) Unclamp the telescope, change face to face right by transiting the telescope.
(ix) Release the horizontal or upper plate clamp and swing left to sight on to the forward object again.
(x) Repeat procedures (v), (vi) and (vii) for a second reading to the forward object on FR.
(xi) Unclamp the telescope, release the horizontal or upper plate clamp, swing left to sight on the RO using the sights or collimator, and repeat procedures (v), (vi) and (vii) again to obtain a second reading on the RO on FR.
(xii) Calculate the mean value of the angle.

Four circle readings should have been read and recorded, two on each face, giving two measures of the angle which may be meaned. If desired, the whole process from (ii) above may be repeated with a new zero value set on the circle. If two zeros are to be used, the first could be made just over 00° and the second just over 90°, and if four zeros are to be used, they could approximate to 0°, 45°, 90° and 135°, being just over in each case. The general aim is to distribute the measurements around the circle to help reduce errors.

Two or more zeros are essential when an accuracy close to the limitations of the equipment is demanded. Thus, when using a one-second theodolite it is necessary to use four or more zeros to ensure that the mean angle is correct to within one second. In addition, it is desirable to use two or more zeros if re-observation would entail a large expenditure of time and effort. In angular observation work generally a disproportionate amount of time is lost in getting to the site, setting up the equipment, etc., while the time spent on observing is minimal. An experienced surveyor would have observed and recorded the four readings above within ten minutes, but it would have taken as long to set up the equipment and still longer if targets for the reference and forward objects had to be established. Travelling time to the site must also be taken into account.

6.6.5 A round of angles

If angles are to be measured from the RO to several forward objects, the above simple reversal procedures can be used, suitably adapted.

With the instrument on face left, after noting the circle reading on the RO, swing right and note the readings on all the forward stations in turn.
Transit the telescope to face right, turn the alidade and sight the last station again and note the reading.
Swing left and note the readings on all the stations in turn, including the RO.

This gives one complete *round of angles*, on *one zero*. For improved accuracy a further round can be taken by transiting to face left, changing the zero setting on the circle to another value, then repeating the operations again. This may be done with as many zeros as desired; *Figure 6.11* shows readings on two zeros.

Horizontal readings at .... Station. A3												
Point	Face Left			Face Right			Mean			Reduced to RO		
B (RO)	00	01	50	180	02	10	00	02	00	00	00	00
C	18	20	30	198	21	10	18	20	50	18	18	50
D	55	18	25	235	19	25	55	18	55	55	16	55
B (RO)	45	15	10	225	16	20	45	15	45	00	00	00
C	63	34	20	243	35	10	63	34	45	18	19	00
D	100	31	00	280	32	00	100	31	30	55	15	45
Final Angles	B-A3-C = 18° 18' 55"						B-A3-D = 55° 16' 20"					

Figure 6.11

6.6.6 Repetition

The *repetition method* is used to obtain a relatively high accuracy measurement of a small horizontal angle. As an example consider stations A and B, the angle they subtend at the instrument station being required. The procedure is as follows:

(i) Set up the instrument, set a small zero on the horizontal circle, point station A and note the reading.
(ii) Release the upper circle, turn and point B, note the reading. The difference between the readings is a measure of the angle.
(iii) With the upper circle still clamped, release the lower clamp and point A again, apply lower clamp (circle still shows the reading obtained when pointed on B).
(iv) Release the upper clamp, turn and point B again.

The current reading on B *minus* the initial zero setting is actually twice the required angle. This process can be repeated, say ten times to give ten times the angle, then the final value on B *minus* the zero setting may be divided by ten to give a measure of the angle. The measurements described have all been taken on FL, but the process can be repeated on FR and swinging left.

An instrument with repetition clamp (circle locking lever) is particularly suitable for this operation, the horizontal circle being alternately clamped down to the base and clamped up to the alidade using the repetition clamp.

6.6.7 Vertical angles

A recommended procedure for vertical angle observations is as follows:

(i) Release all clamps, point the telescope at the target with face left, using the open sights or finder-collimator.
(ii) Apply the vertical circle (telescope) and horizontal circle clamps.
(iii) Focus carefully on the target.
(iv) Bring the intersection of the cross-hairs exactly on to the target, using a horizontal slow-motion screw and the telescope slow-motion screw.
(v) Read and record the vertical circle reading. *Note that if an altitude bubble is fitted, this must be centred before the reading is made.*

The vertical circle in modern instruments is usually graduated from 0° to 360°, with zero at the zenith, and when face left the vertical circle reading is actually a zenith distance or zenith angle. Thus with face left, the vertical angle is equal to *90° minus the circle reading*, but with face right the vertical angle is equal to *the circle reading minus 270°*.

To eliminate errors, a vertical angle should be observed face left and face right and the two values meaned. Horizontal angles are generally required to be of greater precision than vertical angles and it is therefore common practice to observe all the horizontal angles at a station before commencing the vertical angles, in case any settlement of the tripod occurs. On some occasions, however, such as in poor visibility, it may be expedient to observe the vertical angles first, as this gives the surveyor some familiarity with the target direction.

6.7 Adjustments

From time to time it becomes necessary to make adjustments or carry out maintenance on the equipment. Tripod fittings may become loose and require to be tightened up, wear may develop in the foot screws, etc. Many of these details are specific to the individual instrument and the maker's handbook should be referred to for the details of what maintenance may be carried out by the surveyor. Certainly a surveyor working in the less populated areas of the world should include the instrument handbook in the survey stores.

Depending upon the age and type of instrument, there may be up to four permanent adjustments of the theodolite. These are:

(i) plate level axis perpendicular to vertical axis,
(ii) collimation line perpendicular to transit axis, combined with diaphragm orientation,
(iii) transit axis perpendicular to vertical axis, and
(iv) vertical circle index adjustment.

These are described in the following sections.

6.7.1 Plate level perpendicular to vertical axis

To check: Carry out the operations described in Section 6.2.3.1 (i), (ii), (iii) and (iv). If the alidade is now rotated through 180° and the bubble does *not* stay central, then adjustment is required.

To adjust: Rotate the alidade through 180° again to the first position, then using the two foot screws parallel to the plate level, bring the bubble half-way back to centre. Turn the alidade through 90° and again bring the bubble half-way back towards the centre position. The instrument should now be level, but the bubble tube requires to be adjusted.

Return the alidade to the start position, then centre the bubble using the adjusting screws fitted at one end of the bubble casing. Repeat the check and adjust as necessary.

6.7.2 Collimation line perpendicular to transit axis

To check: Set up the instrument in an area of reasonably level ground in the normal way and level up carefully. On face left, aim on a fine target at least 100 m away, carefully bisecting the target with the intersection of the cross-hairs, using the clamp and tangent screw. Book the horizontal circle reading as $R1$.

Unclamp the alidade, transit the telescope, turn through 180° and bisect the target again on face right. Book the horizontal circle reading as $R2$.

If the horizontal collimation adjustment is correct, then $(R2 - 180°) = R1$. If not in adjustment then error $E = ½[(R2 - 180°) - R1]$.

To adjust: Calculate $(R1 + E)$, unclamp the alidade, transit the telescope, turn and point towards the target again, and bring the horizontal circle reading to exactly $(R1 + E)$ using the clamp and tangent screw.

The vertical hair will no longer bisect the target. To complete the adjustment, move the vertical cross-hair to exactly bisect the target using the lateral adjustment screws of the diaphragm. (This must be done carefully to avoid any rotation of the diaphragm.) Repeat the check and adjustment as necessary. *It is preferable to have this adjustment carried out by the manufacturer!*

6.7.3 Vertical cross-hair orientation

Movement of the diaphragm in carrying out the horizontal collimation adjustment may result in a small rotation of the diaphragm in its mounting, so that the vertical hair is not properly vertical.

To check: With the instrument still set up as above, carefully levelled, observe a fine distant target and tilt the telescope up and down and check whether the vertical hair stays on the point, on both faces. If it stays on the point, no adjustment is needed.

To adjust: If the vertical hair strays off the point, the diaphragm must be rotated in its mount to remove the error. Again, this is better carried out by the maker.

6.7.4 Transit axis perpendicular to vertical axis

In older instruments, the position of one end of the transit axis could be adjusted up or down. Today, manufacturers do not provide any adjustment, maintaining that modern production methods make this unnecessary. The effects of any error can be eliminated by meaning face left and face right observations.

6.7.5 Vertical collimation (vertical circle index error)

When the theodolite is correctly levelled up and the collimation line is set horizontal, the vertical circle on most modern instruments should read 90° or 270°, according to which face is being used, although some read 0° or 180°.

To check: Set up the instrument with face left, sight a fine mark on a target about 100 m distant with the horizontal hair, read the vertical circle. Transit the telescope to face right, turn and point the mark again, read the vertical circle. Add the two vertical circle readings together, they should sum to 360° on most instruments. Any difference between the sum and 360° indicates vertical index error.

To adjust: The adjustment can only be carried out by the manufacturer, but pending this the effects of the error can be eliminated by taking the mean of FL and FR readings.

6.7.6 Optical plummet adjustment

When the theodolite is levelled up, the optical plummet's collimation line and the vertical axis of the instrument should coincide.

To check: If the plummet is in the alidade, and thus can be rotated in plan without disturbing the levelling, level up the instrument then place a piece of paper on the ground and carefully mark the point where the plummet collimation line impinges upon the paper. Turn the alidade through 180° and make a second mark. If the two marks do not coincide the plummet requires adjustment to the mean position.

If the plummet is in the tribrach and cannot be rotated without affecting the levelling, then the instrument should be laid on its side on a bench with its base towards a wall and the point where the plummet collimation line hits the wall marked. Turning the tribrach through 180° and repeating the process will give a second mark, and the two should coincide.

To adjust: The maker's handbook should be consulted for guidance for the actual instrument concerned as it may be the diaphragm or the objective lens of the plummet which must be moved.

As a general rule it is strongly recommended that if adjustments are required to optical reading systems or the instrument axes, they should be carried out by an instrument mechanic specializing in optical theodolites. As with a car, a complex modern theodolite should be serviced occasionally.

Chapter 7

Indirect distance measurement

7.1 Introduction

This chapter deals with the indirect measurement of distance, using *optical distance measurement (ODM)* methods or the more modern *electromagnetic distance measurement (EDM)* methods.

In the past special instruments were produced for the optical measurement of distance, including *direct reading tacheometers* or *tachymeters*, for detail survey, and *horizontal wedge* instruments, *subtense bars*, etc., for more accurate distance measurement. 'Tacheometry' is from the Greek *tachus* (swift) and *metron* (measure), and the term was applied collectively to all forms of optical distance measurement, as they were much quicker than traditional direct distance measurement. An early form of English spelling was *tachymeter* and *tachymetry*, now lapsed.

The tacheometers had a typical accuracy in distance measurement of only about 1:500, and are all now obsolete, having been displaced by modern EDM methods and equipment. The wedge instruments could measure distance with relatively high accuracy, but again could not compete with EDM. Subtense bars are still used for some industrial measurement control tasks but will not be covered here.

Although specialist optical distance measurement equipment is no longer made today, occasional tasks may be carried out by the method of *vertical staff tacheometry* or *stadia measurement* using an ordinary theodolite, and it may be a useful standby on occasion. Such tasks might include limited detail survey of natural features or contouring, since both the distance and height of points may be obtained.

EDM can be carried out either using a theodolite with a distance-measuring device attached, or using a *total station*, a special-purpose instrument resembling a theodolite but capable of measuring both angles and distances by electronic methods.

7.2 Optical distance measurement

Vertical staff tacheometry or stadia measurement, the simplest surviving form of ODM, uses an ordinary theodolite or a surveyor's level with horizontal stadia hairs on the diaphragm (as described in Chapter 5), in conjunction with a levelling staff. Before the advent of EDM this method was widely used for providing detail and contours for 1:2500 or 1:5000 scale mapping. At the larger scales, however, the accuracy is adequate only over short distances, and control lines would need to be measured by other methods.

7.2.1 Distance measurement by stadia observations

In the chapter on levelling it was shown how the stadia hairs on the level diaphragm are used to check the centre hair readings and also the distance from the instrument to the staff. Thus, with a horizontal sight line and a vertical staff, as in *Figure 7.1(a), horizontal distance from centre of instrument to face of staff* = 100 × staff intercept, or

$$D = 100s,$$

or

$$D = sk \tag{7.1}$$

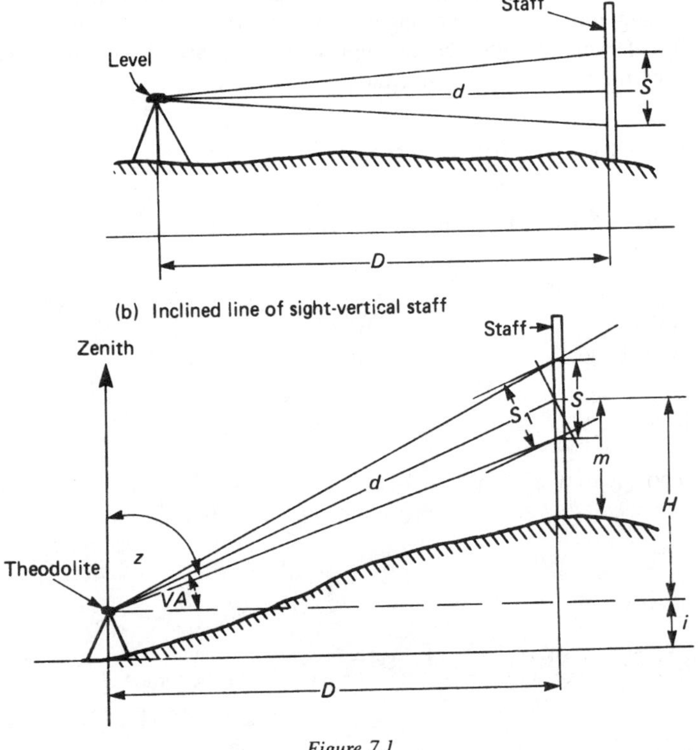

Figure 7.1

where k is the instrument multiplication constant of the instrument, generally 100 today, s is the staff intercept and D is the distance from the centre of the instrument to the face of the staff.

Seventy or eighty years ago or so most theodolites had an additive as well as a multiplication constant and the formula was written

$$D = sk_1 + k_2$$

k_1 and k_2 being the multiplication and addition constants, respectively. Some modern telescopes have an additive constant, but it is normally so small with respect to the accuracy of stadia measurement that it is ignored.

In addition, in older instruments the multiplier was often not exactly 100, but this problem no longer arises unless a reticule has been replaced in a telescope.

In stadia surveys using a theodolite, the line of sight is seldom horizontal, the typical case being rather as in *Figure 7.1(b)*. This shows a theodolite aimed on a staff held at a point which is higher than the instrument station, but it may be re-drawn to cover the case where the telescope is depressed. In the case illustrated, s is the staff intercept, but the inclined distance d is not equal to $100s$. In fact the inclined distance is equal to $100s_1$, and it is necessary to calculate s_1.

It will be observed from the figure that the angle between the faces of the real and imaginary staves s and s_1 is the reduced vertical angle, *VA*. The two small triangles formed by s, s_1 and the lines of sight through the stadia hairs are roughly equal, the larger angle in each triangle approximating to a right angle, one being slightly greater and the other being slightly less than 90°. Taking these as being right angles for practical purposes, then

$$s_1/2 = \cos VA \times s/2, \text{ or } s_1 = s \cos VA$$

From *Figure 7.1* it will be seen that for inclined sights, horizontal distance $D = d \cos VA$, but

$$\begin{aligned} d &= 100s_1 \\ &= 100s \cos VA \end{aligned} \quad (7.2)$$

therefore

$$D = 100s \cos^2 VA$$

or

$$D = sk \cos^2 VA \quad (7.3)$$

If the zenith distance angle z is used, the expression becomes

$$D = sk \sin^2 z \quad (7.4)$$

7.2.2 Heights by stadia measurement

Stadia observations can also be used to determine the height difference between an instrument station and a distant ground point where the vertical staff is held. Referring to *Figure 7.1(b)*, with an inclined sight,

$$\textit{vertical distance } H = d \sin VA$$

therefore,

$$H = 100s \sin VA \cos VA$$

or

$$H = sk \sin VA \cos VA \tag{7.5}$$

If the zenith distance angle z is used, the expression becomes

$$H = sk \sin z \cos z \tag{7.6}$$

(It should be noted that $\sin z \cos z = \frac{1}{2} \sin 2z$, and this form is sometimes preferred for calculation.)

If both D and H are to be calculated, then

$$H = D \tan VA \tag{7.7}$$

or

$$H = D/\tan z \tag{7.8}$$

The value H, of course, is the difference in height between the transit axis of the instrument and the centre hair reading on the staff, while what is required is the height difference between the reduced level of the instrument station and the ground point at the staff. Taking the instrument transit axis height above the station marker as i, and the centre hair reading as m, then in *Figure 7.1(b)* the required height difference is $i + H - m$, or in a more general form, $i \pm H - m$.

Thus, if the instrument station reduced level is RL_{inst}, the reduced level of the staff position is

$$RL_{inst} + i \pm H - m \tag{7.9}$$

The arithmetic is simplified if the centre hair is always pointed at the height of instrument, so that i is kept equal to m.

7.2.3 General

Equations (7.3) and (7.5), or (7.4) and (7.6) are used to deduce the required values for distance and height, and reduction was commonly carried out by the use of special tacheometric tables and/or tacheometric slide rules or diagrams.

This work was very laborious before programmable electronic calculators and computers were developed, but the fieldwork itself is relatively fast.

Sources of error in stadia measurement include:

(i) differential refraction, making s unreliable,
(ii) non-vertical staff, again making s inaccurate,
(iii) equipment faults, errors in k_1 and k_2, not generally significant today, and
(iv) mistakes in reading s and the vertical circle.

If sights are kept short, say not more than 100 m, a staff bubble is used to ensure verticality, grazing rays within 1 m of the ground are avoided, and all readings made carefully, the error in a single distance should be between 1:500 and 1:1000 at the best, and height differences should be accurate to ± 50 mm.

Due to its relatively low accuracy in distance measurement, theodolite tacheometry is not used for the measurement of survey lines in control. The application of theodolite tacheometry to *detail survey by radiation* is outlined briefly in Chapter 10. Note that the method of stadia measurement described is sometimes known as *vertical staff tacheometry* because there is another version of the method in which the graduated staff is not held vertically but rather normal (at 90°) to the instrument sight line.

7.3 Electromagnetic distance measurement

7.3.1 Background

Electromagnetic waves transmit energy through space and matter. Radio waves, infrared, visible light, ultraviolet light, X-rays and gamma rays all travel at a speed of approximately 300 000 km/s in vacuum and slightly less in the atmosphere.

EDM in surveying may be considered to have had its origin in the 1939–45 war development of radio-location or radar. Assuming a fixed speed, a radio signal was 'bounced' off a target and the time delay between the emission of the signal and its return allowed calculation of the target distance. Although accurate enough for its purposes at the time, the method was not sufficiently accurate for survey work.

Independent research in Sweden and South Africa produced the first practical EDM instruments for surveyors. Dr Bergstrand of the Geographical Survey of Sweden, in collaboration with the company AGA, produced the AGA Geodimeter Model 1 instrument in the early 1950s. This was a long-range instrument for distances up to 30 km, using visible light. The instrument was placed at one end of the line to be measured and a mirror placed at the other end, then the signal was transmitted along the line and reflected back to the instrument.

Dr T. L. Wadley of the South African Council for Scientific and Industrial Research at a similar time developed the Tellurometer MRA1, which used radio waves. Two transmitter/receiver instruments were used, a Master and a Remote, at either end of the line to be measured, and equipped with radio communication links.

In each case the instruments gave a measure of the slope distance, which had then to be corrected for slope, temperature, barometric pressure, humidity and the curvature of the earth.

7.3.2 Measurement principle

Although visible light waves, radio waves, etc., all travel at the same speed, their wavelengths and frequencies may vary. *Wavelength* is quoted in *metres*, *frequency* in *cycles per second* or *hertz* (1 cycle/s = 1 hertz), as illustrated in *Figure 7.2*.

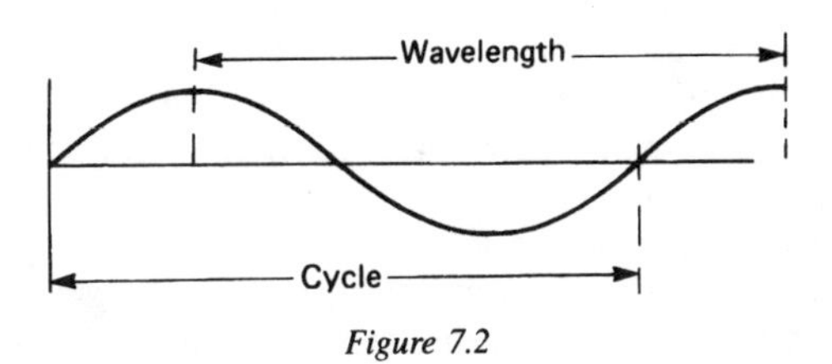

Figure 7.2

Taking the approximate speed of 300 000 km/s, a signal with a wavelength of 1500 m will have a frequency of

$$300\,000 \times 1000/1500 = 200\,000 \text{ cycles/s} = 200\,\text{kHz}$$

VHF radio may use a frequency of 300 MHz, or 300 million cycles/s, giving a wavelength of 1 m.

Modern EDM makes use of these wavelength and frequency characteristics by transmitting signals on more than one frequency, then determining the *phase difference* between the emitted and the returned signal at each frequency, as in *Figure 7.3*.

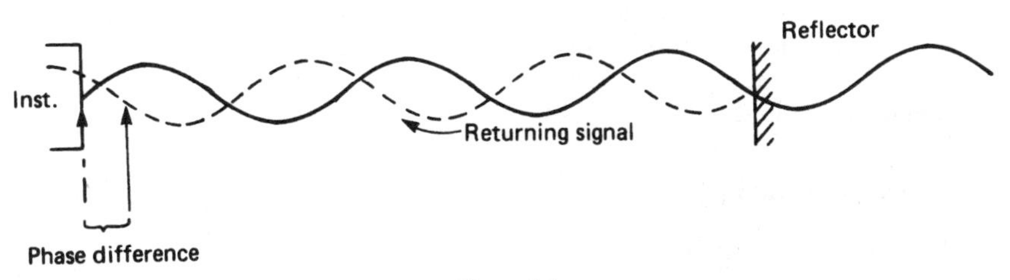

Figure 7.3

In practice, the EDM signal is a *carrier wave*, modulated to produce a measuring frequency and the instrument uses the phase difference of the measuring frequency to determine the slope or slant distance between the instrument and the retro-directional prism reflector placed at the other end of the line being measured. The reflector prism returns the beam along a line parallel to the incident beam. The instrument and reflector must be intervisible, and the range will be affected by mist and rain. Atmospheric conditions generally affect the signals, in particular relative humidity, and grazing rays should be avoided. Modern instruments are powered by clip-on rechargeable NiCad batteries.

7.3.3 Development of modern equipment

The early instruments were medium to long range and would not have been suitable for the class of work covered by this text. By the mid-1960s, however, the traditional survey instrument makers were developing and producing EDM equipment and since then there has been continuous development of new instruments. These were originally made as 'add-on' units attached either to the telescope or to a yoke on the standards of a theodolite, but today both add-on and independent instruments are made. An add-on instrument is commonly termed a *distancer*, while an independent instrument is termed a *total station*, or sometimes an *electronic tacheometer*, and consists of an electronic theodolite with a built-in EDM distancer.

In either form of instrument, the slope distance is measured, and the vertical angle observed, then the horizontal distance and vertical height difference may be calculated. Modern short-range instruments may measure distances from under 1 m up to 3 or 4 km with a single retro-reflector prism at the other end of the line, typically to an accuracy of ±5 mm ±5 parts per million (usually written ppm) or better, using modulated infrared light, and longer ranges can be obtained with more prisms in the target.

Makers typically specify instrument accuracy in the form ± x mm ± y ppm, where x represents the instrument constant and y is the systematic error, varying with the distance being measured.

Today, then, the simplest add-on distancer may give slope distance only, while a total station can display slope and horizontal distances, horizontal and vertical angles and vertical height differences, and even compute and display the rectangular co-ordinate values of the observed points. Calculations which originally took much labour and an A4 page of computations are displayed directly by the instrument, and all the data may be transferred to a computer, or stored for later transfer.

Where an add-on instrument is used, some makers supply a keyboard calculator which can be attached to the theodolite supporting the distancer. The slope angle and horizontal direction may be read manually, then keyed into the calculator keyboard, and the horizontal, vertical and co-ordinate distances displayed as required.

7.3.4 General field use

7.3.4.1 Add-on instruments

To measure a line using an add-on distancer, a theodolite and tripod are carefully set up over the mark at one end of the line being measured, and centred over the mark and levelled up in the usual way, with the distancer attached to the theodolite. (The following notes and illustrations refer to Leica equipment, but these are simply used as typical examples of a coherent range, there are many manufacturers who produce similar ranges of equipment.)

Figure 7.4 shows a Leica Wild DI1600 Distomat attached to a theodolite. At the other end of the line a reflector prism is placed over the other end mark. The prism may be supported on a hand-held rod as in *Figure 7.5*, or placed in a tribrach mounted on a tripod centred over the mark. In either case, the prism mounting may incorporate an aiming target, since the aiming head must be pointed accurately at the prism. No great accuracy is needed in pointing the prism

Figure 7.4

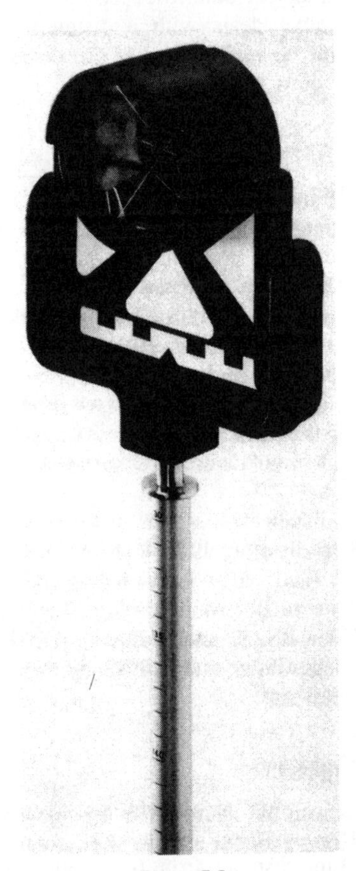

Figure 7.5

at the aiming head, however. Longer measuring ranges may be achieved by using multiple reflectors, up to 11 in the Leica system.

The DI1600 can operate in four distinct measuring modes.

In *standard mode* the measurement is made in a time between 1.5 and 3 s, with a resolution of 1 mm and an accuracy of ±3 mm ± 2 ppm.

In *repeat mode* automatic repeat measurements are made and the mean value displayed. This improves reliability in difficult atmospheric conditions.

Tracking mode is used for setting out. After the initial measurement successive measurements are made every 0.3 s, giving an accuracy of ±10 mm ± 2 ppm. An even faster form of *rapid tracking* gives an accuracy of ±20 mm ± 2 ppm.

With the DI1600 on the T16 the angle of slope must be read manually to calculate the required horizontal and vertical distances, and the instrument and target prism heights above the ground mark must be measured, and the reductions carried out as in optical tacheometry. Alternatively the Leica Wild GTS5 keyboard can be clipped to the side of the theodolite and the angles can be input and the various values calculated directly. If co-ordinate differences for the distant point are to be found, then the horizontal circle reading must be observed and included in the calculations, or input to the attached keyboard.

The DI1600 is classed as a medium range (max 8 km) and medium accuracy instrument, there are a variety of other models.

While the example above deals with a distancer attached to an optical theodolite, the work would be still quicker if the distancer was attached to an electronic theodolite.

7.3.4.2 Total stations

A typical mid-range total station instrument is shown in *Figure 7.6*, the Leica Wild TC1100. This is essentially an electronic theodolite with an EDM distancer built in.

The total station must be set up over one end of the line to be measured and the target prism(s) at the other end as described previously. The instrument must be oriented and pointed manually at the prism, and any required corrections or parameters input to the built-in keyboard, together with any data coding, but thereafter all distance values and angle readings are produced electronically. With many instruments such as this it is only necessary to point then press an 'ALL' key, and the distance and point details are automatically recorded. A typical keyboard is shown in *Figure 7.7*.

The TC1100 measures distances to ±2 mm ± 2 ppm, at a range of 2.5 km with one prism, and can store details of up to 8000 points on a PCMCIA memory card. It has keyboards on both faces, a two-axis liquid compensator and automatic correction of all axis errors, an electronic level, and an eight-line matrix display. Data may be transferred via RS232 and it may be interfaced to data recorders.

Higher accuracy and longer range instruments are made, and also total stations suitable for the construction site.

7.3.4.3 Timed-pulse distancers

These instruments differ from the instruments previously described in that they use an infrared laser carrier wave, but instead of continuous modulation of phase the carrier is pulsed and the distance from the emitter to the target and back to

Figure 7.6

Figure 7.7

the receiver is determined from the time taken for a pulse of infrared light to cover this course.

Figure 7.8 shows the Leica Wild DI3000S, designed for attaching to a theodolite. This can measure up to 9 km with a single prism, and up to 19 km with 11 prisms under good conditions, while up to 1500 m is possible with reflecting tape instead of prisms. This instrument can measure very rapidly, typically to ±3 mm ±1 ppm using a reflector, and a particular advantage is that it can be used with moving targets. It can be used for such varied tasks as directing moving machinery, e.g. tunnelling equipment, deformation measurements, off-shore work, etc.

Figure 7.8

A variation on this, the Wild DIOR3002S, can be used *without reflectors* up to 350 m, depending upon the conditions, and projects a laser dot in the centre of its beam, with an accuracy of 5 to 10 mm. The reflectorless operation is appropriate for the measurement of the surface of cavities, e.g. chambers, tunnels, tanks, and for surfaces such as cooling towers, dams, quarry faces, etc.

7.3.5 Error sources and corrections

7.3.5.1 Atmospheric conditions

EDM results are directly affected by the atmospheric conditions prevailing at the time of measurement and the greater the distance the greater the effect. To correct for these some instruments are fitted with an *atmospheric correction switch*, while others allow corrections to be input directly to the control panel. In either case, the temperature and pressure must be measured carefully and the maker's instruction sheets followed. Corrections are usually entered in 'parts per million'.

7.3.5.2 Prism constant

This is related to the distance between the effective centre of the reflective prism and the plumb point. If the instrument maker's own prisms are used no correction is needed, but if a different make of prism is used a correction will need to be entered into the instrument.

7.3.5.3 Slope

When distances are measured on slope using a total station, the instrument calculates the horizontal distance and height difference directly. If a distancer is used on a theodolite it may be necessary to calculate the slope correction, although some distancers carry out the calculation and display D and H. Where a distancer is used, it is important to ensure that the measured slope line is parallel to the slope angle measured by the theodolite. (This should not arise with a total station as the EDM optics are co-axial with the telescope.)

7.3.5.4 National Grid scale factor and height correction

Horizontal distances may require to be transformed to OS National Grid distances and reduced to sea level, as in Sections 4.4.8 and 4.4.9.

7.3.5.5 Instrumental errors in EDM

EDM measurements are subject to *scale error* (or *frequency drift*), *zero error*, and *cyclic error*.

Scale error is the result of changes in the modulation frequency of the EDM instrument and is proportional to the length of line measured.

Zero error occurs where there are differences in the mechanical, optical and electrical centres of the instrument and reflectors. This error is a constant.

Cyclic error results from variations in the phase shift measurement.

Equipment should be maintained regularly by the maker, but it may be checked for the above errors on site by a *calibration* procedure, although no adjustments to the instrument can be made on site. A more advanced text should be consulted for this procedure.

7.3.5.6 Accuracy of EDM

As shown earlier, the accuracy of EDM equipment is typically stated by manufacturers in the form

$$\pm x \,\text{mm} \pm y \,\text{ppm} = \pm x \,\text{mm} \pm y \,\text{mm/km}$$

The values x and y are both standard errors, thus the standard error of a distance measurement D metres is

$$s_D = \{x^2 + (D \times y \times 10^{-3})^2\}^{1/2}$$

For an instrument quoted as ± 5 mm + 3 ppm, on a distance of 1300 m, then

$$s_D = \{5^2 + (1300 \times 3 \times 10^{-3})^2\}^{1/2} = 6.34 \,\text{mm}$$

It will be evident that except for very long distances the y component can be ignored for practical purposes.

7.4 Applications of EDM

EDM instruments are widely used now for all forms of linear measurement where distances exceed a tape length or so, with increased speed and accuracy. In addition, total station instruments may incorporate software to carry out a variety of specialized tasks, although manufacturers' products vary in their details.

7.4.1 Control

Control layouts which formerly relied primarily on high accuracy angle measurement and triangulation can use trilateration methods with lines measured to very high accuracy, and with much greater speed. Total station software relevant to control surveys may include the following.

(1) *Orientation program.* When set over a station and the co-ordinates of a distant reference station entered, the program can calculate the bearing of the distant station. On sighting that station the program will then set the instrument's horizontal circle to the calculated bearing.
(2) *Tie distance program.* If the instrument is pointed on two distant stations in turn and these are observed in the usual way, this program can determine the length of line and height difference between the points. This may be continued through a series of distant points.
(3) *Resection program.* If observations are made on two distant points of known co-ordinates and height then this program can compute the three-dimensional co-ordinates of the instrument's position (see Section 9.6.2.4).

7.4.2 Setting out

Setting out operations which relied heavily on site grids and offsets from lines now use polar methods and location by co-ordinates, and road curve centrelines can be set out by polars from offset positions rather than by tangential deflection angles along the centreline.

Where an instrument has a 'stake out' program, a required set out distance can be entered into the instrument, then observation made on a reflector at the approximate distance. The instrument will then display what distance the reflector must be moved to reach the required distance, plus or minus. See Chapters 12, 13.

7.4.3 Detail survey

Detail survey by optical tacheometry is now displaced by electronic tacheometry, giving faster and more accurate results, and with the stream of data automatically reduced and stored in a data recorder or transmitted to a computer.

Where a detail survey has been carried out by EDM the resulting computer database of detail points may be used to create a *digital terrain model* as discussed in Section 10.6.

7.4.4 Levelling

Levelling may be carried out by EDM methods, with multiple heights fixed from one instrument station, using the same basic equation (7.9) as shown for vertical staff tacheometry, with i the height of the centre of the transmitter above the station point and m the height of the reflector prism centre above the ground point. Again, it is convenient to keep i and m the same value.

A specialized levelling application is remote elevation measurement. With this program the height of a distant inaccessible object may be obtained, provided it is possible to place a reflector immediately vertically below the distant object. With the reflector in position it is observed in the usual way, giving the point distance and height, then when the telescope is pointed on the high object the program will calculate the height of the object. A typical example might be the determination of the height of overhead high tension electrical cables.

Chapter 8

Levelling applications

8.1 Introduction

This chapter is concerned with the applications of traditional height determination using a surveyor's level, commonly termed *levelling*. Levelling is used to establish the heights of one or more points, located either in a line over the Earth's surface, or in a series of lines, or covering an area. This chapter deals with the common ordinary levelling tasks of

(1) levelling to establish a temporary bench mark (TBM),
(2) levelling for contours, and
(3) levelling for sections and cross-sections,

together with some consideration of precise levelling and river crossing problems. Levelling for setting out is considered in Chapter 12.

8.2 Establishing a TBM

This is done by running a line of *flying levels* (back and foresights only, no intermediate sights) from a bench mark to the proposed TBM, then from the TBM to another bench mark, or alternatively, back to the original bench mark. This procedure allows a check on the calculated level of the TBM.

8.2.1 Operations before starting to level

Before starting to level, obtain the necessary bench mark information, locate the bench marks, select and establish a mark for the TBM. Decide on the route the levelling is to follow and (if considered necessary) check the accuracy of the level to be used.

8.2.2 Levelling the TBM

Set up the level, sight on to the opening (starting) bench mark, read the staff to the nearest 0.001 m, book the reading as a backsight, then check it again. Signal the staff holder to move to the next staff position. (It is important that the surveyor does not leave the level unattended at any time, particularly where there is considerable pedestrian or vehicular traffic, thus the use of hand signals is important.) Sight the staff at the new position, read the staff, book the reading as a foresight, check as before.

Signal the staff holder to remain in position and move the instrument to the next position. Note that at all times either the staff should be in position or the level should be in position – they should never both be off the ground and moving at the same time. When moving the instrument, it may be carried on its tripod with the legs closed and held in a vertical position, but if being transported in a vehicle then the level should be placed in its case and the case held on the lap of one of the passengers.

Set up the level at the new position, sight the staff and take a backsight reading as before, signal the staff holder to move on, then take a foresight reading on the staff and repeat these operations until the TBM is reached. The line of levels must close (finish) with a foresight reading on to the new TBM.

When the TBM has been reached, move the instrument to a new position and repeat the whole process to close a new line of levels on to the second bench mark, or alternatively, back on to the opening BM.

8.2.3 Reducing the levels

The levels should be reduced by the Rise and Fall method, as described in Section 5.9.7, and the line should close within $\pm 10\sqrt{K}$ mm, where K is the total distance levelled in kilometres.

In establishing a TBM the positions selected for the instrument and the staff should be firm and solid, keeping cumulative errors at a minimum and any misclosure should then be distributed equally through the changepoints.

8.3 Contouring plans by level and staff

A *contour* may be defined as a line on a plan or map representing an imaginary line on the ground connecting all adjacent points of equal height. (The best visual contour line is the water's edge of a still lake.) The difference in height between successive contour lines is termed the *contour vertical interval* (VI) and is generally constant throughout one drawing. The shortest horizontal distance between any two contour lines varies with the slope of the ground and is termed the *horizontal equivalent* (HE). The salient features of the ground are readily observed from a study of the contour lines, close contours showing steep slopes, widely separated contours showing gentle slopes and evenly spaced contours a constant slope. Contour lines never cross one another, but several contour lines coming together indicate a cliff or overhang and contours always close on themselves, although they may not do so on the particular map or plan.

Individuals who make infrequent use of contoured maps or plans sometimes have difficulty in deciding whether a particular feature is a ridge or a dry valley, since both appear similar on the drawing, see *Figure 8.1*. The simplest approach is to imagine oneself standing on the ridge or in the valley and looking along the feature, then turn through 90°. An increase in the contour values indicates that

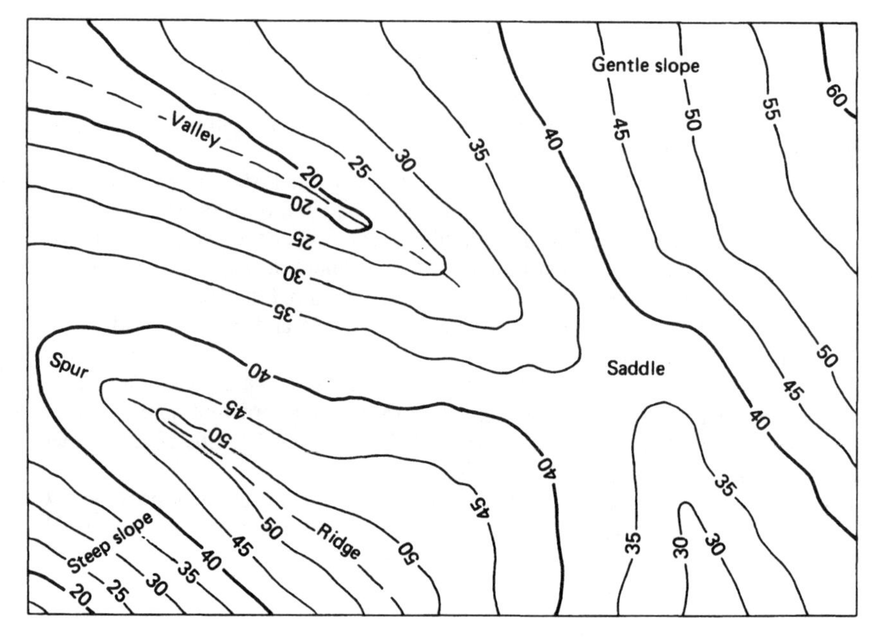

Figure 8.1

one is looking up a hill (in a valley), while a decrease indicates looking down a hill, standing on a ridge.

8.3.1 Uses of contoured maps and plans

A contoured map or plan should give the user a picture of the ground which will enable the relief of the ground to be interpreted. A ground profile, i.e. a section, could be produced along any required line on the plan, thus determining the location of a route to a suitable gradient, or a check on whether distant points are intervisible, or the suitability of an area for a reservoir, or the extent and volume of earthworks. Contoured plans are often useful for the planning of the layout of estates, although on small building sites, etc., contours are not often demanded. For minor road and drain layout, a few spot heights scattered over the area concerned is often sufficient.

8.3.2 Choice of contour interval

This depends primarily upon the scale of the map and the ground relief, together with the purpose of the survey, although time and the finance available will have an influence. On small building sites a VI of 0.2 to 0.5 m is often recommended, but the plans of such sites are often to very large scales. Large estates, housing or industrial, and reservoirs, then a VI of 0.5 to 2.5 m is frequently recommended. For small-scale plans for roads and railways, often 1 to 5 m is used.

As a guide, *Table 8.1* is recommended (note the similarity of the digits in the centre row). On very steep ground the VI can be greater, and conversely smaller on very flat ground.

Table 8.1

Ground relief	*Common plan scales*			
	1:200	*1:500*	*1:1000/1:1250*	*1:2500*
Very flat	0.2 m	0.2 m	0.5 m	1 m
Gentle slopes or undulations	0.2 m	0.5 m	1.0 m	2.5 m
Very steep	0.5 m	1.0 m	2/2.5 m	5 m

8.3.3 Methods of locating contours

All contouring methods involve heighting a number of points, known as spot heights, then deducing the position of the contours from these spot heights. The various methods may be classed as *direct* or *indirect*. Direct or 'contour chasing' methods require the location of spot heights actually on the desired contours, at intervals of from 5 to 50 m apart, depending upon the scale of the plan and the shape of the ground. The spot heights are joined by a smooth curve, similar to connecting the points on a graph. Indirect methods entail locating spot heights in positions such that, for any pair of adjacent 'spots', the ground surface may be considered to be a line of constant slope or gradient from one spot height to the next. Contour positions are then interpolated between the spot heights and again the contour points are joined to form smooth curves. Interpolation is generally carried out by estimation rather than with any great precision, e.g. if two spot heights are A and B, with heights of 34.8 and 35.4 m, respectively, then the 35 m contour would be 2 units from A and 4 units from B and lying on the line connecting A and B. In this case, the contour position is one-third of the way from A towards B, the third being estimated on plan.

The actual method to be used depends upon a number of factors, including whether the task is to contour an existing plan or involves the survey of both detail and heights. Again, the plan scale and the general form of the ground are important.

8.3.3.1 Direct contouring

These methods are seldom used, except in special circumstances. They should be the most accurate, but are generally slow and hence more costly than indirect methods. Direct methods are at their best on scales of 1:500 to 1:1000 and when the slope and shape of the ground are clearly visible to the surveyor.

(i) *Direct contouring by arbitrary selection of spot heights along the contour.* The following example illustrates the method.

Imagine a level set up and levelled from a BM, the collimation height of the instrument being found to be 104.230 m and it is required to locate the 103 m contour line. It will be clear that when the base of the staff is standing on the 103 m contour line, then the centre hair reading on the staff will be 104.230–103.000 = 1.230 m. To locate a point on the contour, the staff holder should hold the staff erect, base on the ground and then be directed up or down hill until the centre hair reads 1.230 m. When the reading is correct, a peg or lath may be placed at the staff position to mark a point on the contour. Repetition of this process, the staff holder moving along the slope between each point fixing, will establish a series of points on the contour. Other contour lines may then be fixed in the same way, with the appropriate centre hair reading for the particular contour being selected. To facilitate the work, a line of levels should be run through or around the site initially, giving a series of TBMs such that, wherever the surveyor sets up, there will be a TBM visible and the collimation height may be readily obtained. Note that the spot heights should extend somewhat beyond the site boundaries, to avoid distortion of the contours at the boundaries. The plan positions of the contour line points must be located thereafter and this can be done by any appropriate detail supply method. If there is no plan, then it is customary to survey both the site and contour points by tape and offset or other detail survey methods. The various contours may be identified by different coloured laths and when the levelling is complete the contour lines may be picked up from the chain lines as other detail.

If a plan already exists, then it may be better to fix the position of the spot heights by bearing and distance from the instrument station. This requires a level fitted with a horizontal circle for the bearings, the distances to the staff being obtained from the stadia hair readings and the instrument stations must be located by measurements from site detail so that they may be plotted on the plan.

(ii) *Direct contouring by sections*. In this method, a base line is laid out on the site approximately parallel to the contours, then a series of ordinates or section lines set out at right angles to the base line. Thereafter, contour points are located on each section line, in the same way as described above, effectively fixing the points where each contour line crosses all the section lines. In some circumstances this approach may be easier and quicker than the basic method.

8.3.3.2 Indirect contouring

These methods are the most popular for area. The level and staff should be used for scales of 1:1000 or larger and trigonometric heighting with theodolite stadia readings at smaller scales. Modern sophisticated EDM equipment can be used at all scales, see Chapters 7 and 10.

(i) *Grid levelling*. This is the most commonly used method when the site is not extensive and has only minor variations in slope. The area is covered by a grid of lines, generally forming squares of 5 or 10 metres or so, levels being taken at the intersections of the grid lines. It is common practice to give the lines in one direction identifying letters (A, B, C, ...) and in the other direction identifying numbers (1, 2, 3, ...), thus any intersection point may be specified, e.g. A4, B6, C3, etc.

The grid spacing will be influenced by the extent of the undulations of the ground, the accuracy required and the need to be able to draw a reasonably accurate smooth curve through the interpolated points. The last point tends to limit the maximum grid spacing on the plan to 25 or 30 mm, that is 5 or 6 m apart on the ground at 1:200 scale, 12 to 15 m at 1:500 and 25 to 30 m at 1:1000 scale. Generally the grid is laid out by tape and optical square, particularly on small sites.

The grid must be tied to a straight line, often one of the measured control lines is used, but any straight part of detail may be suitable. If the offsets are to be kept short, then it is preferable that the grid be based on a straight line through the centre of the site. Often offsets will seem excessive by tape and offset survey standards, but errors which arise should not significantly affect the end product. Positions for spot heights (intersections) may be ground marked with pegs, short pieces of lath, or arrows with strips of bunting tied to the loops. On larger sites, ground marks may be dispensed with by using ranging rods around the perimeter, or pairs of rods on two adjacent sides, the staff holder being responsible for taping distances between the rods to locate spot height positions.

Where there are marked changes of slope lines through the site, it is advisable to attempt to arrange for grid lines to pass along these features. Alternatively, additional spot heights may be needed and means of identifying them. Again, the spot heights should extend beyond the site boundary, but if this cannot be arranged then additional spot heights at the boundaries will be required. The simple method of letter/number for spot identification may not always be convenient. An alternative method is to identify each point as a distance left or right from a known distance along a central grid line, as in *Figure 8.2*. This technique is also used for booking sections and cross-sections.

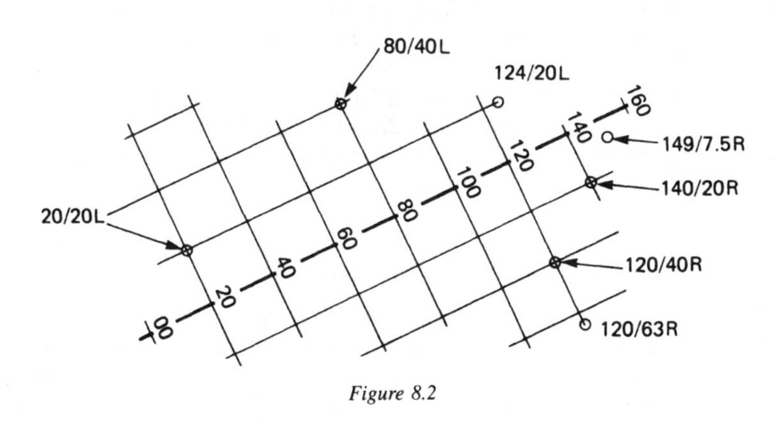

Figure 8.2

(ii) *Contouring by sections at changes of slope*. This method is similar to the second of the direct methods, using a base line and ordinates or sections at right angles to it. In this case, however, the staff holder goes along each ordinate to a change of slope point rather than a contour point, the spot heights being identified as in *Figure 8.2*.

This method is not as popular as simple grid levelling, perhaps because it is not so convenient for volume calculations. In addition, if it is to be economical in time then the decision as to where to take the spot heights must be made by the staff holder and he therefore needs to be experienced.

(iii) *Contouring by spot heights at arbitrary change of slope points*. In this method, spot heights are taken over the whole area of the site at all those points at which the surveyor (or the staff holder) considers there are significant changes of slope. Some positions will be self-evident, e.g. changes of slope at the site boundaries, on ridge or valley lines, at the top of conical hills, at the bottom of hollows, etc. It is more difficult on long convex or concave slopes, but it should be remembered that it is unlikely that spot heights will need to be closer than 6 m apart for 1:500 scale mapping, 12 m at 1:1000 scale, and so on.

The spot height positions will have to be located by the methods of *Figure 8.2*, or by bearing and distance from the instrument position. For this reason, the method is not generally popular with a level unless bearing and distance methods are being used. It is, however, the ideal method for theodolite radiation techniques in conjunction with trigonometric heighting.

8.3.4 Booking and the reduction of levels

The levelling procedure, including the booking, should be similar to that previously described. The 'Remarks' column of the book may need to be modified so that the spot heights can be identified and located. In grid levelling a column could be headed 'Grid point', while in bearing and distance methods columns may be headed 'Hor. Angle', 'Stadias', 'Distance'. When using stadia distances from instrument to staff the distance can be mentally deduced in the field and entered in the Distance column, without writing down the actual stadia readings, but it is better practice to record the stadia hair readings in the book. If sections or ordinates at right angles to a base line are used, then it will be necessary to include extra columns for these.

Reduction should preferably be by the Rise and Fall method, since there are likely to be many intermediate sights. At an intended 0.5 m contour VI, then readings to the nearest 0.005 m and reduced to 0.05 m are more than adequate, and proportionally at other contour intervals.

8.3.5 Interpolation and the plotting of contours

Whichever method has been used for spot heighting in the field, in the office the spot heights are plotted in their correct positions on the plan and the reduced level of the point written alongside, to the limit quoted above. In indirect methods, every adjacent pair of spot heights is examined and, if one is above the level of a required contour line while the other is below, then the actual position of the cut of the contour line is estimated between the spot heights and marked.

Figure 8.3 illustrates a method of interpolation, but it also shows a problem which may arise with a beginner. Here a study of the interpolated points shows that the contours will cross the grid square either at an angle of roughly 45/225° or 135/315°. The square cannot be part of a plane surface, but must contain either a valley running SW–NE or a ridge running NW–SE, and there is no information to indicate which. In practice, a smaller grid could have been used, or a note should have been entered in the field book to the effect that one of the diagonals of the square was a ridge or a valley line.

The interpolated points, or the contour points located by the direct methods should be joined by smooth freehand curves to represent the contour lines. It should be appreciated, however, that the contour line should be as straight as possible between each adjacent pair of common points. Adjacent contour lines are often roughly parallel, and they can only end at a plan edge.

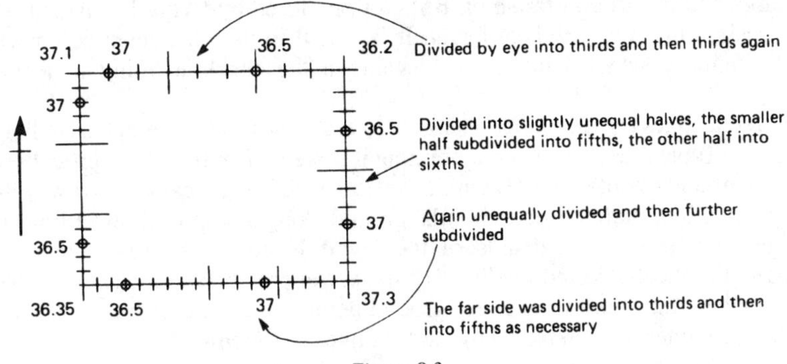

Figure 8.3

If the plotter is satisfied with the pencilled contour lines then they may be inked in. Traditionally contours are coloured brown or orange, but for some reprographic processes they must be in black. To distinguish them from detail and provide contrast, they should be drawn in a narrower gauge line, or in a different form of line. Every fourth or fifth contour line may be emphasized by a wider gauge of line. The height of contour lines is indicated at a gap in the lines, such that they are read looking uphill, and if at all possible they should be capable of being read without having to turn the plan (see *Figure 8.1*).

8.3.6 Other methods of representing relief

There are many methods of representing relief other than by the use of contour lines, the use of slope symbols as in *Figure 2.8* being one. Not many of these other methods are suitable for site plans of proposed construction sites, but some may be useful in special circumstances, e.g. in illustrating the remains of a medieval village, where traditional contours and spot heights would be quite inadequate. For further information, see the Bibliography.

8.4 Sections and cross-sections

In building and engineering work it is often necessary to prepare a profile of the ground along a particular line. Such a profile, termed a *section of levels*, is obtained by running levels along the required line, then plotting the heights and distances on paper to appropriate scales.

8.4.1 Longitudinal sections

A ground profile along the centre or other longitudinal line of an existing or proposed road, railway, pipeline, canal, etc., is termed a *longitudinal section.* Levels are observed along the ground line by series levelling (Section 5.9), generally at a standard interval of horizontal distance such as 20, 25, 30, 50 or 100 m, together with points where the ground slope changes distinctly or natural or artificial features disturb the ground profile. As always, to minimize error, levelling should start and finish on BMs, or be closed by flying levels and back and foresight distances be kept equal. It is useful to use changepoints on well-defined features which can be used again on the check-levelling to localize error.

The reduced levels are plotted using distorted scales so as to emphasize height variations. Typical examples are a horizontal scale of 1:500 with a vertical scale of 1:100, or a horizontal 1:2000 with a vertical 1:200. *Figure 8.4* shows a typical example of a longitudinal section of the ground along a proposed sewer line. The same section also shows details of the invert levels of the sewer manholes proposed and the suggested gradients of the sewer pipe between manholes. Note that a vertical line is drawn at every ground point to the scale height and that the ends of these lines are connected by straight lines, no attempt being made to draw the surface as smooth curves, since the scale distortion makes this pointless.

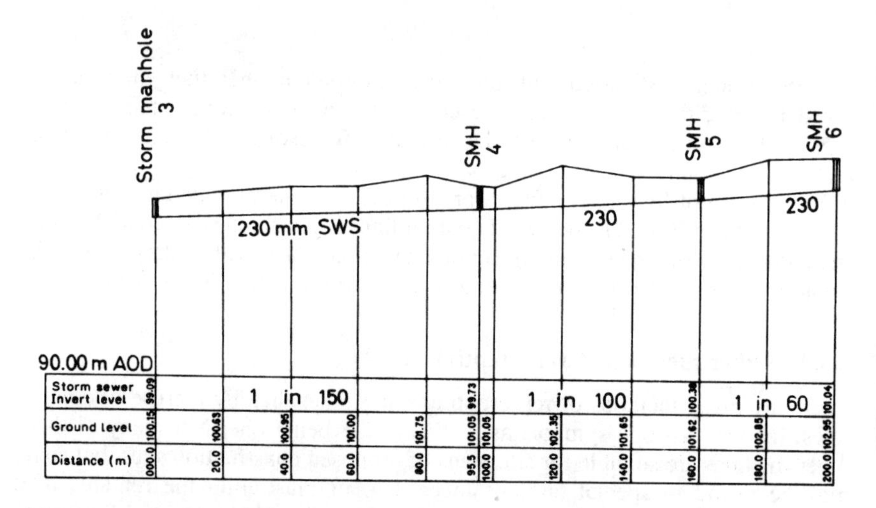

Longitudinal Section: Storm sewer 'A', CHGE. 0 to 2 + 00

Scales: Horizontal 1 : 500 (original plotting scales)
Vertical 1 : 100

(All measurements in metres)

Figure 8.4

All details as to levels and distances are shown in the 'boxes' drawn across the bottom of the section. Sufficient space must be left between the top line of the boxes and the lowest level of the section to allow for extra boxes which may be required for other details such as, for a road, perhaps road formation levels, horizontal curve data, storm sewer details, foul sewer details, cut and fill for excavation, etc.

Note in *Figure 8.4* that the datum height should preferably be a multiple of five metres and that some 50 to 100 mm clearance is reasonable between the top line of the boxes and the lowest level of the section. When booking levels for this type of work it is sensible to have a 'Distance' column within the remarks space in the book.

8.4.2 Cross-sections

Where the proposed construction is of considerable width the longitudinal section information must be supplemented by cross-sections. A *cross-section* is a profile of the ground at right angles to the longitudinal line, serving mainly to allow the calculation of the volumes of earthworks. Cross-sections are not usually taken for pipelines, but are required for roads, railways and canals. Cross-sections must be taken at regular intervals along the centre-line, the same regular points generally as used for the longitudinal section. The distance apart depends upon the nature of the ground, perhaps 20 m on broken ground, or even 100 m on gentle slopes. Occasional sharp changes in ground configuration, such as rock outcrops, may necessitate extra sections at other non-regular points. The centre-line should be pegged at all points where cross-sections are to be taken, the pegs being driven to ground level and marker pegs placed beside them for

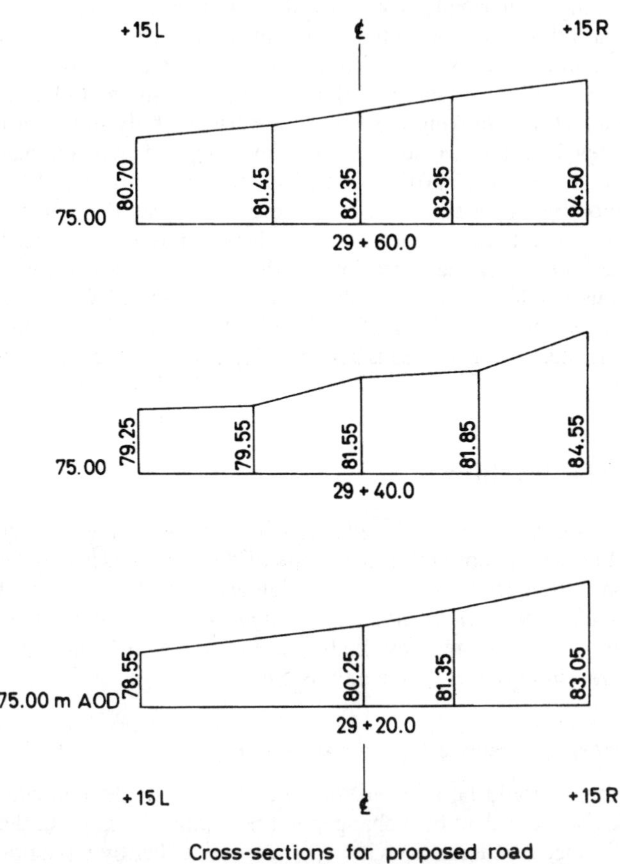

Cross-sections for proposed road
(ground as existing)

Original scales: Horizontal and vertical 1 : 200
(All measurements in metres)

Figure 8.5

identification. The pegs are best placed before the longitudinal section is taken, then the peg levels provide a comparison between long-section and cross-section levels if the two tasks are done separately, cross-section levelling acting as the check-levelling for the long-section levelling. Alternatively, both long-section and cross-sections may be levelled at the same time and checked by flying levels.

A cross-section is identified by the longitudinal section through chainage at its centre and distances on the cross-section are measured and noted as left or right of centre-line.

This is one of the methods recommended for use in indirect contouring. The width of a cross-section is fixed by consideration of the construction width and the width of land reserve available. The distances left and right are measured by synthetic tape, directions usually being judged by eye, like offsets in chaining. Levels are taken at the centre-line, at all changes of slope and at the extreme width of the section. Flat ground may only need three levels, broken ground perhaps twenty or more.

The levelling is normally series-levelling and usually one cross-section is completed at a time, then on to the next, and so on. Very steep side-slopes may need two or more set-ups per section and then it may be faster to take the downhill levels for two or more sections from one set-up and their uphill levels from another set-up. Booking must be done very carefully in this case, to avoid mixing the levels of the two sections. The levelling and chainage measurements may also be successfully carried out by radiation methods as in Chapter 10.

When plotting cross-sections, it is normal to use the same scale both horizontally and vertically. Generally the scale is that used for the verticals on the longitudinal section for the same job. In the two examples quoted earlier, the cross-sections would probably be plotted at 1:100 and 1:200, respectively (see *Figure 8.5*), but note that 29 + 20.0, 29 + 40.0, etc., is the chainage notation system sometimes used for roads and railways, meaning 2920 m and 2940 m, respectively.

8.5 Precise levelling

As stated previously, geodetic levelling such as the heighting and adjustment of large levelling nets is not within the scope of this text, but it may be necessary on occasion to transfer levels with very high accuracy. For such work a *precise tilting* or *precise optical automatic level* with *parallel plate micrometer* should be used, or perhaps one of the new higher accuracy digital levels, in conjunction with *invar levelling staves* (see Sections 5.6 to 5.8).

8.5.1 General procedure for precise levelling

The instrument should be allowed to settle at the local air temperature before commencing work. Select firm changepoint positions, arranged so that back and foresight distances are equal to within 0.5 m or so, either by taping or by pacing the distances out carefully. If the difference of the sum of the lengths of the foresights and the sum of the lengths of the backsights exceeds 1.5 m during the levelling, then future instrument positions must be arranged to reduce the difference. Observing distances should be from about 25 to 40 m and grazing rays within about half a metre of the ground should be avoided due to the effects of variable refraction of the air. The level should be protected against sun and wind by umbrella and windshield, if necessary, but levelling should not be carried out in high winds.

Using two staves, one should be placed at all odd staff points and the other used for all even staff points, to ensure that all observations at one point are made on the same staff. At odd instrument positions, commence observations with a backsight; at even instrument positions commence with a foresight, to help reduce systematic instrument errors. When setting up the instrument at each station, point the telescope at the staff on which the first reading is to be taken, before levelling-up the circular bubble.

For best results, the line should be levelled at least twice; once outwards, once backwards in the opposite direction, and the two sets at different times of day under different atmospheric conditions, using different changepoints. The results of the two levellings will be meaned if their difference is within acceptable limits. Note that although the instrument and staff positions are set out by measurement, the actual distance to the staff from the instrument must be noted at each position using the stadia hairs. The line should be arranged so that there are an even number of instrument set-ups. The actual observing order depends upon whether single-scale or double-scale staves are used.

8.5.2 Using single-scale staves

Referring to *Figure 8.6*, successive instrument stations are numbered 1, 2, 3, etc. It is required to level from BM A to a distant point E. B1 means 'Backsight reading from instrument station 1', while F1 means 'Foresight reading from instrument station 1'.

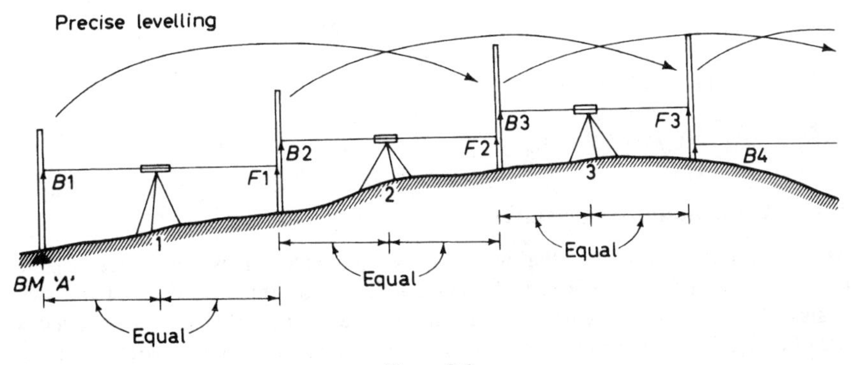

Figure 8.6

Staff readings are made in the sequence:

B1, F1; (from station 1)
F2, B2; (from station 2)
B3, F3; (from station 3), etc.

At each observation, read and book all three hair readings; the central and the two stadia hairs, using the parallel plate micrometer for each in turn, see Section 5.9.4. For example

	Readings to B1	*Reading order*		*Readings to F1*
Centre hair reads	1.895 49 m	2	3	1.539 52 m
Upper stadia hair	2.057 10 m	1	4	1.702 95
Lower stadia hair	1.733 62	1	4	1.376 45
Centre hair reads	1.895 37	6	5	1.539 43
Sum stadia rdgs	3.790 72			3.079 40
Sum centre rdgs	3.790 86			3.078 95
Diff of stadia rdgs (×100)	32.348 m			32.650 m

Note: Throughout this book, *lower hair* or *bottom hair* is taken to mean the stadia hair or line which indicates the smallest reading on the observed staff.

Before proceeding further, the following recommended limitations should be checked:

Centre hair readings on the same staff should agree within 0.000 50 m.
The sums of the readings on to the same staff to agree within 0.001 00 m.
The horizontal distances B1 and F1 to agree within 0.5 m or so.

If these limitations are not achieved, the observations should be repeated as necessary, the objective being to reduce, as much as possible, the effect of cumulative errors.

To compute the total difference in level between A and E, total the backsights, total the foresights and their difference gives twice the required amount. To check the arithmetic, compute also the rise or fall between every pair of changepoints, the difference of the totals must equal the previous figure.

This may be expressed as:

$$H_E - H_A = \Sigma(B - F) = \Sigma_B - \Sigma_F.$$

8.5.3 Using double-scale staves

The procedure is similar to that used with single-scale staves, but instead of the centre hair being read twice, each centre hair is read once on each of the two scales. Both sets of stadia hairs are usually read and booked. The reading checks are obtained by comparing the readings on the two scales, centre readings against centre readings, stadia differences or distances against stadia differences or distances. The same limitations are recommended and if these are not achieved the observations must be repeated, otherwise proceed.

Finally, the difference of elevation of E and A is obtained as before, but two values are obtained from the two sets of scale readings. The two values should agree within a suitable specified tolerance, for example $4\sqrt{K}$ mm in which K is the distance levelled in kilometres. If the line is levelled twice as is normal practice, then the two final results are meaned again.

8.6 Reciprocal levelling

Where it is not possible to equalize the backsight and foresight distances from an instrument position, such as in taking levels across a wide river, the technique of *reciprocal levelling* may be used.

8.6.1 Distances of 60 to 100 m

In this instance, the method is as for ordinary levelling, except that the gap should be observed twice, as in *Figure 8.7*.

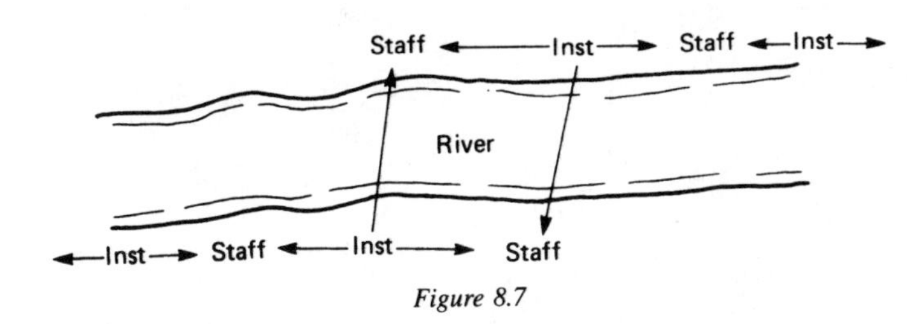

Figure 8.7

8.6.2 Longer distances, ordinary work

In *Figure 8.8* the level is first placed at position A, then at position B, and level staves at positions 1 and 2. The distances from A to 1 and from B to 2 should be equal. From A, readings are taken to 1 and 2, and from B readings are taken to 1 and 2. The readings are noted as A1, A2, B1, B2 and the differences (A2 – A1) and (B2 – B1) calculated. These differences will not be the same, but their mean gives the difference in level between points 1 and 2, provided the atmospheric conditions have not changed between taking the two sets of readings.

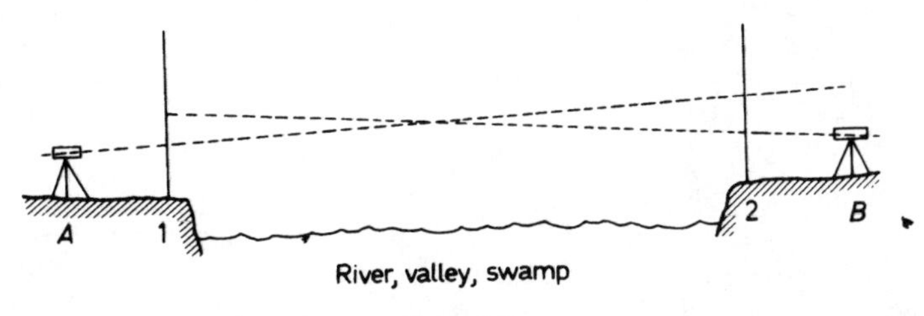

Figure 8.8

Should there be a long time delay in getting the instrument from one side of the river to the other, then it is better to use two levels simultaneously, to overcome the possibility of changing refraction. A target on the staff may be necessary as distances become very large.

8.6.3 Longer distances, precise work

In precise work, the traditional approach was to use a precise level with either a gradienter screw or a bubble scale for distances between 100 and 1500 m. Beyond 1500 m the preferred method was *reciprocal trigonometric heighting* with two theodolites which could be read to decimals of a second, many FL and FR readings being needed. Automatic levels and most modern tilting levels, are not fitted with gradienter screws or bubble scales, thus, for all precise work over 100 m the reciprocal trigonometric heighting technique is recommended.

Chapter 9

Control surveys

9.1 Introduction

Control was referred to briefly in Chapter 1 as being the principle of tying measurements together in such a way as to produce a survey which is accurate in proportion and scale. The form of control to be provided depends on the type of survey and the accuracy demanded.

Where a detail survey is to be carried out over a small area, and the accuracy required is not high, it may be sufficient and economic to carry out the work with steel-taped measurements only. In this case, a series of measured survey lines forming a framework of triangles will be superimposed over the area to be surveyed, forming the control for the survey. The survey lines forming the framework are measured carefully, so that the framework can be reproduced to scale on a plan, and the location of all the detail is fixed by appropriate measurements from the survey line. The taped survey line framework then *controls* the detail measurements. The procedure for carrying out such a control survey is described in Section 9.2.

Where a linear survey is very extensive, the standard of accuracy of ordinary taping may be inadequate to keep the whole survey accurate in proportion and scale, and it then becomes necessary to provide a higher accuracy framework to control the survey. Control frameworks, then, for large or high accuracy surveys, are measured to a higher standard than can be attained in simple linear surveys based on direct linear measurement and they are likely to involve both linear and angular measurement. Control may consist of a *traverse survey*, or a *triangulation scheme* from a base line, or *trilateration*, or even a *triangulateration network* where the last two are combined. The latest method for providing control is GPS using artificial Earth satellites.

Generally the most economical method of providing a higher standard of accuracy of control in terms of time and money is the *traverse survey*. Traverse survey relies on a chain of connected lines rather than a network of triangles, with both lengths and angles measured, and co-ordinates can be computed for the stations at the end of each line. The traverse *control* frame would be plotted first, then any lower accuracy survey lines fitted to the control and then the detail plotted from the survey lines. This demonstrates the survey maxim of *working from the whole to the part*. Traverse survey is described in Section 9.3.

In a very large task, there may be successive levels of control. Thus the Ordnance Survey's primary triangulation of the United Kingdom, measured to the highest standards, is described as 1st order control, the infill between these is 2nd order control, and ultimately fitted to this was the 3rd order work which actually controlled the detail measurements.

This chapter considers the various methods used for the provision of control, some in outline only.

9.2 Linear survey control

If the three sides of a triangle are marked out on the ground and measured using a tape, then the triangle may be plotted to scale on paper using a scale-rule and a pair of compasses. One line can be drawn to scale, then from its ends arcs may be drawn to the scale lengths of the other two sides. If all the measurements are correct, the intersection of the arcs defines the third 'corner' of the triangle.

In tape and offset linear survey, a number of straight lines are laid out in the area to be surveyed, arranged so that they form a framework of triangles, the intersection points being termed *stations*. If all the survey lines forming the triangle sides are measured, then the framework can be reproduced on a drawn plan in the same way as the simple triangle above.

Detail in the area is located by measurements from the lines of the framework. These, in turn, are plotted to scale on the plan until the plan gives a proportional representation of the area surveyed. The survey lines should be run close to the detail to be 'picked up', and extra *detail lines* may be included in the frame where the detail is not close to a line. Detail is picked up by offsets or intersecting ties from the survey lines, as shown in Section 1.6.1 and described in Chapter 10.

The triangle technique is simple and direct, but it has one important defect. This is that should an error be made in measuring the field length of one side of a triangle, then the triangle can probably still be plotted, but it will not be the same triangle as exists upon the ground. Errors in measurement are easily made – misreading figures, tape not straight, etc., therefore it is essential that an extra *check line* be measured in every triangle to act as a check on the measurement of the sides. The check line is not used in plotting the triangle, but *after* plotting – its length is then scaled off the drawing and compared with the measured field length. If the two agree it proves that there is no serious error in the measurement or the plotting of the triangle. *Figure 9.1* shows an example of a very simple framework of survey lines which could be used to survey a field. The pecked check line checks the plotting of both triangles.

When a survey consists of several triangles, they must not be simply 'tacked-on' to one another, since this leads to a gradual build-up and exaggeration of any errors. Instead, one long line (as long as possible) should be placed right through the area of the survey and all the triangles based on this *base line*. The base line ties the whole framework together and prevents any gradual twist or deformation of the framework. It is, of course, the linear survey method of satisfying the rule, *work from the whole to the part*. The 'art' of this basic survey method lies in proper selection of the framework lines.

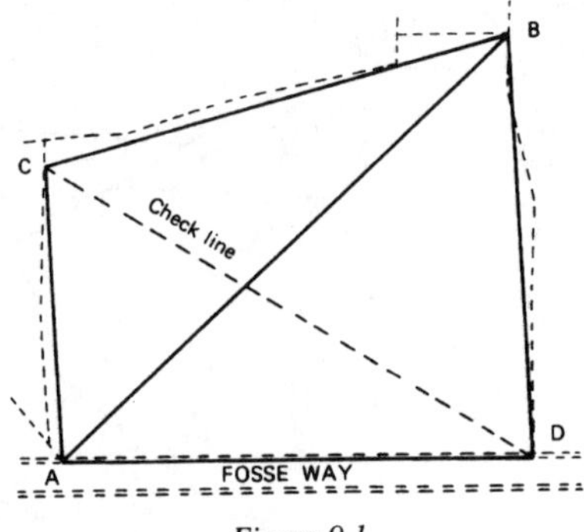

Figure 9.1

9.2.1 Field procedure

9.2.1.1 Selection of survey lines

On arrival on site, the surveyor walks over the entire area to be surveyed to familiarize himself with the general layout and in particular any problems it may present. Time spent on a thorough reconnaissance is never wasted. The survey lines and their layout are decided upon the basis of the principles set out below.

(i) The framework of lines must be primarily triangular. (A simple triangle around the site would be ideal, but this is rarely achieved.) There should be a *base line*, as long as practicable and preferably, but not necessarily, run through the middle of the area, the triangles being based on this base line. Sometimes a triangle must be set up, say to pick up detail, but it cannot be based on the base line. In such a case, the triangle may be erected on a side, or a part of a side, of another triangle, provided that the new triangle is 'tied back' to the base line by a suitable measured check line. When setting up a framework, triangles may overlap one another, and they may share common sides.

(ii) The triangles formed should be as few in number as possible, and they should be *well-conditioned* (i.e. as judged by eye, no angle of a triangle should appear to be less than 30° nor more than 120°).

(iii) Each triangle must be checked, therefore each triangle must be provided with one or more *check lines*.

(iv) Measurements to the detail must be kept short, therefore the survey lines should be positioned close to the detail but not so close as to cause difficulty in measuring the lines. The need to keep the tape line straight should have priority, however, and it is preferable to locate the lines along straight features or along points on the same alignment. For example, the faces of buildings or the tangent points of telegraph lines or the lines of straight kerbs or pavement edges are all suitable, provided that the resulting offset values do not become excessive.

(v) Obstacles to ranging and measuring should be avoided where possible.

(vi) Lines should be positioned over level ground if at all possible.

(vii) The total chainage involved in the survey should be close to the minimum required to achieve the above.

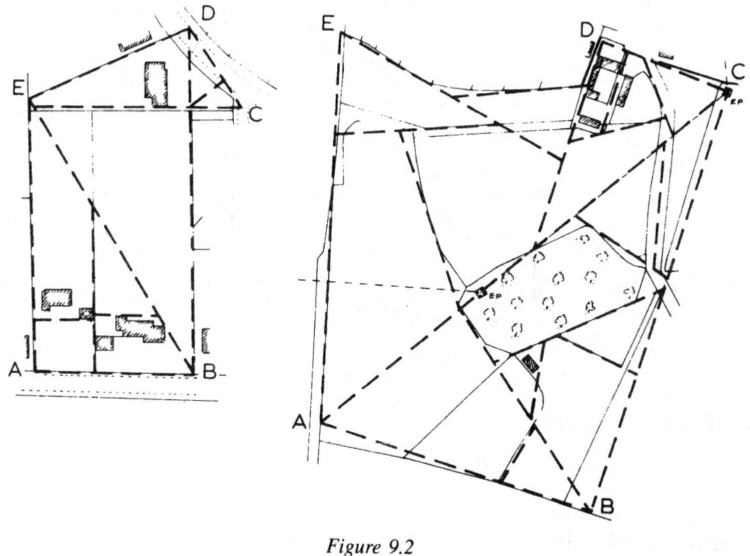

Figure 9.2

Figure 9.2 shows examples of frameworks selected for two different sites. It should be appreciated, however, that except on the simplest of sites, it is unlikely any two surveyors would select exactly the same arrangement of lines.

It will be noted that the survey lines start and finish either at the terminal points of other lines or at some point on another tape line. The terminal points of the former are known as *survey stations*, lettered A, B, C, etc., in *Figure 9.2*. The terminal point of a survey line that ties out at some point along another line is known as a *tie point*.

Stations may be marked in a variety of ways, depending upon the permanency required. Where marks are only required for the day – as typical on a small tape survey – then a ranging rod is simply stuck in the ground at the desired position. On hard ground the rod may be held above the mark by a tripod support. If more permanent marks are required these may be made using wooden pegs or steel pegs surrounded in concrete.

Tie points are indicated by similar marks, normally established during the measuring process, e.g. in the larger survey diagram, the tentative position of all the lines would be decided at the reconnaissance stage, the positions of A, B, C, D and E being confirmed by inserting marks. Later, while measuring the base line AC, markers would be inserted to denote the intended tie positions.

9.2.1.2 *Ranging the lines*

It is generally important that there be no misalignment when measuring a line, otherwise any detail near the erroneous alignment and any tape line tying into it will also be in error. As a consequence, lines of more than one tape or tape length may need to be aligned by *tracing* or ranging before measurement, as described in Chapter 4.

It is recommended that line ranging be carried out if the line exceeds 200 m length, or lies over very broken ground. For lines over 400 m length, the ranging should be by theodolite (see Chapter 6).

9.2.1.3 Measuring the lines

The lines are taped, one at a time, measuring all detail and offsets before moving on to the next line. (The actual measurement of survey lines by tape is covered in Chapter 4, and the measurement of detail is considered in Chapter 10.) The order in which the lines are measured is immaterial, except that unnecessary walking should be avoided.

9.2.1.4 Picking up the detail

Detail in linear surveys is normally supplied by rectangular offsets or ties from the survey lines as they are being taped, so that when one line has been measured both the line measurements and the detail measurements for the line are complete and it is not necessary to go over the line again.

9.2.2 Linear survey booking

Linear survey booking may be carried out either on *booking sheets* or in a field book. The ideal booking sheet or field book for linear survey is A4 sized and of a plain or 5 mm square ruled paper. Linear survey 'chain survey' books as obtainable from stationers or suppliers of office materials are usually small octavo size (between A5 and A6) and characterized by either one red line or two red lines 15 mm apart ruled down the centre of the page. While suited to the beginner, in practice these will be found inadequate since the page size is restrictively small and the position of the red lines may not always be convenient. With A4 paper on a clip-board the surveyor may rule any lines to suit his own convenience.

Methods of booking vary but most surveyors adhere to a small number of basic rules which enable any other surveyor to understand what has been recorded in case the survey has to be completed or plotted by some other surveyor.

Clarity and accuracy of booking are obviously essential. These may be achieved by neatness, taking especial care with one's handwriting, numbering and possible emphasis of detail, i.e. printing the numbers and using a slightly broader gauge of line to represent the detail. The surveyor must learn to do the bookings once only, since hand copying will lead to errors and considerably increase the time taken on the job, a satisfactory presentation and speed will come with practice. Mistakes inevitably occur, but no harm is done if these are carefully corrected – incorrect figures should be cancelled by drawing a single line through them, and revised values printed above or below the original figures, while cancelled detail should be indicated by small crosses on the cancelled lines.

Explanatory notes may be added to assist the plotter and the numbering of pages is helpful and also provides a check as to whether any loose sheets may be missing.

It is unreal to expect the surveyor to be able to keep the pages of a field book clean, since tapes become dirty and there is no way of preventing some of this dirt from being transferred to the book unless an extra (generally uneconomic) assistant is employed.

Where a traditional 'chain survey' field book is used for booking the survey, it is conventional to start at the *back* of the book, working towards the front. A sketch, an outdated plan or a smaller-scale plan attached to the field book will be found useful at the reconnaissance stage, and the proposed stations and survey lines may be drawn on this. The first page of the field notes should include relevant information such as the name of the site or the job, the address, the name of the client if appropriate, the surveyor's name, the date, etc.

A survey line is usually represented by two parallel lines about 15 mm apart, drawn down the centre of the page. The successive distances along the survey line are entered between these parallel lines as running measurements from the commencement of the line, starting at the *bottom* of the column. (It should be noted that these distances along a line are still often termed *chainages*, even though the chain is no longer used.)

The readings should be entered sequentially, so that they read in the direction of travel along the line, and they should be spaced out so as to allow any entry to be amended or an additional entry inserted and no attempt should be made to observe scale. The length of the line is entered in the column on completion of the line, and it is emphasized either by circling it or 'topping and tailing' it, that is, entering the length sideways-on between parallel lines in the direction of measurement, as in *Figure 9.3*. It must be emphasized that only chainages along the survey line are entered between the parallel lines – all other measurements and sketches of detail are placed outside the parallel lines.

If necessary the booking of a survey line may be continued onto the next page or pages, as required. Each line of the survey should be started on a new page, so that there is never detail from two lines on the same page.

A *survey station* is represented by a square box at the end of the chainage column, the station letter (or number) being written within the box, while a tie point is represented by an arrow head with a brief description. Both of these are shown in *Figure 9.3*.

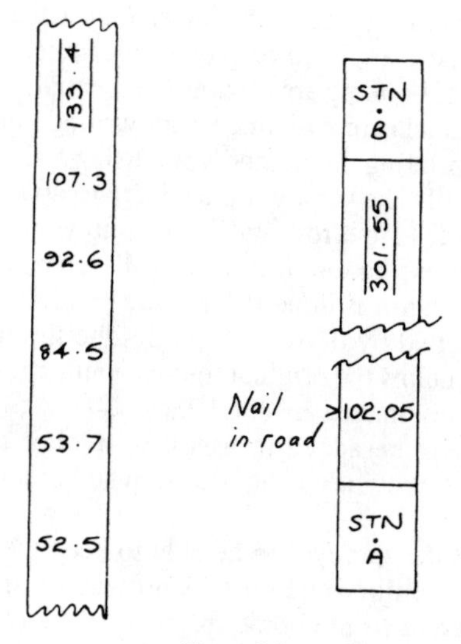

Figure 9.3

Figure 9.4 shows how to represent tie points at the ends of a survey line 61.20 m in length, commencing at tie point 102.05 on the line from A to B and closing at the tie point 81.65 on the line 241.20. It should be noted that the tie point values are 'topped and tailed' for emphasis, and that the previously measured survey lines are repeated as single lines only, known as *skeleton lines*. Only the essential information about these skeleton lines is given, that is to say their length, page numbers and values of the tie points values being recorded and aligned as in the original survey lines. This duplication of information may not be necessary if the site is small and the plotting is to be carried out by the surveyor himself.

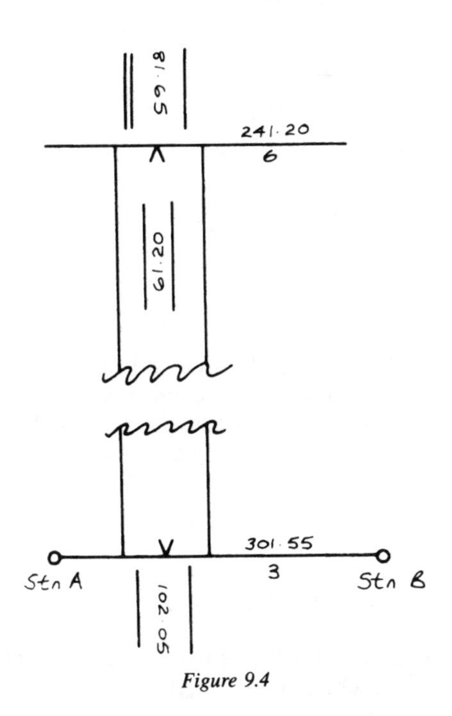

Figure 9.4

9.2.3 Linear survey plotting

In order to avoid duplication, linear survey framework plotting is covered in *Section 10.2.3.2*. Regarding the orientation of the drawing, it is customary to show the direction of north on the plotted plan, and in the UK this can generally be transferred from the area OS map to the plotted plan with sufficient accuracy for the purposes of linear surveys. If this is not practicable it will be necessary to obtain the magnetic bearing of one of the survey lines using a prismatic compass. Many varieties of prismatic compass are in use, a typical example being illustrated in *Figure 9.5*.

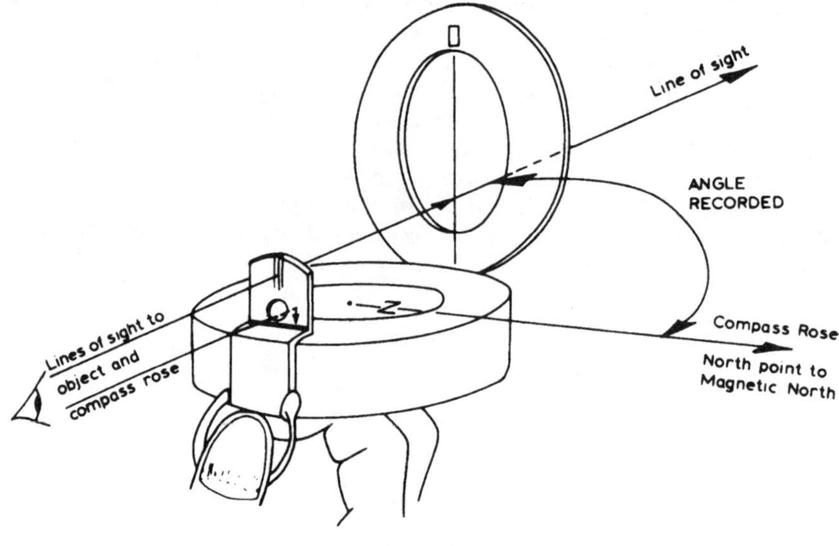

Figure 9.5

9.3 Traverse survey

Traversing is probably the most widely used and flexible method of providing control for site surveys, particularly in urban areas where the formation of triangles is difficult and sometimes impossible. A *traverse* has been defined as a series of connected straight lines whose lengths and bearings can be determined. The lines are known as *legs* and the end points of the lines as *stations*. Typical measurement arrangements are:

(i) The angles between the legs measured by theodolite, with the leg lengths measured by steel tape, the detail probably fixed by offsets and ties from the legs, or
(ii) the angles measured by theodolite and the leg lengths measured by an EDM attachment on the theodolite, detail supplied by radiation from the station points, or
(iii) the survey carried out using a *total station* instrument which can measure both the angles and the distances, and the detail by radiation from the stations.

Apart from fixing detail by measurements from the traverse legs and stations, extra detail lines can be run between points on the traverse legs as required. Traverses may commence and close from either end of a base line, from triangulation or trilateration control points, or from other existing traverse stations. In the survey of small sites, a traverse often consists of a single closed loop of connected straight lines.

The following sections deal with the techniques to be used in providing control for survey areas by the use of traversing, where the traverses will seldom exceed ten kilometres in length.

9.3.1 Types and classification of traverses

Traverses are described as being either *closed* or *open*, a *closed traverse* being one whose start and end points (*terminals*) are fixed co-ordinated points. The observations in a closed traverse can be numerically checked and the results mathematically adjusted, but an open traverse cannot be checked or adjusted.

A traverse which commences and closes on the same station is a *closed loop traverse*, sometimes known as a *ring traverse*. A traverse which commences and closes on two different stations, the positions of both of which are already known (co-ordinates for both stations fixed) is a *closed traverse*, sometimes termed a *link traverse*. *Open traverses* should generally be avoided.

Traverses may be classified by the accuracy attained, typically as precise, semi-precise, or low accuracy. *Precise traversing* today usually involves the use of a 'single second' theodolite in conjunction with EDM equipment. An accuracy considerably greater than 1:10 000 is expected.

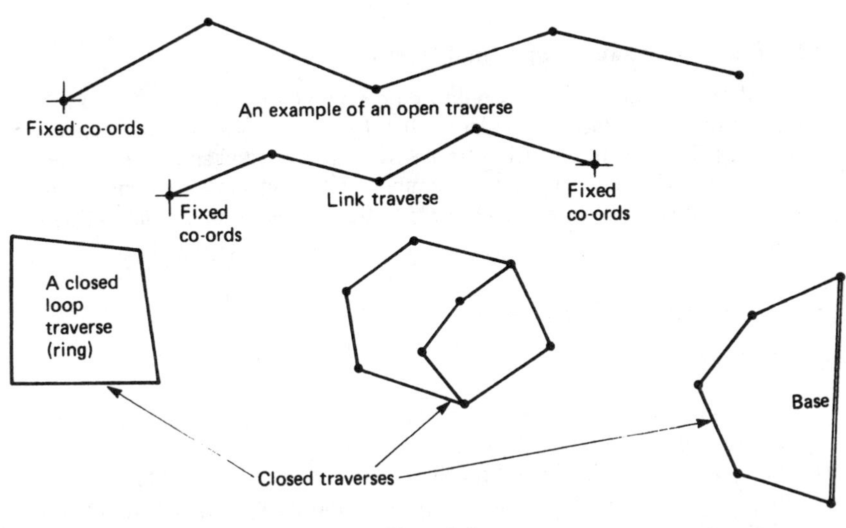

Figure 9.6

Semi-precise describes the majority of traverses. A typical example might include the use of a 20" theodolite with EDM or a steel tape, attaining anything up to 1:10 000 or better.

Low accuracy traverses indicate an accuracy of less than 1:5000, such as small-scale detail or reconnaissance surveys, using a 20" or 1' theodolite and stadia measurement or tape.

At the office stage of the work, traverses may be described as being either *computed* or *plotted*. A *computed* traverse is one in which the field observations and measurements are used to compute rectangular co-ordinates for the stations of the traverse, the traverse then being plotted using the computed co-ordinates. All precise and semi-precise traverses should be computed, since this allows field errors to be located before plotting, mistakes in plotting can be avoided, and the accuracy of the traverse can be calculated. A *plotted* traverse is one which is plotted from the field notes, using a scale and protractor. Low accuracy traverses, perhaps for preliminary reconnaissance, may be plotted in this way.

9.3.2 Appropriate uses of traversing

Traverse survey will be a suitable method if:

(i) control is to be provided for a site survey; or
(ii) tape and offset linear survey techniques are inadequate to meet the needs of the task; or
(iii) the accuracy demanded is greater than can be achieved by tape and offset survey; or
(iv) the introduction of traversing into a tape and offset survey will reduce the costs and/or the duration of the job, by reducing the number of tie and check lines that would normally be needed.

9.3.3 Traverse reconnaissance and layout

On arrival on site, the surveyor should first carry out a thorough examination of the area to determine the best positions for the traverse stations and hence the traverse legs. Time spent on such a 'recce' is seldom wasted and generally leads to a more satisfying end product. A number of factors must be considered in making the choice of station and leg positions, identified in the following sections.

9.3.3.1 Preliminary considerations

These may include:

(i) the need to tie the traverse to higher order control in order to meet the accuracy demanded or help locate any errors;
(ii) the equipment available and its characteristics, e.g. if taping the legs, then they should preferably run over the more level ground, avoiding obstacles;
(iii) the climatic conditions, e.g. mist and haze, may limit leg lengths;
(iv) the quality and quantity of the labour to be used;
(v) the time available for the task;
(vi) the method of supplying detail. If offsets from the legs are used, the legs must run close to the detail. If radiation is used, there must be a clear view of the area to be surveyed from any station.

9.3.3.2 Traverse layout

A *closed loop traverse* should preferably be regular in shape, e.g. a perfect square, hexagon, etc. This is seldom practicable, but large length differences and acute direction changes in consecutive legs should be avoided, in order to simplify the distribution of the small errors which can be expected in the observations and measurements of the traverse.

The *legs* should be as long and as few as possible, bearing in mind the equipment to be used and the climatic conditions. This will usually minimize the observational, computational and plotting time. Legs of less than 20 m in length should be avoided, due to the effects of centring errors on such legs and the distribution of errors through these legs. Legs should be located along the more level or even ground, avoiding obstacles, steep slopes, rough ground and marsh if distances are to be taped on the surface. *Stations* should, if possible, be placed on permanent ground features, e.g. the inside corner of a manhole cover frame, since this simplifies their location. Artificial station markers might include a wooden peg with a nail driven in the top to mark the exact station point, or a steel pin placed in concrete, in either case with possibly a wooden marker peg placed alongside bearing the station identification in waterproof crayon or paint.

Station markers should be protected from damage, either by 'fencing-off' or by suitable location selection. The type of marker depends upon the permanence required. If station markers are likely to be disturbed, they should be *referenced* to other pegs or permanent objects, and the referencing noted in the field book. To *reference a point*, note its distance from three or more permanent marks, thus the point may be relocated later by measurement from the permanent marks.

9.3.3.3 Tying to other control

If a closed loop traverse is inadequate for a survey task, then a network of *secondary traverses* may be a possible alternative. These secondary traverses should all lie within the 'primary' loop traverse. A secondary traverse should be as straight as possible between its opening and closing stations (terminals) and all its legs should lie within the circle which has the line joining the traverse terminals for a diameter, as in *Figure 9.7*.

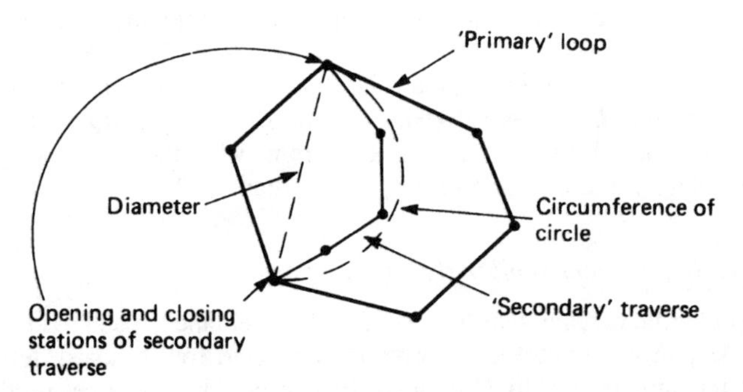

Figure 9.7

Adjacent traverses in a network which run close to one another should, if at all possible, be linked at their mid-points, as in *Figure 9.8*. This is to ensure that the detail from both traverses is compatible, e.g. the north face of a rectangular building could be supplied from one traverse and the south face from the other traverse.

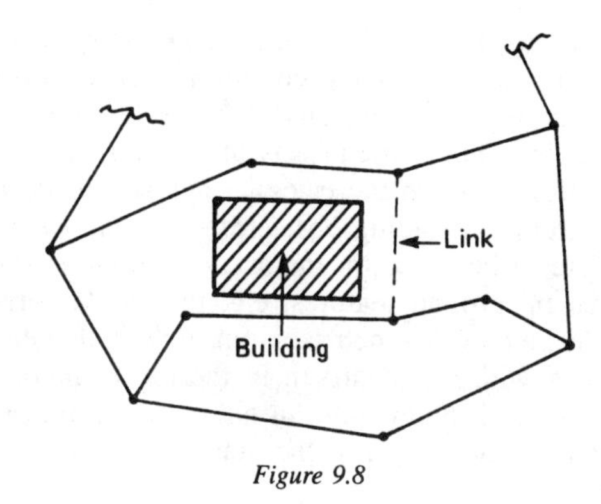

Figure 9.8

Where a traverse is linked to other control, either a primary loop as mentioned above or higher order traverse or triangulation, the terminal points of the traverse would obviously be known co-ordinated points on the existing control. The co-ordinates of the terminals being known, then the traverse angles and the computed traverse co-ordinates may be checked and adjusted. On occasion, when tying to existing control, there may be problems such as an existing point which can be measured to, but which the theodolite cannot be stationed over. This and similar problems are considered in Section 9.6.

9.3.4 Traverse angle and distance measurement

Where a *network of traverses* is required, or a closed loop traverse holds a network of taped survey lines, it is usual to aim for an accuracy of 1:5000 or better.

Setting out traverses, that is to say traverses from which detail is to be set out in preparing for building or engineering work, generally require still higher standards of accuracy. British Standard BS 5964: 1980, *Methods for Setting Out and Measurement of Buildings: Permissible Measuring Deviations*, specified that such traverses should close to within $\pm 2\sqrt{L}$ mm, where L is the length of the traverse in metres, and this is normally better than 1:5000.

9.3.4.1 Setting out and similar traverses

Each *leg* should be measured at least twice. If a steel tape is used readings should be to the nearest millimetre and corrections applied for standard, temperature, slope and tension. If tied to Ordnance Survey control, the measured distance should also be corrected by the OS National Grid local scale factor.

Where a line is measured by steel tape, the two *field measurements* of the line should agree within 2 mm/30 m of tape or part tape length. The tape may not need to be standardized if it is fairly new and undamaged. Long legs over undulating ground should be measured in catenary and corrected for sag.

Where the legs are measured by EDM, undulating ground is no problem so long as the end stations are intervisible, but grazing rays within 1 m of the ground should be avoided.

Horizontal angles should be observed with a theodolite or total station. With an optical instrument the readings should be booked to at least the nearest 20", on face left and face right (the simple reversal technique) and the reduced readings should agree within one minute of arc. Where traverses are of considerable length (say 1500 m or more) the readings should be taken on two or more zeros. On long traverses with few legs this is essential to maintain accuracy, and on long traverses with many legs this may eliminate the need for later re-observation, which could entail considerable expenditure in time and money.

Three tripod traversing is a method used to reduce errors in centring over station points and in bisecting observed targets, and hence improving the accuracy of the angle measurements, particularly when the leg lengths are very short.

Example: With traverse stations A, B, C, and D, it is required to measure the included angle at B. Tripods with theodolite tribrachs attached would be set up over stations A, B and C, and carefully centred over each of the station points. A traverse sighting target would be set in the tribrachs at A and C, and a theodolite placed in the tribrach at B. The included angle ABC would be measured using the theodolite and sighting on the targets at A and C. To measure the angle at C, the tripod and tribrach at A would be moved to D, the theodolite removed from B tribrach and placed in C tribrach and the target at C moved to the tribrach at B. This process can be repeated as necessary, in every case the theodolite being moved forward into a tribrach which has previously contained a sighting target. Despite being termed 'Three tripod traversing' the method is more efficient if five tripods and tribachs are used.

When using a steel tape the *angle of slope* (vertical angle) between the traverse stations and/or changes of slope are often measured by the theodolite, but the Abney level may be used or a level and staff, in all cases for the purpose of reducing the slope leg length to the horizontal. The *vertical angles* are usually observed on completion of the horizontal angles. They are measured in one direction only, but on both left and right faces. Alternatively, they may be read on one face only, but in both directions, i.e. both up and down the slope. The vertical angles are often read and booked to the nearest 20" of arc, but on gentle sloping sites readings to the nearest minute are adequate.

9.3.4.2 Single detail traverse

The linear measurement may be carried out as in tape and offset survey, with distances read and booked to the nearest 0.01 m generally being adequate.

Horizontal angles should be read and booked to the nearest minute of arc on both faces, the reduced readings should agree within two minutes of arc.

Slope correction may be made as in tape and offset survey.

9.3.5 Traverse data recording

Traditional traverse booking may be carried out using loose-leaf booking sheets or a field book, in either case A4 size is recommended. The methods used for booking vary, but most surveyors tend to use one of the styles illustrated below. As with all booking, of course, clarity and accuracy are essential.

9.3.5.1 *Detail traverses*

In taped *detail traversing,* it is common practice to record the horizontal leg lengths, together with any necessary offsets, in the tape survey booking. Accordingly, a single booking sheet can be used, recording all the angular observations in the traverse but with no linear measurements shown, as in *Figure 9.9.*

Inst. stn	Face	Station observed	Horizontal reading	Reduced to RO	Mean	Corrected angle
A	L	RO. Stn. D	00° 00'	° '		
		Stn. B	80 27	80 27	80° 27½'	80° 27'
	R	Stn. B	260 28	80 28		
		RO. Stn. D	180 00			
B	L	RO. Stn. A	03 10			
		Stn. C	99 01	95 51	95 51½	95 51
	R	Stn. C	279 01	95 52		
		RO. Stn. A	183 09			
C	L	RO. Stn. B	50 14			
		Stn. D	136 09	85 55	85 55	85 54
	R	Stn. D	316 09	85 55		
		RO. Stn. B	230 14			
D	L	RO. Stn. C	327 29			
		Stn. A	65 19	97 50	97 49	97 48
	R	Stn. A	245 18	97 48		
		RO. Stn. C	147 30			
				Sum.	360° 03'	360° 00'
				Misclosure	03'	✓

Correction = -¾' per station

(Note each corrected angle has been rounded up or down to the nearest minute)

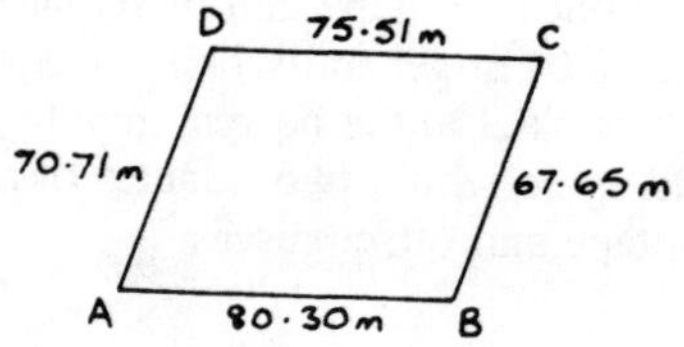

Nurses home traverse
LRI
Surveyed June '80
by C. Freestone

Figure 9.9

The form is fairly self-explanatory, but a few points may be noted:

RO = *reference object*. This is the object on which the theodolite is pointed first when starting to observe a set of directions. The horizontal circle readings are recorded in the order in which they are observed, working downwards from the top of the page. Three different 'zero setting' techniques are illustrated in the figure, although in practice only one, the surveyor's own preference, would be adhered to.

In *Example 1*, at station A, pointing the RO on face left, the circle has been set at 00° 00' 00".

In *Example 2*, at station B, a small angle of 03° 10' has been set on the instrument.

In *Example 3*, at station C, no particular angle was set on, the theodolite was simply pointed at B and the reading noted, and in this case it happened to be 50° 14'. This method was also followed at station D, pointing on C, but in this case the reading is rather large and is likely to make the subsequent arithmetic subtraction work laborious.

Note that at each instrument station the FR readings should differ from the FL readings by approximately 180°. The difference would be exactly 180° if the observations were free from all error or if the errors compensated one another.

The *Reduced to RO* column gives the results of subtracting the RO reading from the forward reading for each consecutive pair of readings. Thus at A, FL is 80° 27' – 00° 00' = 80° 27', FR is 260° 28' – 180° 00' = 80° 28'. At B, FL is 99° 01' – 03° 10' = 95° 51', and so on.

The traverse is a closed loop, and the readings have been booked moving round the stations in an anti-clockwise direction in plan, thus the angles reduced to RO are the internal angles of the figure. Since the internal angles of a polygon sum to $(2n - 4)$ right angles, where n = the number of sides, the angles should sum to 360°. This may be checked on the booking sheet and the meaned values adjusted equally before the traverse computation, if desired. (This check should be carried out before leaving the site, to detect any error.) The traverse could have been carried out in a clockwise direction, then the reduced angles would be the *external* angles of the figure, which should sum to $(2n + 4)$ right angles.

Traverse points		*Horizontal angles*	
Inst.	*Target*	*Face Left* ° ' "	*Face Right* ° ' "
B	A	03 10	183 09
	C	99 01	279 01
Angles =		95 51	95 52
Mean =		95 51 30	

Figure 9.10

Figure 9.10 shows an alternative form of presentation for one 'round' of horizontal angles. The observations at Station B in *Figure 9.9* are illustrated. See also the method illustrated in Section 6.6.5 and *Figure 6.11*.

9.3.5.2 Other theodolite traverses

When a traverse is to be tied to other traverses, or if the individual station details are more complex than with a simple detail traverse (e.g. observations on more than one zero, vertical angles by theodolite and distances measured by steel tape and corrected for temperature, standardization, etc.), then it is often better to have a separate sheet or page for each traverse station's booking. An example is shown in *Figure 9.11*. As an alternative, one book could be used for angular measurement recording and another for linear measurement details.

Again the form illustrated is fairly self-explanatory, but the following points should be noted:

The *horizontal angle readings* are at a station identified as 209, on a traverse known as LRI – TWO. This station is also the opening station of traverse LRI – THREE, which was observed on completion of traverse LRI – TWO. The observations have been taken to station 303 in order to avoid the need to re-occupy station 209 at a later date, i.e. when traverse THREE is commenced. Note that for the first zero, the observer has actually used 00° 00' 00", but the zero for the second round is a completely different, arbitrary setting of 71° 12' 20". The 'Mean' column shows the means of the four readings for each direction.

The *reduced vertical angle* is the angle of slope of the leg (or part) concerned. The calculation depends on the mode of graduation of the vertical circle of the theodolite in use. In this instrument, the vertical circle is designed to read 90° 00' 00" when the telescope is horizontal on face left, and 270° 00' 00" on face right. Thus the reduced vertical angle to station 210 is 90° 00' 20" less 90°, giving a slope of –00° 00' 20" on FL, and on FR 270° 00' 10" less 270°, giving +00° 00' 10", with a mean value of –00° 00' 05".

Note that when the line of sight is nearly horizontal, it often happens that one reduced angle is positive and the other negative. For clarity, the signs have been entered here, but they are not essential – see the second result of 00° 15' 05".

The measured length of the leg from 209 to 208 is quoted as 121.344 m, i.e. 4 × 30 m lengths + the final tape length of 1.344 m, all measured at 3°C. Note that there is a change of slope between stations 209 and 303. Distance measurements to the nearest 0.25 m are usually adequate for the length from the station to the change of slope point, 47.5 m in the example. Remember, however, that the total length should be recorded to the nearest millimetre. Note also that the leg 209 to 303 is still booked to the nearest millimetre. Had 21.8 been booked, the individual doing the computation would be uncertain as to whether the distance had been measured to three decimal places.

9.3.6 Traverse computation

The minimum equipment for traverse computation today is a calculator with trigonometric functions, including polar–rectangular and rectangular–polar conversion, and a square root key. If several traverses are to be computed, a programmable calculator is to be preferred, ideally with program cards and a

page 17

Traverse LRI - TWO	Instrument Wild T16
Date 20 Nov '79 Weather Fair	Observer Greasby
Horizontal angles at station 209	Booker Hemingsley

Station observed	Face / Zero	Horizontal reading	Reduced to RO	Mean
RO. Stn 208	L/1	00° 00' 00"	° ' "	° ' "
210		179 25 40	179 25 40	179 25 38
303		247 31 50	247 31 50	247 31 55
303	R/1	67 32 00	247 31 40	
210		359 26 00	179 25 40	
RO. Stn 208		180 00 20		
RO. Stn 208	L/2	71 12 20		
210		250 37 40	179 25 20	
303		318 44 10	247 31 50	
303	R/2	138 44 20	247 32 20	
210		70 37 50	179 25 50	
RO. Stn 208		251 12 00		

Vertical angles at station 209

Stn. obs'd	Face	Vertical rdg.	Reduced angle	Mean
210	L	90° 00' 20"	-00° 00' 20"	-00° 00' 05"
	R	270 00 10	+00 00 10	
C/S I	L	90 15 20	00 15 20	00 15 05
	R	269 45 10	00 14 50	

Observer Greasby
Booker Hemingsley
Date 20 Nov '79

Surface taped distances – Tape No. 5 Length of tape 30 m

Line	No. of full tape lengths laid	Final tape rdg.	Total measured length	Temp °C	Remarks, sketch or change of slope diagram
209-208	✓✓✓✓	1.344	121.344	3	
209-210	✓✓✓✓	1.239	121.239	3	
209-303	✓✓✓	21.800	111.800	3	Booker Hemingsley Date 23 Jan '80

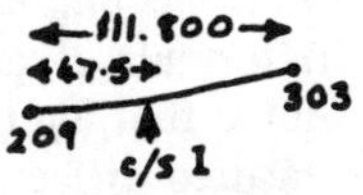

Figure 9.11

printer, or better still a microcomputer with suitable software and peripherals.

Before the actual calculation of co-ordinates from the observed angles and measured distances, the steps outlined in the next three sections may require to be carried out.

9.3.6.1 Reduction of measured lengths

The *field measured leg lengths* must be reduced to the *corrected horizontal equivalents*, by applying corrections for slope, temperature, standardization, etc., as covered earlier. *Figure 9.12* illustrates part of a typical 'Reduction of the measured length' form which could be used with a calculator. The following points should be noted:

REDUCTION OF THE MEASURED LENGTHS

~~Base~~/ Traverse LRI - FIVE

Calculated by R Jones Date 3 Mar '80

Correction factors for:

STANDARD		TEMPERATURE		PROJECTION	
Tape No.	Factor	°C	Factor	N.G Eastings	Factor
(5)	0.99990	11	0.99990	~~450 000~~ 460 000 ~~470 000~~	~~0.99963~~ 0.99965 ~~0.99966~~
(7)	0.99987	14	0.99993		

Leg	Measured	Corrected for: Standard	Temperature	Projection	~~Cos~~. VA for Slope	Reduced Horizontal Distance
505 -506	41.800 41.798	(7) (5) 41.7942	11°C 14°C 41.7906	41.7760	01°53′28″	41.753
506 -507	45.850 45.848	(7) (5) 45.8437	11°C 14°C 45.8398	45.8238	00°58′52″	45.817
507 -602	36.810 36.815	(7) (5) 36.8083	11°C 14°C 36.8051	36.7923	00°24′58″	36.791

Figure 9.12

Correction factors for standardization and temperature have been obtained from the formulae in Sections 4.4.2 and 4.4.4, respectively.

The OS projection correction factor may be calculated to five decimal places using the formula given earlier in Section 4.4.8.

Taking the first leg (505–506) as an example, it will be seen that the leg has been measured twice (41.800 and 41.798). These measurements have been meaned, then the result multiplied, in turn, by the correction factors for *standardization* (giving 41.7942), for *temperature* (giving 41.7906), then by the *OS scale factor* (giving 41.7760), and finally by the cosine of the vertical angle to give the *reduced horizontal distance* (here 41.753 m).

9.3.6.2 *The traverse abstract*

The *horizontal angles* at each traverse station, measured clockwise from the back station to the forward station (i.e. the angle included between the pair of legs meeting at the station, hence known as *included angles*) are readily obtained from the field bookings, but if there are any complications it may be better to prepare an *abstract of the angles*. The complications referred to may result from the use of more than one booking sheet for a station, the employment of two surveyors on the same task, or difficulties in tying the traverse to other traverses.

A part of such an abstract is shown in *Figure 9.13*. The entries under '1st zero' are the horizontal angular readings reduced to RO, the 2nd and 3rd zero columns indicating seconds only. In this case, the reduced angles on 2nd and 3rd zero differ from the 1st zero values only in the number of seconds in each, but if there were a difference in the numbers of minutes also, then the minutes values would also be shown.

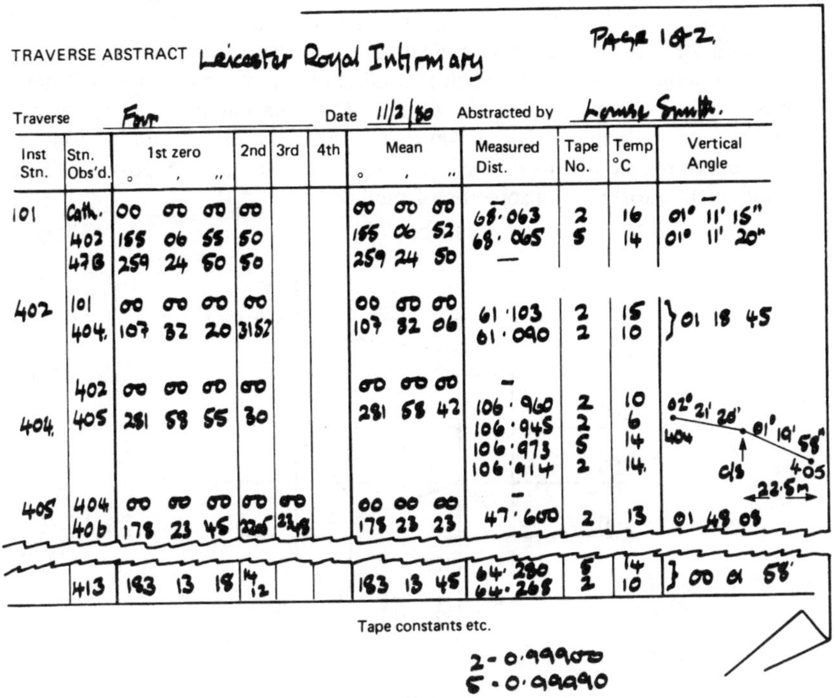

TRAVERSE ABSTRACT Leicester Royal Infirmary PAGE 102.

Traverse Four Date 11/3/80 Abstracted by Louise Smith.

Inst Stn.	Stn. Obs'd.	1st zero ° ′ ″	2nd	3rd	4th	Mean ° ′ ″	Measured Dist.	Tape No.	Temp °C	Vertical Angle
101	Cath.	00 00 00	00			00 00 00				
	402	165 06 55	50			165 06 52	68·063 68·065	2 5	16 14	01° 11′ 15″ 01° 11′ 20″
	47B	259 24 50	50			259 24 50	—			
402	101	00 00 00	00			00 00 00	61·103 61·090	2 2	15 10	} 01 18 45
	404	107 32 20	31 52			107 32 06				
404	402	00 00 00	00			00 00 00	—			
	405	281 58 55	30			281 58 42	106·960 106·945 106·973 106·914	2 2 5 2	10 6 14 14	02° 21′ 20″ 01° 19′ 58″ 404 c/s 405 22·5m
405	404	00 00 00	00	00		00 00 00	—			
	406	178 23 45	22 45	23 48		178 23 23	47·600	2	13	01 48 08
	413	183 13 18	14 12			183 13 45	64·280 64·268	5 2	14 10	} 00 01 58

Tape constants etc.

2 - 0·99900
5 - 0·99990

Figure 9.13

On the traverse booking sheet, *Figure 9.11*, the booker meaned both zeros and this meaning of each zero is a preferable alternative if an abstract is to be used. Note that additional readings may be taken and entered when the horizontal angle readings and/or the measured distances do not conform to the job specification.

9.3.6.3 *Allocation of opening co-ordinates and bearing*

Before starting to compute the traverse, the co-ordinates of the first or start station of the traverse (the *opening co-ordinates*) and the bearing of the first leg (*opening bearing* or *initial bearing*) must be known. If the traverse commences on a previously co-ordinated point, such as an OS or other control point, then the opening co-ordinates can be obtained and the opening bearing may be deduced. (Section 9.6.2 deals with a variety of problems which may arise when tying a traverse to existing control.)

If the traverse also closes on a previously co-ordinated point, then it will be possible to check and adjust both the angles of the traverse and the calculation of the co-ordinates. Where the traverse is independent of existing control, but is to be plotted by co-ordinates, then the surveyor must decide what values to use for the opening co-ordinates, and what opening bearing to use.

Methods commonly used to select *independent opening co-ordinates* are as follows:

(i) Take the first station as simply 0.000 m East, 0.000 m North. The disadvantage of this method is that it is very likely that some stations will have negative co-ordinates, and matters are simpler for the beginner if all co-ordinates are arranged to be positive.

(ii) Select opening co-ordinates such that all stations in the traverse will have positive co-ordinate values. In *Figure 9.14*, Station A has been given opening co-ordinates of 100.000 m E, 100.000 m N, thus all the stations lie in the north-east quadrant and all co-ordinate values are positive. This removes the arithmetic problems of method (i).

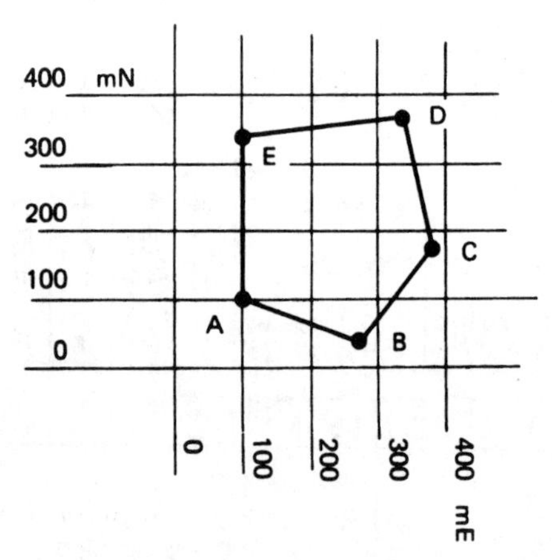

Figure 9.14

(iii) Scale the co-ordinate values for the start station from an existing gridded map, e.g. an OS map, and use the rounded scaled figures. For example, scaled OS values for the point might be 410895.2 m E, 286116.6 m N, then the surveyor might decide to use 895.000 m E and 6117.000 m N for his opening co-ordinates. This has the advantage of giving the client some compatibility between the OS map and the new plan or map.

The *opening co-ordinates* fix one point of the survey on the grid, while the *opening bearing* decides the orientation of the survey. Opening bearings often used for independent traverses are as follows:

(i) Grid north for the particular grid in use.
(ii) The compass bearing of the opening or closing leg of the traverse, possibly corrected for *magnetic variation*. (*Variation*, also known as *declination*, is the horizontal angle between true north and magnetic north at a place, and it may be east or west of true north. Variation differs from place to place, and is subject to small annual and diurnal changes, e.g. in 1991 in eastern England it was about 4° W, changing by about 0.5° E in the next four years, while in the south-west it was about 9° W and changing by about 0.5° E in the next three years. Some local and national map sheets give this information.)
(iii) The approximate bearing of the first leg, as plotted roughly on an existing map and measured by protractor, or by scaling co-ordinates and calculating an approximate grid bearing. Again, this gives some compatibility between the existing map and the new plan or map.

9.3.7 Calculation of traverse bearings

The *whole circle bearings* of all the legs of the traverse are required before the co-ordinates of the stations can be computed. The process of calculating the bearings from the observed angular data is often termed 'working through the bearings'.

9.3.7.1 Forward and back bearings

At any traverse station, the next station ahead is termed the *forward station*, and the preceding station is termed the *back station*. In a closed loop traverse, every station has a back and a forward station, of course, but in a link traverse the back station from the first station will generally be an *opening RO*, reference object, from which the opening bearing will be determined. Similarly, in a link traverse, the forward station from the last traverse station will be a *closing RO.*

At any given station, the whole circle bearing of the leg to the back station is termed the *back bearing* at the given station and the whole circle bearing of the leg to the forward station is the *forward bearing* at the given station.

Figure 9.15 shows station C, and the back bearing at C is the whole circle bearing of the line from C to B (the back station), while the forward bearing is the whole circle bearing of the line from C to D (the forward station). It will be evident that:

> the back bearing at C (bearing of line C → B) + the angle observed at station C (∠BCD) = the forward bearing at C (bearing of line C → D).

If the bearing C → B was deduced as above, then addition of the observed clockwise angle at C gives the forward bearing C → D.

Considering station D, the back bearing D → C is equal to the forward bearing C → D plus or minus 180°, then the process above may be repeated to obtain the forward bearing D → E, and so on.

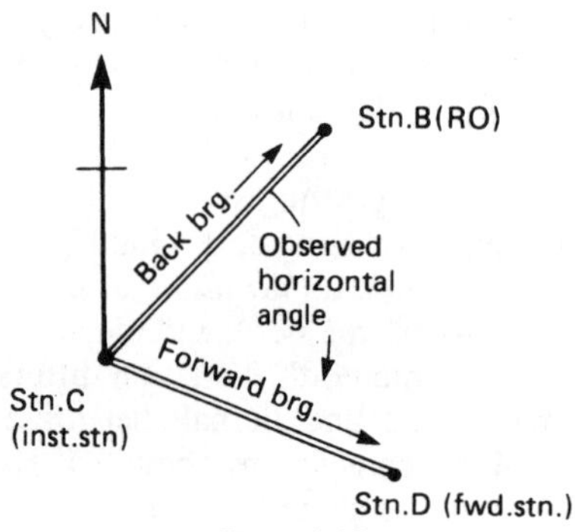

Figure 9.15

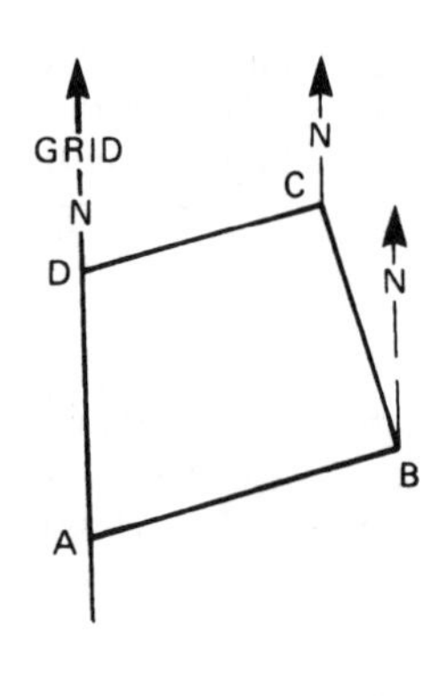

Back brg.	A → D
+∠ obs'd.	+A
= fwd. brg.	A → B
	± 180
back brg.	B → A
+∠ obs'd.	+B
= fwd. brg.	B → C
	± 180
back brg.	C → B
+∠ obs'd.	+C
= fwd.brg.	C → D
	± 180
back brg.	D → C
+∠ obs'd.	+D
= fwd. brg.	D → A
	± 180
back brg.	A → D

Note the repeating patterns:

back brg. + ∠ obs'd = forward bearing

and the grouping of the station letters

AAA, BBB, CCC, etc.

If the observed angles entered are the corrected ones, then the first line should equal the last line as is shown below

A → D	00°	00′
+ ∠A	80	27
A → B	80	27
B → A	260	27
+ ∠B	95	51
B → C	356	18
C → B	176	18
+ ∠C	85	54
C → D	262	12
D → C	82	12
+ ∠D	97	48
D → A	180	00
A → D	00	00
A → D	00	00

(as above)

Figure 9.16

9.3.7.2 *Closed loop traverse*

Figure 9.16 illustrates the closed loop traverse booked in *Figure 9.9* (repeated opposite). The surveyor has decided to take the closing bearing of the traverse (leg A → D) as grid north, then the bearings of the traverse are deduced as shown. Note that in the numeric example the ± 180° has been

Inst. stn	Face	Station observed	Horizontal reading	Reduced to RO	Mean	Corrected angle
A	L	RO. Stn. D	00° 00′			
		Stn. B	80 27	80° 27′	80° 27½′	80° 27′
	R	Stn. B	260 28	80 28		
		RO. Stn. D	180 00			
B	L	RO. Stn. A	03 10			
		Stn. C	99 01	95 51	95 51½	95 51
	R	Stn. C	279 01	95 52		
		RO. Stn. A	183 09			
C	L	RO. Stn. B	50 14			
		Stn. D	136 09	85 55	85 55	85 54
	R	Stn. D	316 09	85 55		
		RO. Stn. B	230 14			
D	L	RO. Stn. C	327 29			
		Stn. A	65 19	97 50	97 49	97 48
	R	Stn. A	245 18	97 48		
		RO. Stn. C	147 30			
				Sum.	360° 03′	360° 00′
				Misclosure	03′	✓
			Correction = -¾′ per station			
		(Note each corrected angle has been rounded up or down to the nearest minute)				

Figure 9.9

done, but the figures have not been written in. The user may prefer to show the figures.

The rule as to whether to add or subtract the 180° is:

if the forward bearing is less than 180, add 180, if the forward bearing is greater than 180, deduct 180, to obtain the back bearing at the next station.

On occasion the summation will give a value greater than 360°, in this case 360° must be subtracted to get the correct value.

The final figures should not be blindly accepted – a small sketch of the traverse should be drawn, to help check that the values determined for the bearings are reasonable.

9.3.7.3 Closed traverse tied to other control

Figure 9.17 illustrates a traverse entitled 'Gateway Traverse', commencing on station 01 at one end of a base line and closing on to station 12 in another traverse called 'Base Traverse'. In this case, the calculations for the Base Line and for the Base Traverse will have to be carried out before the Gateway Traverse can be computed, since the base line and base traverse will fix the co-ordinate values for the start and end stations (01 and 12) of the Gateway Traverse.

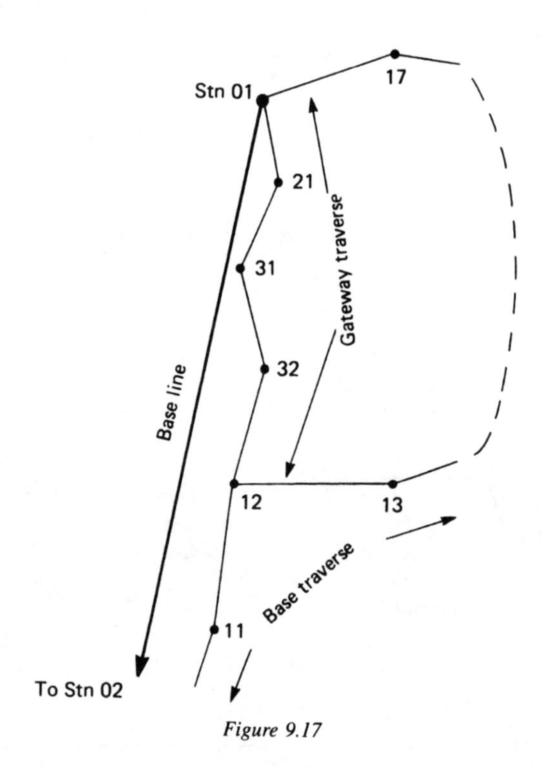

Figure 9.17

(i) *Opening bearings*: Given that the co-ordinates of stations 01 and 02 have already been calculated, then the bearing of the line 01 → 02 can be calculated, in this case it is 180° 00' 00", and this, together with the angle observations at station 01, will give a bearing for the first line of the traverse, 01 → 21, the 'opening' or 'initial' bearing.
Similarly, however, the co-ordinates of stations 01 and 17 will allow the calculation of the bearing of the line 01 → 17, found to be 63° 13' 08" here,

and from this another value for the opening bearing of 01 → 21 may be deduced. Note that the method of calculating bearings from co-ordinates is covered in Section 1.11.

It might seem that only **EITHER** the bearing 01 → 02 **OR** the bearing 01 → 17 need be used to find the bearing of line 01 → 21, but the calculation of two values for the bearing, from different data, provides a check for mistakes. Generally, where two or more opening bearings are available, then either the mean value will be used or the value which is considered to be the most reliable. Note that bearings to distant points are considered usually to be more reliable than bearings to nearby points.

(ii) *Closing bearings*: When an opening bearing for the traverse has been decided, the bearings of the traverse will be worked through, and to provide a check on these calculated bearings, a 'closing bearing' must be determined in a very similar way to the above. Here, using the co-ordinates of stations 12, 11 and 13, closing bearings were calculated for lines 12 → 11 and 12 → 13, and these were 185° 12' 40" and 94° 03' 32", respectively. Working through the traverse bearings will give bearings for these lines also and they may be compared with the co-ordinate-based bearings above. The comparison will determine the angular misclosure of the traverse.

(iii) *Working through the Gateway Traverse*: An abstract of the observed horizontal angles for the Gateway Traverse is shown in *Figure 9.18*. Working through the bearings produces the following result:

Opening bearing	01→02	180° 00′ 00″	... from co-ordinates
	+∠01	347 00 05	... from traverse abstract
	01→21	167 00 05	... 180 + 347 = 527 527 − 360 = 167
Opening bearing	01→17	63 13 08	... from calculated bearings
	+∠01	103 47 02	... from traverse abstract 347° 00′ 05″–243° 13′ 03″
(Check only)	01→21	167 00 10	... Check reading only, 01→02 is longest leg
Back bearing	21→01	347 00 05	
	+∠21	195 12 58	
	21→31	182 13 03	... 347 + 195 = 542, 542 − 360 = 182
	31→21	02 13 03	
	+∠31	190 46 48	
	31→32	192 59 51	
	32→31	12 59 51	
	+∠32	154 14 45	
	32→12	167 14 36	
	12→32	347 14 36	
	+∠12	106 47 18	... the clockwise angle at 12 between 32 and 13
	12→13	94 01 54	... again −360°
Closing bearing	12→13	94 03 32	... from co-ordinates
	Misclosure	−01′ 38″	
	12→32	347 14 36	... repeated from above
	+∠12	197 56 25	... the clockwise angle at 12 between 32 and 11
	12→11	185 11 01	... again −360°
Closing bearing	12→11	185 12 40	... from co-ordinates
	Misclosure	−01′ 39″	

TRAVERSE ABSTRACT *City Campus*
Traverse *Gateway* Date *16/6/83* Abstracted by *R Drewery*

Inst Stn.	Stn. Obs'd.	1st zero ° ′ ″	2nd	3rd	4th	Mean ° ′ ″	Measured Dist.	Tape No.	Temp °C	Vertical Angle
01	02	00 00 00	00	00	00	00 00 00				
	17	243 13 00	00	10	00	243 13 03				
	21	347 00 00	00	10	10	347 00 05				
21	01	00 00 00	00	00	00	00 00 00				
	31	195 13 00	00	10	13/40	195 12 58				
31	21	00 00 00	00	00	00	00 00 00				
	32	190 46 60	50	50	30	190 46 48				
32	31	00 00 00	00	00	00	00 00 00				
	12	154 14 50	40	40	50	164 14 45				
12	32	00 00 00	00	00	00	00 00 00				
	13	106 47 20	00	20	30	106 47 18				
	11	197 56 10	10	40	40	197 56 25				

Figure 9.18

Note that the comparison of the two bearings for line 12 → 13 shows a misclosure of –01′38″, while comparison of the two bearings for line 12 → 11 give a misclosure of – 01′39″. Where two or more misclosures are calculated, the surveyor must normally decide whether to accept one of them as the closing error or take their mean. In this case, the two are practically identical.

For an accuracy of 1:2000, a closing error of $\sqrt{n}$ minutes is acceptable, where n = number of traverse stations, and for 1:5000 then $30\sqrt{n}$ seconds may be satisfactory. If the traverse is short these values may be exceeded provided the co-ordinate misclosure is acceptable, i.e. traverse co-ordinates closing to within 1 part in 2000 or 1 in 5000, respectively (see Section 9.3.10).

In the setting out of buildings, BS 5964:1980 suggested an angular misclosure limit of $\pm\ 0.135/\sqrt{L}°$ where L is the traverse length in metres. The Gateway misclosures of 98″ and 99″ are approximately equivalent to $44\sqrt{5}$″, which is a little in excess of the technical instructions issued to the surveyor. However, the traverse was short, roughly 200 m and only four legs, and hence it was decided that the result was acceptable.

The two misclosures here are along closing ROs of approximately equal length, therefore they could be meaned as being equally reliable. The good agreement is due to the very small angular error of only 5 seconds in the base traverse.

9.3.8 Adjustment of traverse bearings

9.3.8.1 Link traverse

Having computed the *bearings misclosure*, this must be distributed through the bearings of the Gateway Traverse. Normally the bearings error is divided by the number of stations in the traverse, including the terminals (01 and 12), and the result is the angular adjustment to be applied at each station, with reversed sign, rounded to the nearest second. Application of this to the Gateway Traverse gives:

	Leg	Deduced bearing*	Correction	Corrected bearing	
	01–21	167° 00′ 05″	+19″	167° 00′ 24″	
	21–31	182 13 03	+39	182 13 42	
	31–32	192 59 51	+59	193 00 50	
	32–12	167 14 36	+79	167 15 55	
	12–13	94 01 54	+98	94 03 32	from co-ordinates
or	12–11	185 11 01	+99	185 12 40	from co-ordinates

*from Section 9.3.7.

Leg 01 → 21 was corrected by +19", leg 21 → 31 by +19+20, leg 31 → 32 by +19+20+20, and so on, until the correct closing bearings were reached. Note the arbitrary alternation of +19" and +20", to avoid using decimals of a second.

In effect, the application of +19" on bearing 01 → 21 changes ALL the bearings by +19", then a further application of +20" at bearing 21 → 31 changes that and all subsequent bearings by an additional +20", and so on. The tabular method is the easier to follow in practice.

It should be noted that when a traverse has a mix of long and short legs, it is customary to apply a greater adjustment at angles observed between two short legs than at angles between two long legs, the former can absorb greater adjustment than the latter without displacing the traverse.

The methods of correcting bearings or angles before calculating co-ordinates are based on the traditional belief that the angular measurement is more accurate than the linear measurement in a traverse. Where modern high accuracy EDM methods of distance measurement are used, some surveyors consider that, provided the angular misclosure is seen to be acceptable, then the linear measurement is likely to be as accurate as the angles and they do not adjust the angles. Instead, they leave the bearings as observed, use them in the co-ordinate calculations, then adjust the resulting co-ordinates. Whichever methods of adjustment are used, of course, the results should be similar within a given specification.

9.3.8.2 Closed loop traverse

Section 9.3.7.2 illustrated how to work through the bearings of a closed loop traverse where the figure angles had been previously adjusted to the theoretical total of $(2n - 4)$ or $(2n + 4)$ right angles. It is not essential to adjust the angles first and had this not been done then the method of Section 9.3.8.1 could be used.

9.3.9 Calculation of traverse co-ordinates

9.3.9.1 The traverse computation sheet

The various calculations involved in a traverse are best done together, on a single computation sheet, even though they have been treated separately up to this point. *Figure 9.19* shows a typical computation sheet, for use with a calculator, with all the polar co-ordinate data for the Gateway Traverse ready for conversion to rectangular co-ordinates. There is no standard form in use, but this version is suited for use with scientific calculators and it does fit on to an A4 sheet, ideal for binding or photocopying.

Note that the computed rectangular co-ordinates have already been entered here.

TRAVERSE COMPUTATION SHEET (Polar to rectangular)

Traverse ....Gateway...... Calculated by ....R. Drewery....... Date ..17 Jun '83..

Total length ..207·601 m.. Co-ordinate misclosure .... Accuracy 1/.........

POLAR CO-ORDINATES;					RECTANGULAR CO-ORDINATES		
Legs and Stn's	Bearings and Angles	Cn	Corrected Bearings	Distance m	E.	N.	St'n
01-02	180 00 00						
+∠01	347 00 05	+			500·000	500·000	01
01-21	167 00 05	19"	167° 00' 24"	39·754	508·938	461·264	
01-17	63 13 08	} Check only					
+∠01	103 47 02						
01-21	167 00 10						
21-01	347 00 05						
+∠21	195 12 58	+					
21-31	182 13 03	39"	182° 13' 42"	32·488	507·675	428·500	
31-21	02 13 03						
+∠31	190 46 48	+					
31-32	192 59 51	59"	193° 00' 50"	77·578	490·205	353·215	
32-31	12 59 51						
+∠32	154 14 45	+					
32-12	167 14 36	79"	167° 15' 55"	57·781	502·943	296·855	
12-32	347 14 36						
+∠12	106 47 18	+			502·948	296·843	12
12-13	94 01 54	98"	94° 03' 32"				
12-13 (overleaf)	94 03 32						
Misclosure	−01' 38"						
12-32	347 14 36						
+∠12	197 56 25	+					
12-11	185 11 01	99"	185° 12' 40"				
12-11 (overleaf)	185 12 40						
Misclosure	−01' 39"						

Figure 9.19

9.3.9.2 *Partial co-ordinates and co-ordinates*

The method of computing the partial co-ordinates of each line, using the sine/cosine of its accepted bearing from North and the accepted leg length, has been described in Section 1.11. For any given station point, its co-ordinates are the *co-ordinates* of the *preceding station* plus the *partial co-ordinates* of the *leg from the preceding station.*

Thus in the Gateway Traverse shown in *Figure 9.17*, the easting co-ordinate of station 21 is equal to the easting co-ordinate of station 01 plus (distance 01 → 21 × sin brg 01 → 21), or

$$\begin{aligned}\text{easting } E_{21} &= E_{01} + (\text{distance } 01 \rightarrow 21 \times \sin \text{brg } 01 \rightarrow 21),\\ \text{easting } E_{31} &= E_{21} + (\text{distance } 21 \rightarrow 31 \times \sin \text{brg } 21 \rightarrow 31),\\ \text{easting } E_{32} &= E_{31} + (\text{distance } 31 \rightarrow 32 \times \sin \text{brg } 31 \rightarrow 32), \text{ etc.,}\end{aligned}$$

and similarly with the northing values, but using cosine instead of sine.

Full advantage of the calculator should be taken by using the store(s) to eliminate the need to write down and re-enter intermediate answers. As an example, the E_{01} and N_{01} values may be keyed into stores 1 and 2, respectively, then the bearing and distance for 01 → 21 entered and converted to rectangulars using the P → R key. The ΔN_{01-21} should then be added to store 2 and the ΔE_{01-21} added to store 1, thus the co-ordinates of station 21 are obtained and these values may be written down on the traverse computation sheet and left in the stores until superseded by the co-ordinates of station 31. The process may be continued in this way until the closing station is reached. (It is important to note that, when using the P → R key, that many calculators display northings before eastings.)

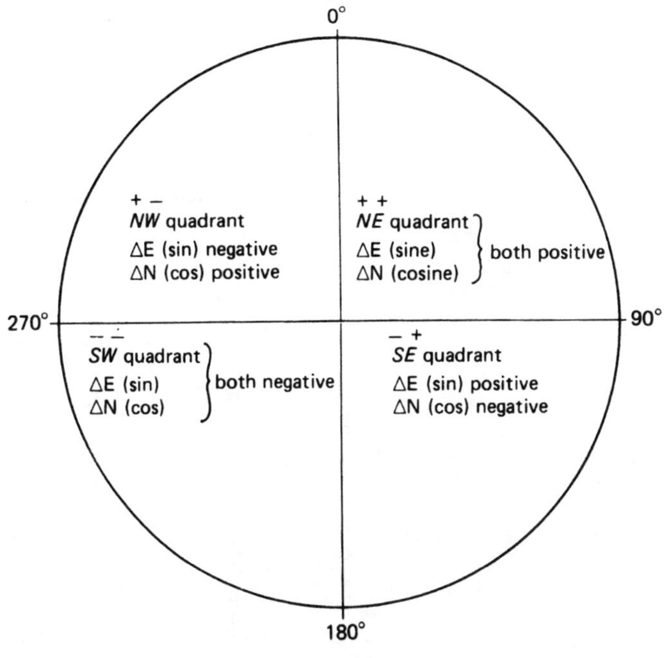

Figure 9.20

The calculated co-ordinates are entered on the computation sheet on the same line as the corrected bearings and distances, as shown in *Figure 9.19*, together with the co-ordinates of the opening and closing stations 01 and 12, on the lines 01 and 12.

The ΔEs and ΔNs used to calculate the co-ordinates may be positive or negative, and most calculators will automatically display and use the correct sign in the calculation of co-ordinates. If this is not so, then it should be remembered that the use of the P $\rightarrow$ R key or the sine and cosine functions is to determine the ΔE and ΔN of a traverse leg. Thus for a traverse leg from station A to station B, it is required to calculate how far station B is east of station A, and how far station B is north of station A. If station B does not lie to the north and east of station B, then one or more of the partial co-ordinates must be negative. The appropriate signs for the quadrants are shown in *Figure 9.20*.

9.3.10 Traverse misclosure and accuracy

The final pair of co-ordinates in *Figure 9.19* are the *accepted co-ordinates* of station 12, while the penultimate pair are the co-ordinates of the same station as calculated from the traverse data. In theory, these pairs should be the same and any difference indicates the *co-ordinate misclosure* of the traverse. Subtracting the accepted values from the calculated values gives, in this case, misclosures of –5 mm east and +12 mm north. The *linear misclosure* of the traverse is the actual linear distance between the plan point defined by the accepted co-ordinates and the plan point defined by the calculated co-ordinates. This is equal to the length of the hypotenuse of the right-angled triangle formed by these errors, equal to $(e^2_E + e^2_N)^{1/2}$ where e_E and e_N are the errors in eastings and northings, respectively. In this case, 13 mm, as shown in *Figure 9.21*.

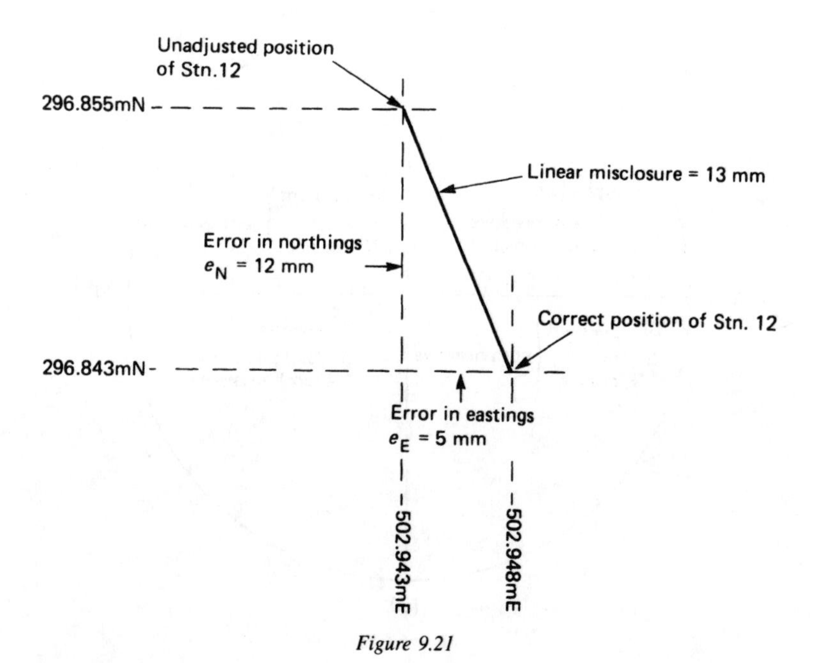

Figure 9.21

Acceptable standards of accuracy are often specified in terms of a stated maximum permissible linear misclosure, such as 50 mm or 200 mm. Another method is to define the accuracy of the traverse as the ratio of the linear misclosure to the total traverse length, in this case 0.013/207.601, which is approximately 1/16 000, or 1 in 16 000. (Note that in the case of a closed loop this form is merely an indication of the consistency of the work and not strictly a measure of absolute accuracy.)

The British Standard referred to previously recommended that traverses should close to within $\pm 2\sqrt{L}$ mm, where L is the total length of the traverse in metres. In the Gateway Traverse, this gives 28.8 mm, and the actual linear misclosure is well within this limit. In specifying permissible linear misclosures, limits may be variously encountered stated in the forms K mm, or KL mm, or $K\sqrt{L}$ mm, or $(k + KL)$ mm, where k and K are suitable factors and L is the traverse length in metres. In the Gateway Traverse, the technical specification required that the linear misclosure should not exceed 0.05 m.

9.3.11 Adjustment to traverse co-ordinates

If the traverse misclosure exceeds the permissible limit, it may be necessary to re-measure the traverse and to re-compute. If the accuracy is deemed to be acceptable, then the co-ordinate misclosures should be distributed through the traverse, *with reversed sign*, so that all station co-ordinates are adjusted and the calculated co-ordinates of the end station match the accepted co-ordinate values.

There are a variety of methods of correcting co-ordinates, some perhaps more mathematically respectable than others, but there is no method which guarantees that the adjusted co-ordinates are the correct values. A very widely used rule in practice is *Bowditch's method*, which has no great theoretical validity but is easy to apply.

9.3.11.1 Bowditch's method

In this method, the partial co-ordinate for each leg is corrected by an amount which is proportional to the length of the leg. In fact, *the partial easting* for each *leg* is adjusted by

$$-e_E \times \frac{\text{length of leg}}{\text{total length of traverse}}$$

or, for easy calculation,

$$\frac{\text{total misclosure in eastings}}{\text{total length of traverse}} \times \text{length of the traverse leg}$$

Similarly, the *partial northing* for each *leg* is adjusted by

$$\frac{\text{total misclosure in northings}}{\text{total length of traverse}} \times \text{length of the traverse leg}$$

Applying this to Gateway Traverse, for each leg:

$$\text{adjustment to } \Delta E = (+0.005/207.601) \times \text{length of leg}$$
$$\text{adjustment to } \Delta N = (-0.012/207.601) \times \text{length of leg,}$$

then

$$
\begin{aligned}
\text{leg } 01 \rightarrow 21 \ (39.754\,\text{m}) \text{ adjustment} &= +0.001\,\text{m E}, -0.002\,\text{m N} \\
\text{leg } 21 \rightarrow 31 \ (32.488\,\text{m}) \text{ adjustment} &= +0.001\,\text{m E}, -0.002\,\text{m N} \\
\text{leg } 31 \rightarrow 32 \ (77.578\,\text{m}) \text{ adjustment} &= +0.002\,\text{m E}, -0.005\,\text{m N} \\
\text{leg } 32 \rightarrow 12 \ (57.781\,\text{m}) \text{ adjustment} &= \underline{+0.001\,\text{m E}}, \underline{-0.003\,\text{m N}} \\
\text{Sum} &= +0.005\,\text{m}, \quad -0.012\,\text{m}
\end{aligned}
$$

Note that in practice, where the error is small as in this example, simple calculation by inspection might well be adequate.

The adjustments calculated above are those for the individual partial co-ordinates of the legs, but it is the co-ordinate values which must be adjusted and the adjustments must be progressively summed and applied to the co-ordinates. Thus, the adjustments to the co-ordinate eastings are +0.001, then +0.002 (+0.001 + 0.001), then +0.004 (+0.001 + 0.001 + 0.002), and finally +0.005 (+0.001 + 0.001 + 0.002 + 0.001).

Similarly the co-ordinate northings adjustments are –0.002, –0.004, –0.009 and –0.012. The layout of the co-ordinate adjustment is shown in *Figure 9.22(a)*.

ır to rectangular)
... R Drewey ... Date ... 17 June '83 ...
ıisclosure ... 13 mm ... Accuracy 1/ 16000 ...

Distance m	RECTANGULAR CO-ORDINATES E.	N.	St'n
	500·000	500·00	01
39·754	508·938	461·264	
	(001) +·001	(002) –·002	
	508·939	461·262	21
32·488	507·675	428·800	
	(001) +·002	(002) –·004	
	507·677	428·796	31
77·578	490·205	353·215	
	(002) +·004	(005) –·009	
	490·209	353·206	32
57·781	502·943	296·855	
	(001) +·005	(003) –·012	
	502·948	296·843	12
Misclosure	–·005	+·012	

SHEET (Polar to rectangular)
Calculated by ... R Drewey ... Date ... 17 June '83 ...
Co-ordinate misclosure ... 49 mm ... Accuracy 1/ 4300 ...

ꓤDINATES Uncorrected Bearings	Distance m	RECTANGULAR CO-ORDINATES E.	N.	St'n
		500·000	500·00	01
167°00'05"	39·754	508·942	461·265	
		(009) –·009	(002) –·002	
		508·933	461·263	21
182°13'03"	32·488	507·685	428·801	
		(007) –·016	(002) –·004	
		507·669	428·797	31
192°59'51"	77·578	490·237	353·211	
		(018) –·034	(005) –·009	
		490·203	353·202	32
167°14'36"	57·781	502·995	296·856	
		(013) –·047	(004) –·013	
		502·948	296·843	12
	Misclosure	+·047	+·013	

Figure 9.22

In Section 9.3.8.1 it was suggested that on some occasions it may be considered appropriate NOT to adjust the traverse angles, but rather to use the unadjusted angles in the co-ordinate computation and then apply adjustments to the computed co-ordinates only. The effect of this procedure may be observed in *Figure 9.22(b)*, and it will be noted that the maximum difference between the results is 8 mm in the eastings at station 31, an amount which is negligible in this particular task. Since the traverse runs roughly north to south, it is probable that the misclosure in eastings is primarily due to angular error.

The calculation of the closed loop traverse co-ordinates, using the same methods, is shown in *Figure 9.23*.

TRAVERSE COMPUTATION SHEET (Polar to rectangular)

Traverse ..Nurses Home LR1..... Calculated by ....C. Freestone............... Date ..July '80..........

Total length ...294·37m............ Co-ordinate misclosure ...73 mm............... Accuracy 1/.4000.........

POLAR CO-ORDINATES					RECTANGULAR CO-ORDINATES		
Legs and Stn's	Bearings and Angles	Cn	Corrected Bearings	Distance m	E.	N.	St'n
A-D	00°00'						
+∠A	80 27				100·000	100·000	A
A-B	80 27		80°27'	80·30	179·187	113·322	
B-A	260 27				(001) +·001	(020) −·020	
+∠B	95 51				179·19	113·30	B
B-C	356 18		356°18'	67·85	174·809	181·031	
C-B	176 18				(000) +·001	(017) −·037	
+∠C	85 54				174·81	180·99	C
C-D	262 12		262°12'	75·51	99·997	170·783	
D-C	82 12				(001) +·002	(019) −·056	
+∠D	97 48				100·00	170·73	D
D-A	180 00		180°00'	70·71	99·997	100·073	
A-D	00 00				(001) +·003	(017) −·073	
(as above)	00 00				100·00	100·00	A
	✓			Misclosure	−·003	+·073	

Figure 9.23

9.3.11.2 *Transit method*

In this method, the partial easting for each leg is adjusted by

$$-e_E \times \frac{\Delta E \text{ for the traverse leg}}{\text{absolute } \Sigma\Delta E \text{ for the traverse}}$$

and similarly for the partial northing, substituting ΔN and $\Sigma\Delta N$.

9.3.11.3 *Equal adjustment*

For EDM traverses with leg lengths of the order of 200 m, it can be shown that the errors in the leg lengths will all be of approximately the same magnitude, hence it will be acceptable to apply the same correction to each line. Thus the correction to each partial easting will be

$$-e_E/n$$

where n is the number of lines. For each partial northing, $-e_N/n$.

9.3.11.4 Least squares

For the surveys covered by this text, Bowditch's method will be adequate, but in very high accuracy work the *method of least squares* would be used. This is now feasible using commercially available software on a PC, but it will not be considered further here.

9.3.12 Locating mistakes in traverse calculations

9.3.12.1 Error in angle

If the bearings have been worked through and the angular misclosure is excessive, the following procedures may help locate an incorrect angle.

(i) Check all arithmetic on the traverse computation sheet.
(ii) Check the angles transferred to the computation sheet from the field book/booking sheet.
(iii) Check the reductions in the field book/booking sheet, particularly angles close to 90/180°. An obtuse angle may have been entered instead of the reflex, or the FR to the forward station may have been subtracted from the FL to the RO, etc.
(iv) If the error is large, say 5° or more, it may be located by *either*:
  (a) roughly plotting the stations on an existing map (if any) then measuring the angles by protractor and comparing with the observed values, or
  (b) plotting the traverse to some scale, plotting the linear misclosure, bisecting this line, then raising a perpendicular at the bisection. If there is a single large error the perpendicular will pass through or close to the station in error.
(v) Work through the bearings from each end of the traverse in turn, compute uncorrected co-ordinates for the two traverses. Where the co-ordinates are the same or nearly so may indicate the station in error. If no stations appear to coincide there is more than one angular error.
(vi) If no mistake can be found in the calculations, re-observe the station thought to be in error, otherwise re-observe all stations.

9.3.12.2 Error in co-ordinate calculations

If the angles have been accepted and adjusted, yet the linear misclosure is excessive, the following procedures may help locate the error.

(i) Calculate the bearing and length of the misclosure. Any leg with a similar bearing (or ±180) is suspect and the error in the leg would approximate to the misclosure length. If many legs are all on the same approximate bearing, they may all need to be re-measured.
(ii) If the misclosure bearing is not similar to any leg bearing it indicates more than one error or more likely an error in the calculations. Depending upon the method of calculation, the error might be:
  incorrect data entry to calculator,
  sine and cosine interchanged,
  + and – interchanged,
  ΔE and ΔN interchanged,

incorrect addition or subtraction,
error in reduction of measured lengths,
error in abstraction from field book/booking sheet.

(iii) Compare the distances on a rough plot as in (4)(a) above.
(iv) Re-measure the legs as necessary.

Mistakes may only be avoided by the use of suitable working procedures. In the calculations, the best way to detect error is to have independent calculations by two different people, each abstracting the information independently from the field book/booking sheet.

9.3.13 Traverse plotting

The easiest and quickest method of plotting a traverse is, of course, by computer linked to a plotter. Failing access to these facilities and the necessary software, the manual plotting of a co-ordinated traverse is actually simple and fast. A grid of mutually perpendicular and parallel lines is drawn on the plan, at intervals of about 200 mm, actual distances depending upon the scale of the plan. The grid should not be drawn by tee-square and set-square, these being highly suspect generally, but with a steel straightedge with scale, beam compasses, a good quality pencil finely sharpened and a suitable piece of drawing material. When lightly drawn-in with fine pencil lines the grid intersections should be marked with blue ink crosses for permanence and co-ordinate values assigned to the lines and marked against each. *Figure 9.24* illustrates the grid construction method.

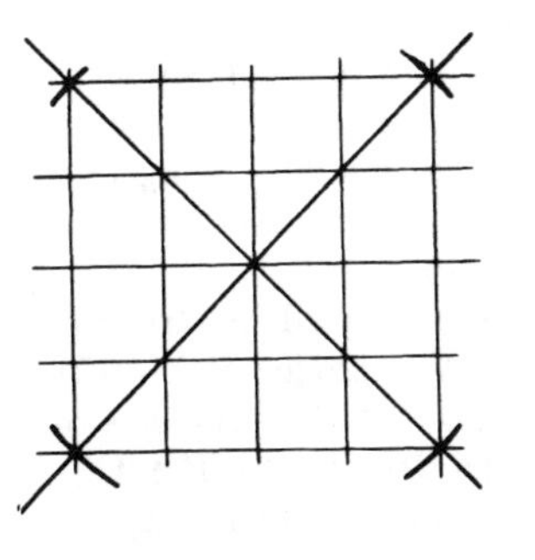

(i) Draw oblique lines in the approximate position of the diagonals
(ii) From the intersection of the oblique lines describe four arcs of equal radius cutting these lines
(iii) Join the cuts so formed, to form a perfect rectangle
(iv) Check with scale that opposite sides are equal
(v) Subdivide as necessary and construct squares
(vi) Check at random the lengths of the diagonals of some of the squares and rectangles
(vii) Erase oblique lines

Figure 9.24

The best equipment for manual grid construction and plotting of traverses is the rectangular co-ordinatograph, a precision instrument fitted to a flat rectangular table top. It has a scale along one edge and a second scale at right angles may be moved along the first scale. The second scale carries a microscope or plotting tool such as a pen, pencil or pricker. The microscope/plotting tool may be moved over the table surface, the two scales giving easting and northing co-ordinate values for precision plotting of the traverse stations.

When plotting by hand with a drawn grid, each survey station is plotted as follows:

> Each ordinate (northing) is plotted twice, once on each of the outside ordinates of the relevant grid square containing the point. The straightedge is laid to connect the plotted positions and a visual check made that the straightedge appears parallel to the lateral grid lines. If it appears parallel, a fine pencil line is drawn to connect the plotted positions. The appropriate easting ordinate is plotted in a similar manner, then the station position marked and circled lightly in pencil, annotating with station name or number. All stations are located on the grid in this way, then the leg lengths scaled off as a check. If all checks, then the station points may be pricked through and any unnecessary construction lines erased.

Detail, if any, is plotted in accordance with Chapter 10.

9.4 Triangulation and trilateration

A triangulation survey consists of a network of triangles in which one side length and all the angles are measured, the lengths of all the other sides being computed without further measurement. The single measured line is the *base line* of the network, it defines the scale of the network and it must be measured with very high accuracy. A *check base line* can be measured at the end of the net and its measured and computed lengths compared as a check on the work. Lower-order survey work can be tied to the triangulation network which then serves as the control for the survey.

Triangulation was formerly used for control frameworks for geodetic surveys and surveys covering very large areas, due to the difficulty in measuring direct distance before the development of EDM, but it is generally being replaced by EDM traverse or trilateration, or even global positioning system (GPS) survey today.

The OS triangulation of Great Britain was organized as 1st order, with triangles of 30 to 100 km side, then 2nd order with triangles of 15–30 km side, and finally 3rd order, with sides of 1 to 15 km. The detail survey was originally by linear methods, controlled by the 3rd order trig work. The OS is now establishing a network of points fixed by GPS survey which may eventually replace the old networks. Much of the detail survey in national surveys today is carried out by air photo survey.

9.4.1 Triangulation figures

Typical arrangements of triangulation networks are shown in *Figure 9.25*. For a large area survey these were generally arranged in chains or belts, or as a complete net over an area.

For lower-order work a simple chain of triangles is suitable, with the triangles as near equilateral as possible. Braced quadrilaterals of four corner stations and observed diagonals are the best form for high accuracy, since there are more conditions to be met in the adjustment of the figure. Quadrilaterals should not be

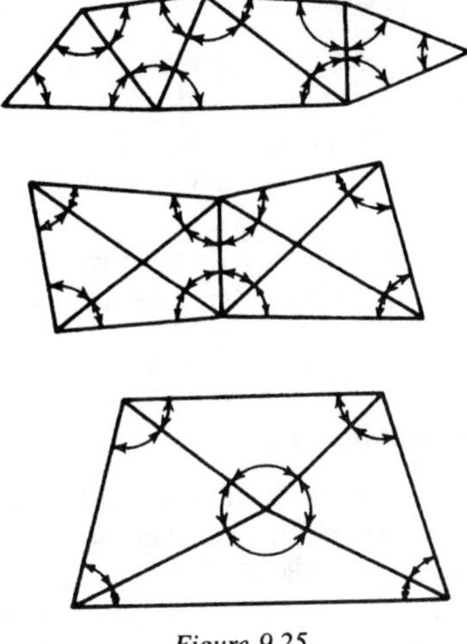

Figure 9.25

long and narrow. Centre point figures lie between the two extremes and are frequently necessary. As a general rule, all triangle angles should lie between 30° and 150°.

9.4.2 Base line measurement

The traditional geodetic base line measurement for triangulation was by *Invar tape* in catenary, with all possible refinements, to achieve accuracy at 1:500 000 or better. For both higher and lower order work this is now replaced by EDM methods.

The base line length was usually much less than typical triangle side lengths, and the base was extended by triangulation until one of the main triangle sides could be computed. In the example in *Figure 9.26*, the measured base is *ab*, extended to *c* and *d* and then to *e* and *f* and finally to the triangulation stations *A* and *B*.

In minor triangulation it is sometimes possible to make one of the triangle sides act as the base line for the network, and this is to be preferred to using any shorter base line.

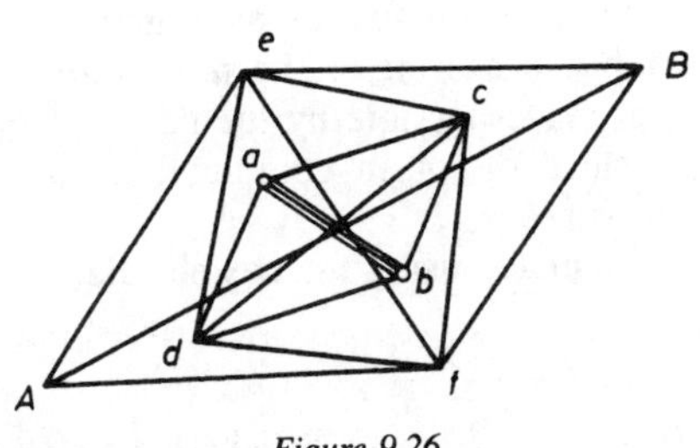

Figure 9.26

9.4.3 Horizontal angle measurement

Horizontal angles are measured as described in Section 6.6.5 for a *round of angles*. The number of rounds and zeros to be used, and the choice of instrument, depend on the accuracy demanded. As an example, the OS 1st order triangulation used 16 rounds, 32 measures of each angle, while the 3rd order used 4 rounds, 8 measures.

Errors in angular measurement have a direct effect on the fixing of the relative positions of points in the net while the base measurement affects the *scale* only.

9.4.4 Triangulation computation

After determining the length of the base line the triangulation figures must be adjusted before the triangles are computed. The method of adjusting the figures depends upon the nature and accuracy of the triangulation concerned. For the highest accuracy the *least squares method* should be used, but a less rigorous alternative is the *method of equal shifts*.

Least squares software on a PC is more widely used today. A more advanced text should be consulted for details of both these methods. For low-order work, with simple figures, it may be adequate to merely adjust the angles of each triangle to 180°. (Add arithmetic means of the three angles, apply one-third of the closing error to each angle before solving the triangle.)

After adjustment, the triangles are solved by the sine rule – $a/\sin A = b/\sin B = c/\sin C$. If the known side is a, then:

$$b = a \sin B \operatorname{cosec} A; \; c = a \sin C \operatorname{cosec} A$$

If a simple figure adjustment to 180° has been made, several values may be computed for any one line. The several values should be meaned before calculating co-ordinates.

9.4.5 Co-ordinates computation

If the co-ordinates of one point, such as an end of the base line, are known, and its orientation, then all other co-ordinates can be computed in the same way as in traversing, using the line lengths and bearings. If known point co-ordinates are not available, arbitrary values may be assumed for a survey of limited extent.

9.4.6 Trilateration

This method relies on setting up networks of triangles, similar to those shown above for triangulation, but measuring all the *side lengths* rather than the angles. This technique has been made possible by the development of EDM, and may quite often be used instead of, or in combination with, traverse survey for engineering and similar surveys.

The angles may be calculated using the cosine rule:

$$\cos A = (b^2 + c^2 - a^2)/2\,bc, \text{ or}$$

$$\sin A = (2/bc)\sqrt{\{s(s-a)(s-b)(s-c)\}}, \text{ where } s = \tfrac{1}{2}(a+b+c)$$

When the angles have been determined then the line bearings, co-ordinate differences and station co-ordinates can be calculated. Again, adjustment may be carried out using appropriate software.

9.4.7 Triangulateration

This is a triangular network again, but with both angular measurement by theodolite and linear measurement by EDM. This system also must be adjusted by least squares by computer.

9.5 Global positioning systems surveying

As has been shown, control has traditionally been provided either by a network of lines between station points, as in triangulation and trilateration, or by running traverses between points of known position. A quite new approach to the provision of control has resulted from the development of artificial satellites in fixed orbits around the Earth. Global positioning systems (GPS) can make use of such satellites to provide the user of the system with accurate three-dimensional position information of points on the surface of the Earth.

The US Department of Defense (USDoD) manages a system termed Navstar (NAVigation System with Time And Ranging) in which 24 artificial satellites have been established in orbits around the Earth, at a height of approximately 20 000 km. The orbits of these satellites are so arranged that throughout the 24 hours of the day at least four satellites can be observed simultaneously from any point on the surface of the Earth, and a user equipped with a suitable receiver may obtain three-dimensional co-ordinates from these to fix his position, regardless of weather conditions.

Each satellite transmits signals in the form of two binary timing codes – C/A (coarse acquisition) and P (precise), modulated on two L-band frequencies L1 and L2, with details of the satellite's orbital data, the almanac of the other satellites, etc. The user's receiver generates a replica of the C/A code, then correlates the generated and incoming code to determine the travel time of the signal and thus obtains a *pseudo-range* between the satellite and the receiver. The range is termed a *pseudo-range* because of the effects of receiver clock inaccuracies, refraction, signal noise and reflections, etc. Simultaneous pseudo-ranges to three satellites will determine latitude, longitude and height, just as in a resection operation, and a simultaneous pseudo-range to a fourth satellite will resolve any major errors caused by receiver-clock inaccuracy. Although an overall accuracy of 10 m should be possible, for reasons of United States national security, the USDoD has implemented accuracy denial systems termed Selective Availability (SA) and anti-spoofing (AS) in an attempt to prevent non-approved civilians obtaining high accuracy fixes.

9.5.1 Differential GPS

Higher accuracy results may be obtained using differential GPS (DGPS). This involves using two receivers, one of them being located on a point of known position. Both receivers calculate their positions from the observed satellite data, then that on the known position can determine the errors in position or pseudo-ranges. Corrections for these errors may be transmitted by radio or modem, or recorded for subsequent post processing, and used to correct the position of the receiver on the unknown point observing the same satellites. Relative accuracies between the two receivers of 2–5 m can be achieved, but this, of course, is not adequate for most survey tasks.

9.5.2 Phase measurement

For the high accuracies appropriate to survey work it is necessary to combine the techniques of DGPS with the carrier phase measurement methods used in EDM. These methods typically remove the codes from the L1 and L2 signals and use the unmodulated carrier waves. Unfortunately, with GPS phase measurement, the process is complicated by the clock errors inherent in the receiver and in each satellite, so in order to remove ambiguities two receivers simultaneously measure phase data from the same satellites. Difference equations of the combined data are used to resolve the point co-ordinates by means of sophisticated computer software. The latest GPS equipment uses these methods to give high accuracy position fixes from a variety of different operational modes, some using post processing and others giving real time results.

9.5.3 GPS measuring modes

9.5.3.1 Static positioning

This was the original high-precision method used in DGPS, with two receivers at different positions collecting data over an extended period, used to measure long lines of over 20 km. This method gives accuracy of about 5 mm + 1 ppm, it is suitable for geodetic control, for the classical baseline measurement, movement monitoring, etc.

9.5.3.2 Rapid Static surveying

In this method, one stationary receiver acts as the *reference station* while the other, the *roving receiver*, is moved from point to point in a survey. The computation methods used in this system only require the rover to occupy each point for 5 to 10 minutes, and give an accuracy of 5 to 10 mm + 1 ppm. This is suitable for any task where many points have to be located quickly, such as local control surveys with lines up to 15 km, engineering survey, detail survey, etc. It is not necessary to maintain lock on the satellites while moving the rover, and it may be switched off while moving.

9.5.3.3 Stop and go surveying

Here, the reference receiver stays in place while the roving receiver is moved from point to point, mounted on a hand-held pole. At the first point, the rover's position is fixed by a rapid-static fix or it is placed on a known point, then it is moved and held briefly on each required point. Accuracy is of the order of 1 to 2 cm + 1 ppm. This method is fast and economical for detail surveys and similar work. The rover may be carried by a vehicle or by hand, but phase lock must be maintained on four satellites at all times. The method is not appropriate in areas where the signals may be shaded, e.g. wooded areas, tall buildings in the vicinity, etc.

9.5.3.4 Re-occupation surveying

This method is similar to Rapid Static, with the rover occupying each required point for a few minutes, but the points are all re-occupied between an hour and two hours later. The two sets of data are combined by software to give accuracies of 5 to 10 mm + 1 ppm. Like Rapid Static, lock need not be maintained and the receiver can be switched off while moving, but the method can be used when conditions would be unsuitable for Rapid Static, e.g. when only three satellites are visible.

9.5.3.5 Kinematic methods

In these methods the roving receiver moves continuously, either mounted on a vehicle or a hand-held pole, and measurements are made at specific time intervals, e.g. 1 second, while lock to four satellites is maintained. Accuracy of 1 to 2 cm + 1 ppm can be attained, and the methods are suited to determining the trajectory of moving objects, surveying road centrelines and cross-sections, hydrographic surveys, etc.

9.5.4 Co-ordinates and heights from GPS

The co-ordinates output from a GPS survey will be in the form of Cartesian or geodetic co-ordinates in the satellite datum – WGS84 – and must be transformed into OSGB36 or whatever local grid is in use. GPS *heights* are related to the ellipsoid (the mathematical representation of the shape of the earth) while traditional spirit levelling relates points to mean sea level. There is no constant difference to allow a general height transformation, and specialist procedures are required to convert GPS heights to heights above mean sea level. GPS heighting can achieve accuracies of two–three times plan accuracy.

9.5.5 Equipment for real-time GPS surveying

A typical set of equipment from the Wild GPS System 300 from Leica is illustrated. *Figure 9.27* shows a reference station setup, with an SR399 GPS sensor mounted on a tripod, together with the CR344 GPS controller, radio modem and battery. The rover shown in *Figure 9.28* is using an SR399 sensor with external antenna, mounted on a hand-held pole, together with a CR344 GPS controller and modem.

These GPS sensors are nine-channel dual-frequency, they can track nine satellites simultaneously and measure on both L1 and L2 signals. The CR344 controller controls the field operation of the sensor, it has a large LCD display and alphanumeric keyboard, a menu-driven interface, it connects to a radio modem, and it can run the RT-SKI (real time static kinematic) software used for real-time applications.

In real-time GPS survey the data from the reference station is transmitted to the rover continuously via the modem. In post-processing GPS, for control

Figure 9.27

Figure 9.28

surveys of the highest accuracy, no data link is used, as the data is recorded for the subsequent post-processing using a PC and SKI (static kinematic) software. PCMCIA cards are used for data storage.

9.5.6 GPS surveying applications

The principal applications of GPS to date have been in control surveys, particularly in improving national networks and carrying out precise engineering surveys over large areas. As an example of the former, the OS is using the method to create a new geodetic network, with the new survey stations located in easily accessible points. The old 'trig' stations were often very difficult of access.

GPS has been used in the control of the Channel Tunnel works, in combination with other methods. Although GPS can be used instead of traverse survey methods, and may even be used for detail surveys, the equipment costs are still high for these applications.

Advantages of GPS include the fact that work can be carried on regardless of the weather conditions, and it is not necessary for stations to be intervisible, and even 'one man surveying' is possible. In all other methods of survey the various control points must be intervisible, and this has meant, in the past, that survey stations were often located on high points difficult of access. The advantages of kinematic GPS for hydrographic survey, with the receiver mounted on a boat, will be evident.

Small hand-held versions of GPS receivers are available for navigational use on water and in cross-country journeys in poorly mapped areas such as desert and bush.

9.6 Co-ordinate problems

A wide variety of co-ordinate problems may arise in traversing, and also in other surveying applications, notably in setting out. Computer methods are available for most of these, but this section outlines the basic methods of dealing with some common problems.

9.6.1 Plane rectangular co-ordinate transformation

Where a co-ordinate grid has been superimposed on a site, while a different grid has been used for the original survey, it may be necessary to convert the co-ordinates of points from one grid to the other. In order to carry out such a transformation the co-ordinates of two points must be known, on both grid systems.

Given the two sets of co-ordinates for the common points, then the shift, swing and scale may be calculated. The *shift* is the displacement of the grid origins with respect to one another, in eastings and northings. *Swing* is the angular difference between the two grid norths and *scale* is the ratio between the two lengths of a line common to both grid systems. When these have been calculated then the remaining co-ordinated points on one grid may be transformed to the other grid.

The formulae required are:

$$E''p = e + s(E'p \cos\gamma - N'p \sin\gamma) \ldots$$

$$N''p = n + s(N'p \cos\gamma + E'p \sin\gamma) \ldots$$

where $E'p$ and $N'p$ are the original co-ordinates of a point, $E''p$ and $N''p$ are the transformed co-ordinates of the same point, e and n are the easting and northing shifts, respectively, s = scale and γ = swing (see *Figure 9.29*).

The scale is calculated as $s = d''/d'$, where d' and d'' are the original and transformed distances between two points common to both grid systems.
The swing is calculated as

$$\gamma = \beta' - \beta''$$

where β' and β'' are the original and transformed bearings of the line between the same two common points.

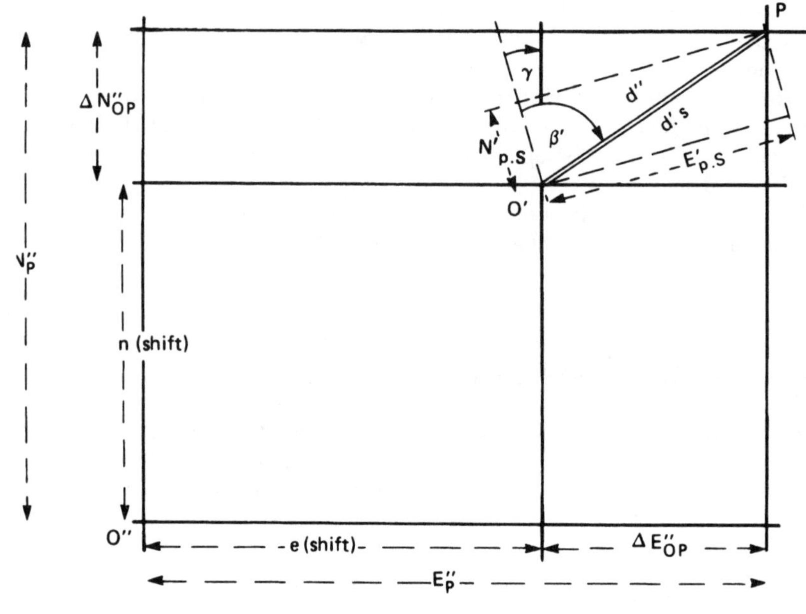

Figure 9.29

The shift constants are calculated from

$$e = E''p - s(E'p \cos \gamma + N'p \sin \gamma) \ldots$$

$$n = E''p - s(N'p \cos \gamma - E'p \sin \gamma) \ldots$$

A more advanced text should be consulted for the derivation of these formulae, which are of particular value in rectangular grid setting out.

9.6.2 Problems in linking to control

Problems frequently arise in linking traverses and other co-ordinated surveys to existing control points. The following sections outline some possible solutions.

9.6.2.1 The 'trilateration' problem

Figure 9.30 illustrates a traverse with stations 1, 2, 3, . . ., to be linked to existing co-ordinated control. It is required to find the opening co-ordinates of station 1 and the opening bearing of the first leg 1–2. Two control points A and B are in the vicinity (the triangle AB1 must be well-conditioned), and the distances 1–A and 1–B and the angle α have been measured, but the theodolite cannot be located over points A or B. Provided 1–A and 1–B are at least as long as the average traverse leg then the triangle AB1 may be solved, the opening co-ordinates of station 1 computed and the opening bearing 1–2 deduced.

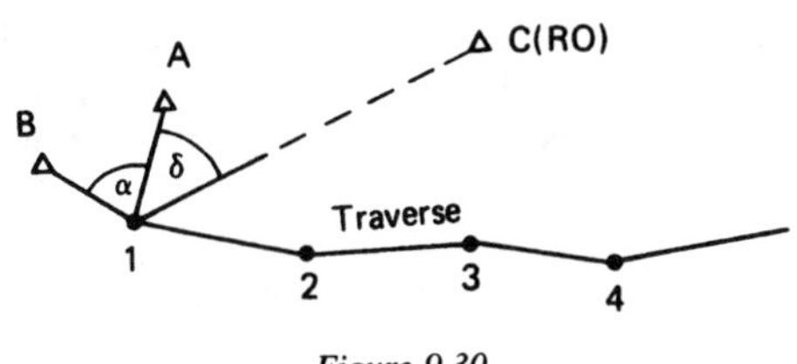

Figure 9.30

If the lines 1–A and 1–B are shorter than the average traverse leg, then an additional angle δ must be observed to a distant RO, a control point C, then the problem may be solved as follows:

(i) Calculate the bearing and length of line A–B from co-ordinates, using R → P.
(ii) Using the cosine rule, calculate the angles A and B in triangle AB1.
(iii) Check that the angles at A, B and 1 in triangle AB1 sum to 180° ± an acceptable tolerance – *do not adjust the angles.*
(iv) Apply the calculated angles to the bearing AB to obtain bearings A1 and B1.
(v) Calculate and check the opening co-ordinates of 1 from A and from B, by bearing and distance, P → R.
(vi) Calculate the bearing of line 1–C from co-ordinates, using R → P.
(vii) Taking C as the RO, determine the opening bearing of leg 1–2 as usual.
(viii) Compute the traverse as normal.

9.6.2.2 The 'satellite' problem

Figure 9.31 illustrates a situation with traverse 1, 2, 3, ... to be linked to a co-ordinated control point C over which the theodolite cannot be located. There is a distant control point A which will serve as the RO. The theodolite is placed at a point S as close as possible to C, forming a 'satellite station' to C. The bearing and length of line C–1 are required in order to find the co-ordinates of 1 and the bearing of leg 1–2.

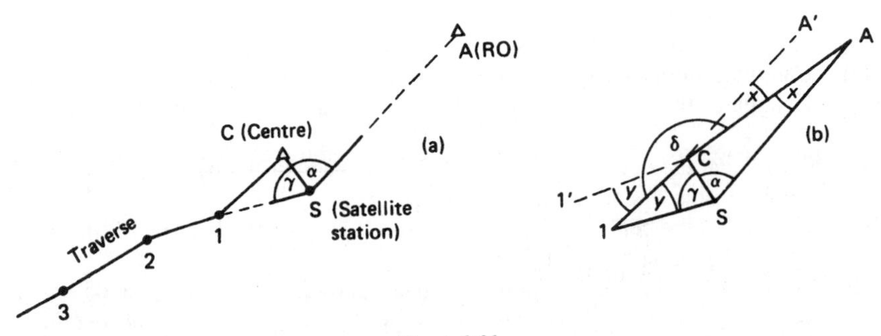

Figure 9.31

To solve this problem, the lines C–1 and C–S are measured, and the angles α and γ observed, then the angle δ (equal to $\alpha + \gamma + x + y$) may be deduced. Note that in *Figure 9.31(b)* the pecked lines C–A' and C–1' are parallel to S–A and S–1, respectively. The procedure then is:

(i) Calculate the bearing and length of the line C–A from co-ordinates, R $\rightarrow$ P.
(ii) Using the sine rule in triangles CAS and C1S, determine the angles x and y, respectively.
(iii) Obtain δ in this example $= \alpha + \gamma + x + y$.
(iv) Apply the angle δ to the bearing CA to obtain the bearing C1 and thus the bearing 1C.
(v) Taking C as the RO calculate the bearing of leg 1–2 as usual.
(vi) Calculate the co-ordinates of 1 from C by bearing and distance, P $\rightarrow$ R.
(vii) Compute the traverse as normal.

9.6.2.3 *The 'auxiliary station' problem*

Figure 9.32 illustrates a similar problem to the last, with a traverse 1, 2, 3, . . . to be linked to control point B which cannot be occupied by the theodolite. An auxiliary station A is located, the distance A–B measured, distant control points C and D will be used as RO1 and RO2, and the angles α, β and γ measured at A. The solution is as follows:

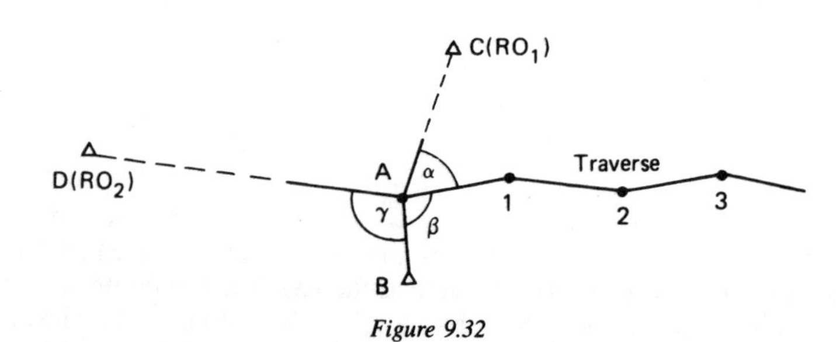

Figure 9.32

(i) Calculate the bearing and length of line B–C from co-ordinates, R $\rightarrow$ P.
(ii) Using the sine rule in triangle ABC, determine the angle at C and hence deduce the angle at B.
(iii) Apply the angle at B to the bearing BC to obtain the bearing BA.
(iv) Calculate the co-ordinates of A from B by bearing and distance, P $\rightarrow$ R.
(v) Calculate the bearings of lines A–C and A–D from co-ordinates, R $\rightarrow$ P.
(vi) Check that the bearing of line C–A from (v) equals the bearing of line C–B from (i) plus the angle at C from (ii).
(vii) Check that the bearing of line A–D from (v) equals the bearing of line A–C from (v) plus angles $\alpha + \beta + \gamma$.
(viii) On the basis of the comparisons made above, decide whether or not to accept the calculated co-ordinates of A. A small change in the length of line A–B, perhaps within its measuring accuracy, would alter the co-

ordinates of A and perhaps improve the comparison between the two calculated bearings for the line A–D. Repeat as needed.

(ix) Taking C as RO1 and D as RO2, determine the bearing of line A–1.

(x) Compute the traverse as normal.

9.6.2.4 The 'resection' problem

Figure 9.33 illustrates a station R (the 'resected point') which is to be co-ordinated from angular observations to four existing distant control points A, B, C and D. In *Figure 9.33*, a circle is drawn through the points A, C and R, and this 'danger circle' must not pass through or be close to the point B, otherwise no solution is possible.

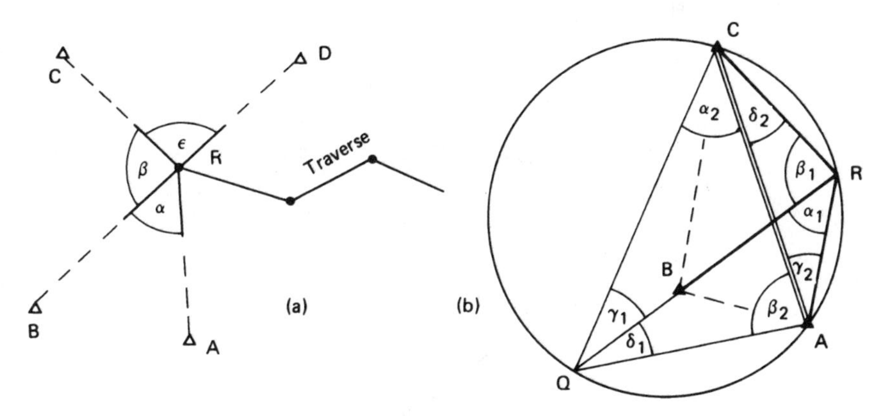

Figure 9.33

The line R–B is extended to meet the circumference at Q here, but note that Q might be on the line B–R or on the line B–R extended. The angles α_1 and β_1 are observed and they are equal to the angles α_2 and β_2 respectively. (Angles in the same segment of a circle are equal.)

The 'Collins Point' method of solution is as follows:

(i) Calculate the bearing and length of the line A–C from co-ordinates, R $\rightarrow$ P.

(ii) Deduce the angle at Q in triangle ACQ ($Q = 180 - \alpha - \beta$).

(iii) Using the sine rule, calculate the lengths of Q–A and Q–C.

(iv) Determine the bearings AQ and CQ.

(v) Calculate the co-ordinates of Q from A by bearing and distance, P $\rightarrow$ R.

(vi) Calculate the bearing of line Q–B from co-ordinates, R $\rightarrow$ P.

(vii) Deduce the angles γ_1 and δ_1 from the bearings QA and QB, and QB and QC, respectively, and hence the angles γ_2 and δ_2.

(viii) Check that $\gamma_1 + \delta_1$ = angle AQC.

(ix) Using the sine rule in triangle ARC, calculate the lengths of the lines A–R and C–R.

(x) Deduce the bearings AR and CR from the bearing AC calculated in (i) and the angles γ_2 and δ_2.

(xi) Calculate the co-ordinates of R from A by bearing and distance and check from C, P → R.
(xii) Repeat using co-ordinated points ABD, ACD and BCD and obtain a mean value for station R.

9.6.2.5 *The 'intersection' problem*

Figure 9.34 illustrates this problem. Two co-ordinated points A and B are available, providing a base line A–B whose length and bearing may be determined from co-ordinates in the usual way, R → P.

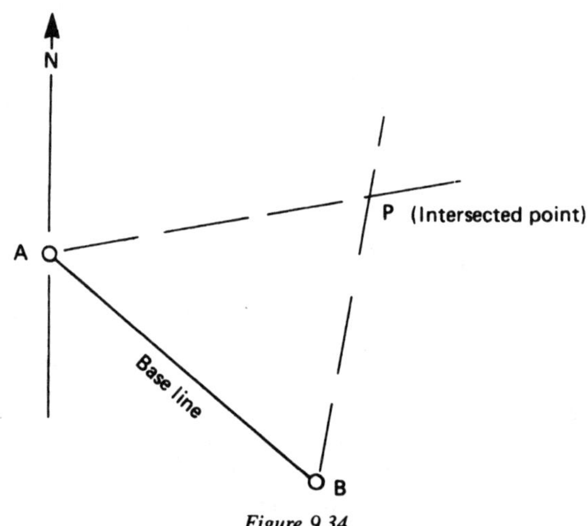

Figure 9.34

A third point P is to be fixed, and this may be done by establishing the rays from A and B through P. The point P may be fixed by either:

(i) observing the angles PAB and PBA, or
(ii) measuring the distances A–P and B–P,

then solving the triangle APB and deducing co-ordinates for P. In either case, a third observation is advisable in order to check the quality of the fix at P. Formula methods are available for this problem but it is probably safer to work from first principles.

Chapter 10

Detail survey

10.1 Introduction

This chapter deals with the various methods of 'supplying' or 'picking up' detail in surveys, assuming that survey control has been provided by one of the methods covered in Chapter 9. In a linear survey, where the control is provided by a triangle-based framework of lines measured by steel tape, the detail will be supplied by tape and offset methods, and these are covered in Section 10.2. However, tape and offset methods may be used to pick up detail from *any* measured survey line, not merely from lines in linear surveys.

Where the survey control is by traverse, triangulation or trilateration, the detail will mainly be supplied by radiation from the instrument stations, i.e. by measured bearings and distances. Most detail surveys by radiation today use EDM equipment, and this is covered in some detail in Section 10.3. Detail survey by optical distance measurement (ODM) radiation methods is much less common now, but ODM radiation with vertical staff tacheometry is outlined briefly in Section 10.4, while Section 10.5 touches on radiation using the theodolite and steel tape.

10.2 Tape and offset detail survey

Detail in linear surveys is primarily supplied by rectangular offsets from the survey lines, and the method may be used to pick up detail in any other survey where taped lines have been run between the control points or control lines. However, it is not always feasible to use offsets and alternatives are sometimes needed. In areas of 'close' detail, in fact, considerable flexibility in the methods used is desirable.

In order to save time it is 'normal' practice to measure the taped survey lines and pick up the detail at the same time. Thus when the tape is stretched out along the survey line, all detail near the tape is picked up before the tape is pulled on again.

The equipment used in tape and offset survey has already been described in Chapter 4, but it should be noted that while the survey line is measured with a steel tape, offsets and ties are generally measured with a synthetic tape. A steel tape may be used if higher accuracy is demanded.

The limits of measurement in a linear survey have traditionally been dictated by the scale to which the survey is to be plotted. Thus assuming a draughtsman can plot to 0.2 mm, at a scale of 1:1000, then the field measurements should be to the nearest 0.0002 × 1000, i.e. 0.2 m. In practice it may actually be easier to simply take all measurements to 0.01 m, regardless of the proposed scale, and the draughtsman may mentally abbreviate the measurements as required. The actual limit of measurement should be specified in the job brief.

10.2.1 Offsets and running offsets

An offset has been defined as a short measurement taken (raised) at right angles from the survey line to the point of detail to be surveyed. Offsets are then said to be *raised* and *measured.*

Running offsets are offsets measured to two or more points of detail along the same perpendicular, as for example to both banks of a stream (*Figure 4.30*). Such offsets are called 'running' because the measurements are what are known in surveying as *running measurements*, that is to say successive measurements recorded along a line from some common point. *Figure 10.1* illustrates such a set of measurements, as do the measurements along a survey line itself. These forms of measurement are widely used in survey.

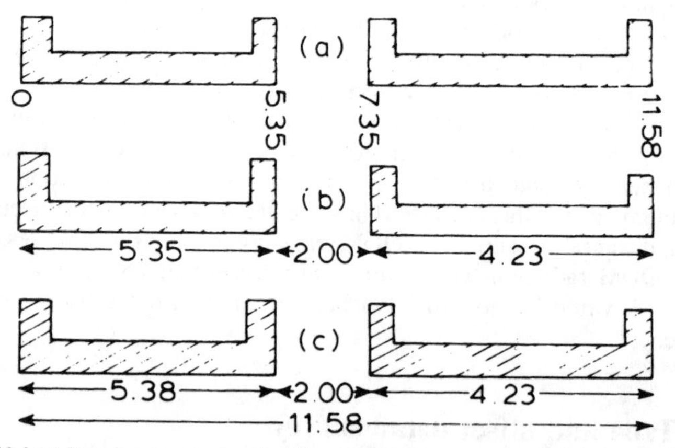

Figure 10.1 (a) Running measurements, (b) separate measurements, (c) separate and overall

With experience it will be appreciated that running measurements, including running offsets, provide the quickest and most accurate method of measuring to a number of points on the same straight line. If 'separate' measurements are made along a line there will clearly be a greater error in the line length than if running measurements are used, where the last measurement is an overall length.

10.2.1.1 Recommended limitations

Offsets to points of detail such as corners, junctions and ends of detail should not normally exceed 8 m in length for scales of 1:500, 1:1000 and 1:1250, nor should they exceed 16 m for scales of 1:2000 and 1:2500.

Similarly, offsets to curving and indefinite detail such as meandering streams and hedges should not exceed 16 m in length for the larger scales or 20 m for the smaller scales.

These limitations are imposed by the need to ensure that the survey measurements are compatible with the maximum accuracy which may be demanded in the plotting, and the need to avoid long offsets which would require much additional care and time to maintain the possible accuracy.

10.2.1.2 Raising and measuring an offset

To raise an offset, the assistant places the zero end of the tape at ground level on the point of detail to be surveyed. The surveyor then holds the tape horizontal and taut at the approximate perpendicular point on the tape lying on the survey line where the offset is to be measured. Up to an offset length of 8 m, the site surveyor should have no problem in judging by eye the right angle required between the tape and the survey line correct to within 0.05 m. If the surveyor gently swings the taut tape from side to side in a horizontal plane like a slow moving pendulum it may assist his judgement of the right angle and location of the correct point on the survey line. Offsets up to 4 m in length are often more quickly raised and measured with a 2 m ranging rod rather than the tape.

When the line of sight between the survey line tape and the detail point is obstructed, or the offset is more than 8 m in length, it is recommended that the right angle be raised with the optical square rather than by eye. The optical square is described in Section 4.2.6.

10.2.1.3 Ties

Ties are measurements taken from two or three different points on a survey line to a common point of detail which is to be surveyed. Ties are used when it is considered that the length of a simple offset would be excessive, or when it is not possible to raise a right angle between the survey line and the point of detail (*Figure 10.2*). Though two measurements are usually sufficient over short distances, three may be preferred when long tape lengths are needed.

A point of detail to be fixed by ties should lie within 20 m of the survey line, unless a steel tape is used for the measurements. The length of the base of the triangle formed on the survey line should be the longest side length, so that a good 'cut' may be ensured when the point of detail is plotted, *Figure 10.3*.

10.2.1.4 Straights

A *straight* is the extended alignment of any straight feature, including features such as the side of a building, a straight wall, etc. The intersection of a survey line and a straight is known as the *point of intercept*. The straight may be useful in areas of 'close' detail and in the unfortunate circumstance of the surveyor not having an assistant. The straight may be 'measured' or 'unmeasured' (*Figure 10.4*).

It must be appreciated that a straight cannot be plotted from a recorded straight measurement unless the other end of the straight is also surveyed by an offset, another straight, etc. To minimize plotting errors which may arise due to possible misalignment of the survey lines, the angle at the point of intercept between a straight and the survey line should not be less than 40°.

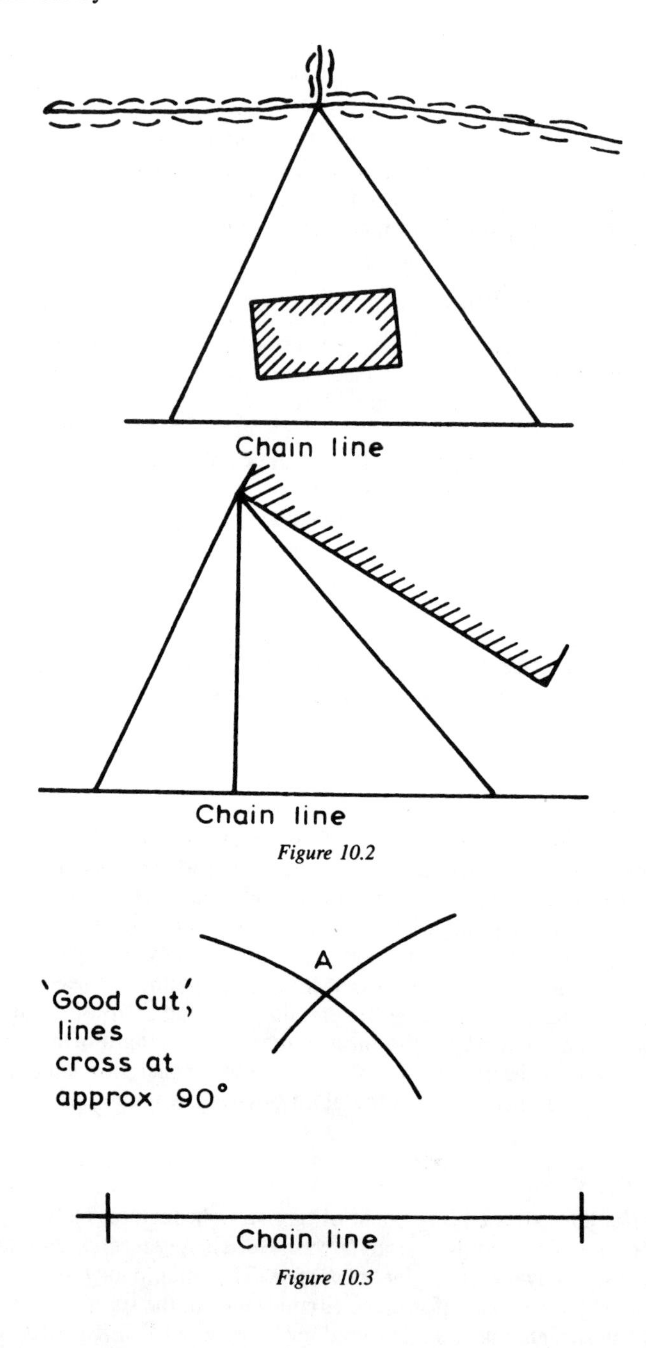

Figure 10.2

Figure 10.3

Measured straights should not exceed 20 m in length, this meaning the distance along the straight between the point of intercept and the detail. Unmeasured straights may be of any length, provided that there is no doubt in the mind of the surveyor as to where the straight cuts the survey line.

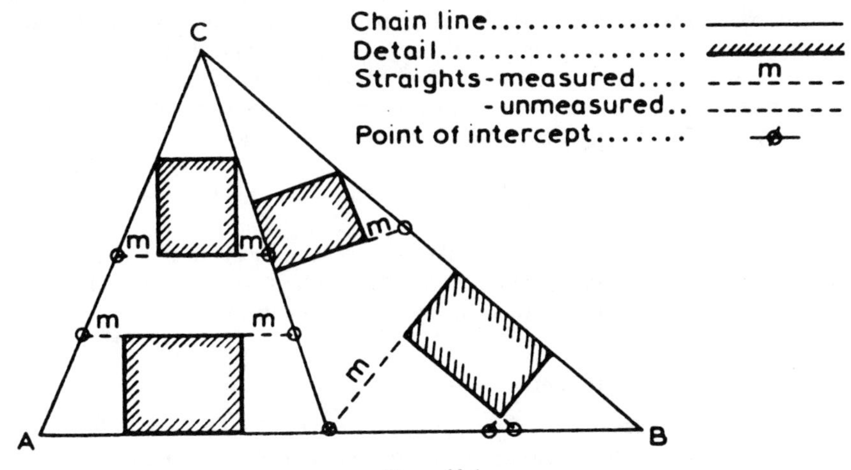

Figure 10.4

It is often preferable that the survey line be arranged so that one end of a straight detail is close to the survey line, say within 20 m, but this will depend upon the length of the detail and the clarity with which the straight can be defined.

10.2.1.5 Plus measurements

Plus measurements are measurements made at right angles to a length of straight detail, often known as an *offsetted base*, which has been surveyed by offsets, straights, etc. Plus measurements are used to enable rectangular shaped buildings to be supplied (*Figure 10.5*).

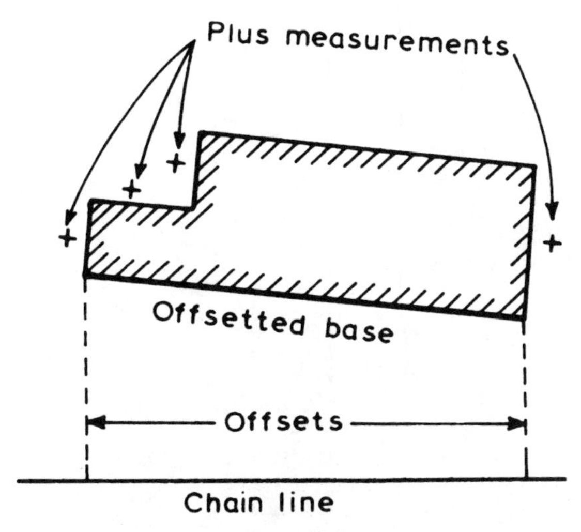

Figure 10.5

There should not be more than three plus measurements from either end of an offsetting base (*Figure 10.5*, left-hand side). The total length of the sum of the plus measurements at either end of an offsetted base should not exceed the length of the base, or 20 m, whichever is the least. The recommendations are intended to ensure that the plotting shall be up to an adequate standard.

10.2.2 Booking tape and offset detail

Tape and offset measurements are booked on booking sheets or in a field book, as described for linear survey taped control lines in Section 9.2.2. The following sections describe the booking of each type of measurement.

10.2.2.1 Offset measurements

If a point of detail lies to the left of the chain line (i.e. on the surveyor's left as he stands on the start point and looks towards the far end of the line) then when booking the measurement, the value of the offset to the detail is written to the left of and on the same line as, the appropriate chainage figure, but outside and as close as possible to the chainage column (*Figure 10.6*). Where detail lies to the right of the survey line then similarly the offset value and the detail are shown to the right of the chainage column in the book.

To avoid ambiguity in linking the value and its detail, the offset value is written close to the chainage column, and the detail is drawn close to the offset value. Mistakes in interpretation are less likely if adequate spacing is allowed

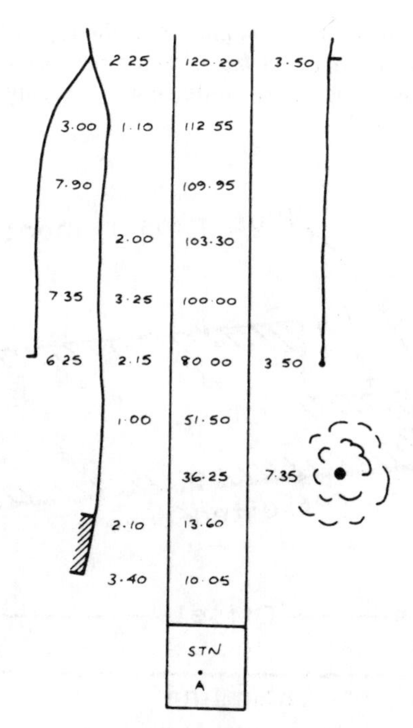

Figure 10.6

between the chainage values, and some surveyors find squared paper useful for showing the relative alignments.

All detail is assumed to be straight between consecutive offsets to the same detail, however much the freehand line drawn to represent the detail may meander, *unless* the surveyor states otherwise in the field bookings. In other words the bookings are diagrammatic only. When taking offsets to curving detail, the usual practice is for the surveyor to raise and measure them at such intervals that for all practical purposes the curve between any two offsets is assumed to be a straight line.

It must be emphasized that no line is drawn between the survey line chainage point and the offset point it relates to (see *Figure 10.6*).

10.2.2.2 *Running offsets*

These are also depicted in *Figure 10.6*, and this illustrates the need to ensure that each offset value is shown close to the point of detail to which it refers.

10.2.2.3 *Common alignments*

In urban areas it is common for a number of buildings to lie on the same alignment. Where such an alignment of buildings is to be surveyed by taping, then it is not necessary to raise and measure offsets to every building corner on the alignment, as this practice is time-consuming and may lead to poor representation of the detail on the plotted plan. Every survey measure and every plotted point will have some accidental error, so that if such a row of buildings is each surveyed separately then, while each building should be correct on the

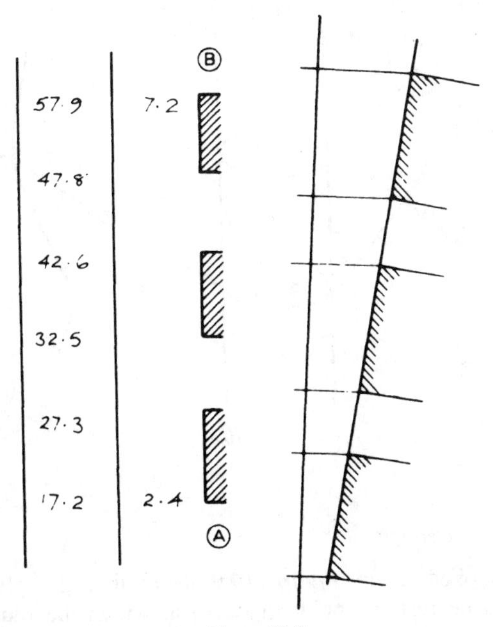

Figure 10.7

plotted plan, the final alignment as a whole will not look precisely correct. This may be overcome by raising and measuring the corners at the ends of the alignment but only *raising* the others (*Figure 10.7*).

Note that the letters A and B circled on the example bookings indicate to the plotter that the detail is straight between the offsets at A and B.

10.2.2.4 Ties

These may be booked as pecked lines, as in *Figure 10.8*, or as light lines with an arrow point at each end, in either case with the measured lengths placed within the lines.

10.2.2.5 Straights

An *unmeasured straight* is booked as a pecked line between the detail and the point of intercept, with the chainage value at the point of intercept circled. A *measured straight* is booked also as a pecked line, but with its measured length circled and placed within the pecked line.

These arrangements of circles are used to avoid any possible confusion between offsets, ties and straights, since in practice it is possible for a single chainage value to refer to all three (*Figure 10.8*).

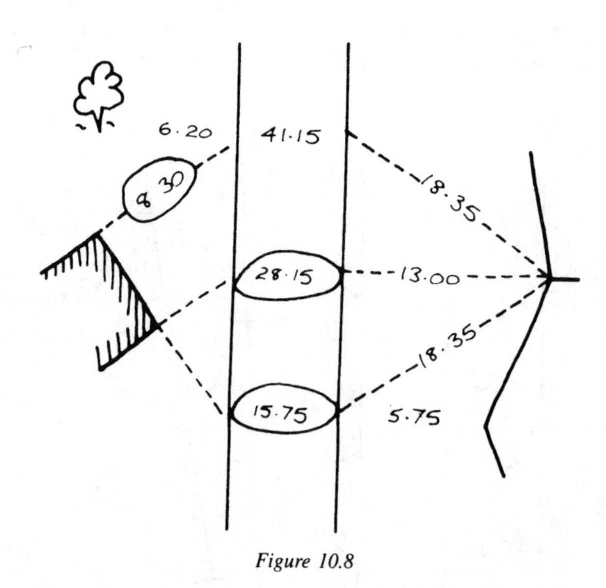

Figure 10.8

10.2.2.6 Plus measurements

These may be booked as in *Figure 10.9*. Note that the plus sign should always be entered nearest the point of detail at which the plus measurement is taken.

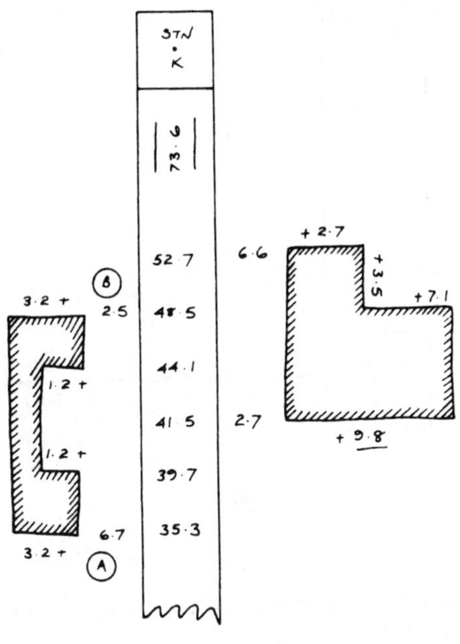

Figure 10.9

10.2.2.7 General

In tape and offset linear survey it will often be found that detail connects between the survey lines, and if the surveyor is not going to carry out the plotting then it is essential, particularly in larger surveys, that this detail be cross-referenced.

Where a point of detail supplied from one survey line connects through to a point of detail supplied from a later survey line, then the former point may be indicated by adding an arrowhead to the line of detail in the field bookings (*Figure 10.10*). The latter point is cross-referenced by the use of a skeleton line

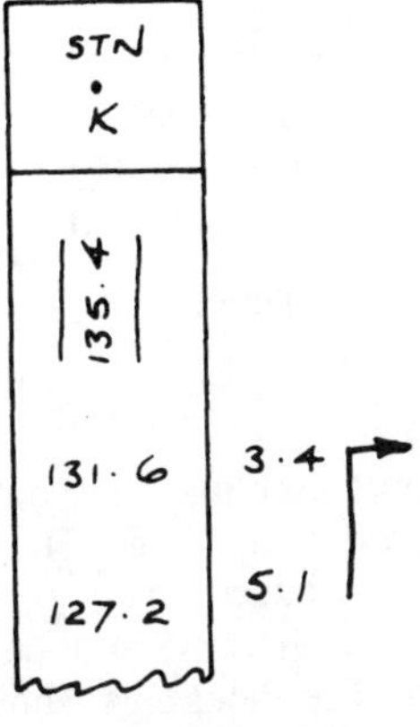

Figure 10.10

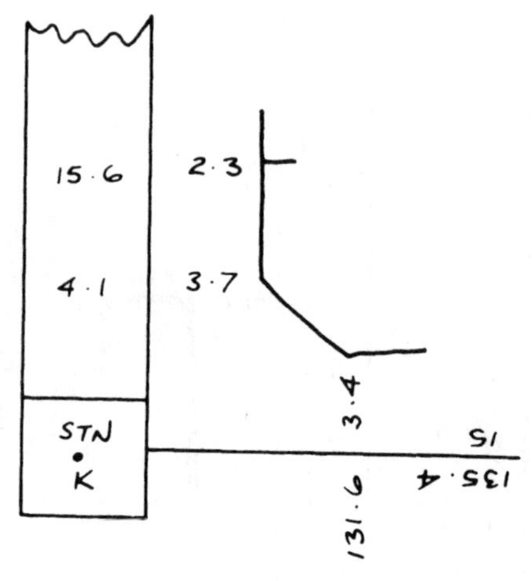

Figure 10.11

giving its length and page number, and repeating the former point's chainage and offset values (*Figure 10.11*).

To indicate where a survey line runs along the line of a kerb, or the face of a building, wall, fence, etc., the line representing the detail should be drawn in a much broader gauge of line where it is contiguous with the survey line (*Figure 10.12*).

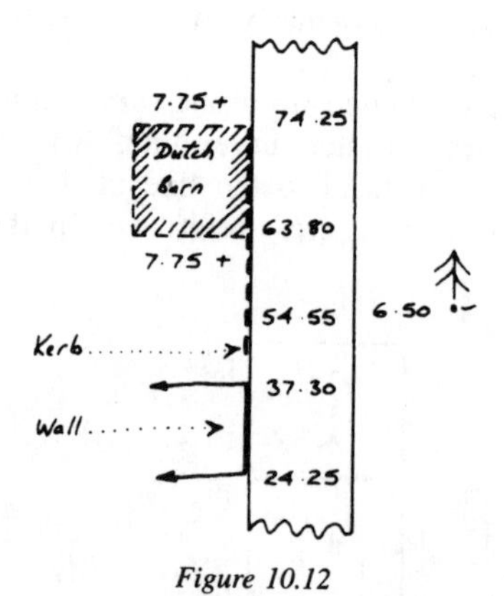

Figure 10.12

The above rules cover most contingencies on site, but if it is considered that at any time there may be some doubt in the mind of the plotter, then an explanatory note should be added. Always when booking, consider the plotting; remember that ideally there should be no need for the plotter to refer back to the surveyor. The bookings should be complete, clear and accurate.

10.2.3 Office procedure and plotting

Tape and offset linear survey *plotting* requires no great skill, although the draughtsman must be accurate and meticulous. Graphic communication skills are of importance in the presentation of the finished work, however.

10.2.3.1 Preliminary considerations

(i) *Choice of drawing material*. The choice of material for the plan or map will depend upon the purpose and use of the plotted survey plan. The material may need to be durable, to be transparent or translucent for purposes of copying and tracing and it may need dimensional stability. The need to take ink and pencil is evident and on occasion it may be necessary for the material to take colour washes. Also, the ability to withstand frequent erasures is generally desirable.

There are three materials in common use – cartridge paper, tracing paper and polyester-based translucent films.

If *cartridge paper* is used, then the heavier and higher quality grades are the more suitable, but a sample should be tested for ability to take ink and colour washes and to withstand erasing. Cartridge paper is not dimensionally stable, and perfect erasure of inks is not possible.

Tracing paper is widely used, and again the heavier grades are better, but it is not dimensionally stable and becomes brittle with age.

Polyester film is strong, durable and dimensionally stable, probably the best material to use but it is more expensive than either cartridge or tracing paper. There is a slight tendency for ink to spread on film and pencils must either be of a hard grade or have plastic-based leads. Both ink and pencil can be erased.

(ii) *Sheet size*. Sheet size is obviously linked to the chosen scale and the area to be surveyed and on a large survey it may be desirable to have two or more sheets. If copies are required then the reprographic facilities available must be considered. Sheet size will also depend upon the amount of marginal and border information required to be shown, for example a title, a north point, a legend, road destinations, etc.

(iii) *Layout on the drawing material*. Having decided on the material and the sheet size to be used, there are three other matters to be considered before plotting can be commenced. These are the orientation and position of the survey on the sheet and the border and marginal information to be provided. When plotted and penned, the drawing should present a balanced appearance pleasing to the eye and much of this will be achieved by the quality of the line work and the lettering. Often, north is placed towards the top of the sheet, but this is not essential.

10.2.3.2 Plotting a linear survey framework

This section describes the procedure in plotting a linear survey control framework such as is detailed in Section 9.2. The equipment required includes a scale, steel straight edge (metric if graduated), set squares, pair of compasses, beam compass, pencil, pencil sharpener and eraser. The pencil should be one which will produce a fine clean line, of even width and density, which will not smudge but may be erased without undue damage to the drawing surface. This means a good quality 'hard' pencil sharpened to a fine point.

Plotting should commence with the drawing of the base line, in a suitable position on the drawing material and to the appropriate scale, together with any relevant tie points. Stations and tie points may be lightly encircled, with their values written adjacent to the line. It should be remembered that all this may have to be erased later.

The largest and best-conditioned triangle tied to the base should be plotted next by using the compasses to describe intersecting scaled arcs of the recorded distances. The survey lines are then drawn in with the straight edge and their lengths checked with the scale. It is most important to ensure that the drawn lines pass exactly through the plotted points – beginners often have initial difficulty in doing this. Next, plot and check any lengths within the plotted triangle and add the relevant chain lengths and tie-point values. The remaining survey lines are plotted in a similar manner.

For checking the lengths of lines, the limiting tolerance usually accepted is 1:500, although occasionally 1:1000 is required. Further, no line should be rejected if the difference between the measured and the plotted lengths is less than three times the recommended limits of measurement of the scale of plotting. Thus, no lines with the following errors should be rejected:

0.15 m at scale 1:500
0.30 m at scale 1:1000
0.75 m at scale 1:2500

Should a line fail to plot within these limits, and the misclosure is large, then the line in error and/or its tie points must be checked and re-measured as necessary. On the other hand, if the misclosure in a line is small, then a judicious adjustment of the surrounding lines may enable it to be fitted into the plot within the above limits of tolerance. That is to say, the network holding the line in error may be expanded or contracted so as to bring all the lines within the limits of tolerance.

Where a line has an acceptable misclosure, then the plotted value of any tie point on that line is to be equated proportionally.

10.2.3.3 Plotting tape and offset detail

All the detail from one survey line should be plotted, completed and draw in, then the next line, and so on until the detail plotting is finished. The chainage values should be plotted as short ticks. The offsets may be raised using a set square and straight edge, the scale being used to plot the offset values. As an alternative to this, specially designed *offset scales* are faster if available. A pair of compasses is required to plot the ties. The straights, whether measured or unmeasured, cannot be plotted until the far end of the straight has been located on some other survey line.

Detail points are then connected as indicated in the field bookings, using the straight edge. Plus measurements may be plotted using the set square, straight edge and scale in a manner similar to the plotting of offsets.

Where a line has an acceptable misclosure, the chainage values as recorded should be increased or decreased proportionally.

Where the acceptable misclosure is at the minimum value, such as 0.15 m at 1:500 scale, 0.3 m at 1:1000 scale, etc., then the misclosure may be 'lost' at the terminal points of the line without adjusting intermediate chainages.

10.2.3.4 Completing the plan

The penning-in of detail can be carried out as plotting proceeds, possibly line by line but preferably on completion of the pencil work.

When all detail and symbols have been inserted, any necessary notes should be added, such as direction of roads 'From Leicester ... To Hinckley', names of buildings and areas, etc.

In positioning the name of a feature on a map or plan there should be no possibility of confusion as to which feature the name refers to, nor should the name obscure any other part of the drawing.

Names identifying linear features such as roads and rivers should be aligned within or alongside the feature, spaced so as to indicate the feature's extent. Names identifying areas, such as field and wood names, should be placed centrally within the area and with sufficient spread to give an indication of the extent of the area.

A simple north point must always be drawn, and a drawn scale should be placed at the bottom of the sheet. It is traditional that the drawn scale be as long as the longest line in the survey, but this is not essential. The drawn scale must have its description clearly marked, e.g. 'Scale–1:500, metres'.

The usual title, firm's name, job number, and the names of persons responsible for surveying, plotting and tracing the work, date, etc., should be added. Dry transfer hatchings, shadings and textures may be added if necessary, depending upon the requirements specified. Finally, it must be remembered that the plan has to communicate its contents adequately and accurately to the user.

10.3 EDM radiation detail survey

This section deals with detail survey by radiation methods, using short-range EDM equipment, as employed in large-scale topographical mapping. Section 10.4 outlines the procedures involved in detail survey using vertical staff tacheometry or stadia survey. Much of the content of this section applies also to the field procedures summarized in Section 10.4.

10.3.1 Basis of the method

Detail and spot heights are supplied by linear and angular measurements taken at an instrument station to all the points of detail and spot heights required to be surveyed in the vicinity of the station. To do this, a reference direction is first established at the station, then the angle between the reference direction and the line to a detail point is measured, together with the distance from the instrument to the detail point, the procedure being repeated for all the detail points and spot heights.

The detail and spot heights are plotted by the use of a protractor and scale, or using a polar co-ordinatograph, or by converting the polar co-ordinates (bearing and distance) to rectangular co-ordinates and plotting these on a rectangular grid.

If available, a rectangular co-ordinatograph may be used to plot rectangular co-ordinates, or they may be plotted by computer. Where the area to be surveyed is large, many instrument stations may be needed and these may be connected together by traversing, triangulation, trilateration, or an appropriate combination of these methods.

10.3.2 Advantages and disadvantages

The primary advantage of the method is that using one set of equipment, at one instrument station, all the detail, height and control measurements in the area may be observed and measured, thus each instrument station need only be occupied once. The principal disadvantages of the method in its early days were the initial capital cost of the equipment and its bulk and weight, but each new model today seems to be lighter, smaller and cheaper than its predecessor. The equipment can be operated by a minimum team of two, but a team of at least three is generally more economical.

On a small survey which only requires to show detail at a scale of 1:500 or smaller, it may be sufficient to simply use tape and offset survey methods, unless there are obstacles to taping and/or traverse control is demanded. Many large-scale surveys are carried out for 1:200 scale plans, but at this scale tape and offset survey is not generally of sufficient accuracy and radiation techniques are recommended at this scale. Where a site survey requires spot heights and/or contours to be shown, it will normally take fewer man hours to carry out the task by radiation methods.

The basic instrument for EDM radiation detail survey is either a *theodolite plus EDM distancer attachment* or a *total station instrument*. Whichever is used, the ancillary equipment will include the necessary reflector prisms with targets, prism poles, tripods and tribrachs, and adaptors to fit the prisms and targets to tripods and prism poles. Batteries are also required and perhaps a charging unit, depending upon the circumstances.

Station marking equipment will be required, as detailed earlier, and data recording equipment such as a portable PC or a data recorder.

10.3.3 Preliminary considerations

A number of considerations must be settled before the fieldwork can be commenced.

10.3.3.1 Suitability of radiation methods

It may be that other methods are more suited to parts of the task. On occasion it may be found that tape and offset survey techniques are suited to part of a survey which is principally being carried out by radiation methods. Where building elevations are concerned, photogrammetric methods may be appropriate and similarly in quarries, etc. Photogrammetry is not covered in this text.

10.3.3.2 The survey scale

The scale of the survey will affect the limits of the linear and angular measurement, the maximum length of traverse which may be used and the equipment required. If the scale has not already been agreed upon, then its choice will be affected by the following factors:

The amount of detail and other information to be shown.
The area to be covered.
The size of the completed plan or map and its convenience to the user.
The purpose of the plan or map.
The accuracies required.

10.3.3.3 The need for additional control

If a site is extensive, it may be necessary to provide additional control in order to achieve a desired accuracy. Additional control measurements may be needed to tie the survey to national or municipal control. As a general guide, it is recommended that an EDM traverse for 1:500 scale mapping should not exceed 20 legs or 2000 m in length and proportionally at the other scales in use. For example, 1:200 scale, 8 legs and 800 m in length, but 1:1000 scale mapping, 40 legs and 4000 m in length.

10.3.3.4 Sight length limitations

The sight length available will affect the layout of stations and of traverses. Generally the EDM equipment itself does not impose limitations, rather it is the survey scale and the atmospheric and environmental conditions. Signalling arrangements are important and visibility, which is affected by the presence of mist, smoke, haze, etc. The angular accuracy of the theodolite may impose a limitation, thus an arc of 10 mm is subtended by an angle of 20" at a distance of 100 m. In built-up areas, stations may need to be not more than 60 to 80 m or so apart to ensure that all detail can be surveyed accurately.

10.3.4 Field procedure

On arrival on site and before making any observations, the surveyor should carry out the usual thorough reconnaissance of the survey area. He should make himself familiar with the general layout and, in particular, identify any likely problems.

10.3.4.1 Selection and marking of instrument stations

Any instrument station should provide a clear view of the area to be surveyed from it. Where the ground is flat, station points should be selected on elevated positions if at all possible, e.g. low bare hills in open countryside, low flat roofs, raised platforms etc., in built-up areas. If the survey is to be controlled by more than two instrument stations then the stations must be tied together and this generally means that at least two other instrument stations must be visible from any instrument position.

The stations should normally be tied together by traversing, either a single loop traverse, or possibly a network of traverses. On a long narrow site they may be tied to a base line or to some other form of control such as the OS National Grid. Triangulation and trilateration may also be considered as possible solutions on occasion, see Chapter 9.

Stations may be marked with a nail, wooden peg, rivet, etc., depending upon the ground conditions, but it is preferable to use an existing man-made feature if possible. Such features include a point on a manhole cover, corner of a drain gulley grating, etc., these being longer lasting than artificial marks and easier to re-locate if observations must be repeated or revision is needed. If centring tripods are used, the station mark should be centre punched to facilitate the positioning of the centring rod.

10.3.4.2 Station point sketches

When the selection and positioning of a station is complete, it is traditional to draw a sketch of the area showing the detail to be surveyed. Normally a separate sketch is drawn for each station and in a large survey department these will be produced to a rigid set of rules similar to those referred to in tape and offset survey, so that all members of the department will fully understand the content of the sketches. Smaller organizations tend not to be so rigid.

Where the survey area is small, a single sketch may be used for the whole site. If the detail to be portrayed is simple, e.g. rows of hedges or a series of spot heights, then the sketch may be dispensed with and a simple annotation placed in the booking sheet adjacent to the relevant field measurements, a sketch only being added for intricate detail. Any sketch produced must be clear and reasonably accurate, the objectives of the sketch being:

(i) To make clear to the plotter how the surveyed detail is linked to the observations, how it is to be shown on the plan and to what other detail it is connected.
(ii) To indicate any taped supplementary measurements such as plus measurements required to complete the plotting of a rectangular feature.
(iii) To indicate any names and their positioning on the drawing.
(iv) To indicate the possibility of any gross errors.
(v) To assist in the interpolation of contours.
(vi) To indicate to the reflector holder the position of the points of detail and the approximate positions of spot heights.
(vii) To ensure that no point to be surveyed is forgotten by the observing team.
(viii) To show the position of the instrument station and reference object in relation to the detail.
(ix) To show what detail, if any, has been surveyed from more than one instrument station.
(x) To indicate detail points lying on the same alignment.

10.3.4.3 Coding systems

Despite the usefulness of the traditional sketch as described above, today the sketch is often dispensed with and instead a system of short standard alphanumeric codes is recorded in the field sheet or recorded by electronic data recorders.

Where total stations are used with data recorders, a coding system must be used for recording detail. In the remainder of this chapter it will generally be assumed that data will be recorded by codes and that the traditional sketch will not be used.

10.3.4.4 Setting up the instrument

As with all specialist equipment, the instrument should be set up in accordance with the manufacturer's instructions. The accuracy of centring over the station mark required is dependent upon the accuracy demanded of the observations, whether they are only for the supply of detail and spot heights or for control observations for computation. If spot heights are to be observed, then the height of the centre of the instrument above the station mark must be measured and recorded. If a centring tripod is used, with central graduated plumbing rod, the instrument is readily set up at a convenient working height.

10.3.4.5 Selection of local reference object (LRO)

The *reference object (RO)* at a traverse station is usually the back object, but if a large number of detail points are to be recorded at a station it will be advantageous to select a *local reference object (LRO)*. The LRO provides a rapid means of checking the orientation of the instrument at any time during the round of observations at the station. The LRO may be any well-defined object, a point of detail which must be surveyed, or even the RO proper. The latter, however, is often inconspicuous or unavailable once the tripod has been removed from the back station, hence the recommendation to use an LRO. If an LRO is selected, it should be observed at least at the commencement and at the end of the round of observations.

10.3.5 EDM radiation survey observations

10.3.5.1 Control

The *control observations* required at a station should preferably be made before the observations to points of detail and spot heights, to avoid the danger that they may be forgotten. Control observations should be made in accordance with the relevant recommendations of Chapters 6 and 9, but it should be noted that with some EDM equipment it is not possible to measure on both faces. An alternative is to read on the one face but on at least two zeros. The LRO, if any, should be observed at the same time as the control observations are made.

10.3.5.2 Detail and spot heights

Detail points and spot heights should be observed on completion of the control observations. They may be observed in any order, but clearly a systematic approach is desirable. It is expedient to observe the detail and spot heights in strings, a string being a series of related points such as the line of a hedge, a kerb line, a row of trees, or the face of a building, etc.

With the most basic forms of EDM equipment, one vertical angle, one slope distance and one horizontal angle to each point of detail or spot height is normally all that is necessary. The LRO should be re-observed on completion of the detail and spot height observations.

10.3.5.3 General

A survey team should consist of a minimum of two, the senior being the team leader responsible for the survey. The team leader decides which detail is to be surveyed and the positions for spot heights and holds the reflector, and also records all the relevant string information except the angles and distances taken at the instrument station. The assistant observes and records the angular and linear measurements taken at the instrument station to the reflector positions.

If separate booking sheets are used at a station, they should be fastened together at the end of the day's work. Alternatively, telephone communication may be used to transfer all information on to one booking sheet, or enter all data in a data recorder. Again, data may be coded into a total station, along with the automatically recorded observations. Where there are three members in a team, the third member can position a second reflector under the direction of the team leader.

10.3.5.4 The reflector

The prism reflector must be placed at each point of detail such that its face is normal to the line of sight from the instrument. It may be tripod-mounted, or fixed to a prism pole, or even hand-held on occasion. Note that where a prism pole with circular bubble is used, keeping the bubble central is more important than accurate aiming of the reflector.

For control observations the reflector will normally be placed on the back and forward stations and possibly on other station points. For these observations it should preferably be centred over the particular station mark on a tripod.

For detail and spot height observations, the reflector should be placed on or over the actual point. Sometimes this is impractical and then it should be placed, as convenient, either in front of or behind the point on the line from the instrument, the displacement being noted. Again, if this cannot be done, it may be held to the left or right of the point and the left or right offset noted.

As discussed in Chapter 7, some EDM instruments can operate without a special reflector, but these are not common and work methods would have to be modified appropriately.

10.3.6 Alternative measurement techniques

Sometimes detail may be inaccessible, or invisible from the instrument station and additional measurements may be needed which cannot be obtained by radiation methods. The following alternative methods may be used, but their use should be limited in order to avoid confusion for the plotter:

(i) Plus measurements to define rectangular detail.
(ii) Detail fixing by intersection.
(iii) Widths of parallel features.
(iv) Tape and offset survey methods.
(v) The use of a cul-de-sac station.

A *cul-de-sac station* is one from which only one other instrument station is visible, sometimes known as the *parent station*. An example might be a station set up solely to measure the detail in an enclosed yard. The location fix of a cul-de-sac station is obviously weak, since it cannot be linked to other stations for a check, hence it should not be used unless there is no economical alternative. A possible checking procedure is to survey a common point of detail from both the parent and the cul-de-sac station, and co-ordinate plot the detail from both stations.

10.3.7 Communication between team leader and observer

A good communication system between these individuals is necessary in order that each may be aware of the other's requirements. If telephones are not available then a simple hand signalling system can be effective. Over long distances, a white handkerchief or a coloured flag may be of assistance. Any such system must be simple, well understood and unambiguous.

The team leader needs to be able to indicate to the observer that the reflector is in position for observing. The observer must be able to indicate:

Measurements complete, move on.
Come to me.
Go back to previous point.
Reflector obscured.
Raise reflector.
etc.

If separate booking sheets are being used, then it is advisable for the two to check with each other, say at every tenth reading and at the end of observations to ensure that no reflector points have been missed and the reflector point numbers on both sheets are the same.

10.3.8 Recording observed data

Where the field observations are being stored in a data recorder no field book or booking sheets are required, since the controlling software will guide the observer as to the input of information. In the absence of a data recorder the observations must be recorded in suitable booking sheets.

Table 10.1 gives a very simple example – the top row of the table gives the identity of the instrument position, the reduced level of the station point, the instrument's height above the mark (this is the value i) and the usual information as to job, location, surveyor, date, etc. It may also include the station point co-ordinates. Each line of the bookings refers to one observed point, and here

TABLE 10.1 Sample EDM tacheometry readings and reductions

	Inst. Station B.					*Date:*	*21/7/94*	*Job:*	*Shek Kip Mei Woh Chai St*
	Red. Level 135.243		*Inst.Ht 1.53 m*						
Point Obs'd	*H.Angle* ° ′ ″	*Zen. Angle* ° ′ ″	*Reflec.* h_r	*Slope* d	*Dist.* D	*Height* H	Δh $i \pm H - h_r$	*Reduced level*	*Remarks or symbol*
Stn.A	00.00.00	(LRO)							
Stn.C	145.25.20								
Point.1	20.30.00	79.00.00	1.530	44.655	43.835	+8.521	+8.521	143.764	*Crnr Bdg*
Point.2	43.20.00	81.20.00	1.530	49.151	48.624	+7.180	+7.180	142.423	*Top Bank*
Point.3	80.30.00	93.00.00	2.530	114.633	114.476	– 5.999	– 6.999	128.244	*Ed. Rd*
1	2	3	4	5	6	7	8	9	10

column 1 gives the identity of the point observed, column 2 the bearing to the point, column 3 the vertical angle as the zenith distance to the collimation line pointed on the reflector target, column 4 the reflector height h_r, column 5 the slope distance d. Columns 6 and 7 contain the calculated values of the horizontal distance D and the height H, column 8 contains the height difference Δh and column 9 the calculated reduced level of the point. In this case the final column has a brief description of the detail point observed.

The actual data to be recorded depends upon the equipment being used, thus for the simplest and most basic EDM 'add-on' the observed measurements for a point of detail would be as shown above, but more sophisticated instruments may directly determine the required values, and also compute point co-ordinates.

In practice, where hand booking is used, it is recommended that two booking sheets be used, one by the observer and the other by the team leader. If telephones are used, then the data may be combined on to a single sheet in the field.

If electronic data recording is used, then the data must be coded to suit the particular equipment in use. The coding shown in the example booking sheets may be suitable, some coding systems allowing numeric codes only and some accepting alphanumeric codes.

The essentials of all bookings are clarity and accuracy, so that the field bookings may be understood by all members of the team. Alphanumeric characters should be printed, corrections should be made by ruling a single line through the characters and printing the correct entries above or below the original. No overwriting should be permitted. Booking sheets may show:

The name of the task and traverse.
Team names, leader, observer, etc.
Survey date, weather.
Identity of equipment in use (theodolite, EDM, reflectors, etc.) and any additive constants or similar values.
Station number, description, height, instrument height.
Control observations; stations observed, vertical angles, slope distances, horizontal angular readings.
Detail and spot height observations; point number, other points connected to the point, line or area symbol to be used by the plotter.

Also generally:

Symbol identification annotations.
Annotations to assist interpretation of contours (ridge or valley lines etc.).
Corrections to be applied when reflector is not on the point being surveyed.
Measurements and type, if any, for the alternative techniques set out in Section 10.3.6.

Figure 10.13 illustrates an observer's booking form, designed for an instrument in which the distances could be read only on face right. The upper part of the form is for station data common to all observations, the next part is for control observations and the lower part is for detail and spot height readings.

Figure 10.14 illustrates a form which may be used by the team leader. This indicates how detail is to be portrayed, how it is connected and the supplementary measurements. Note that the data will be coded and coding may

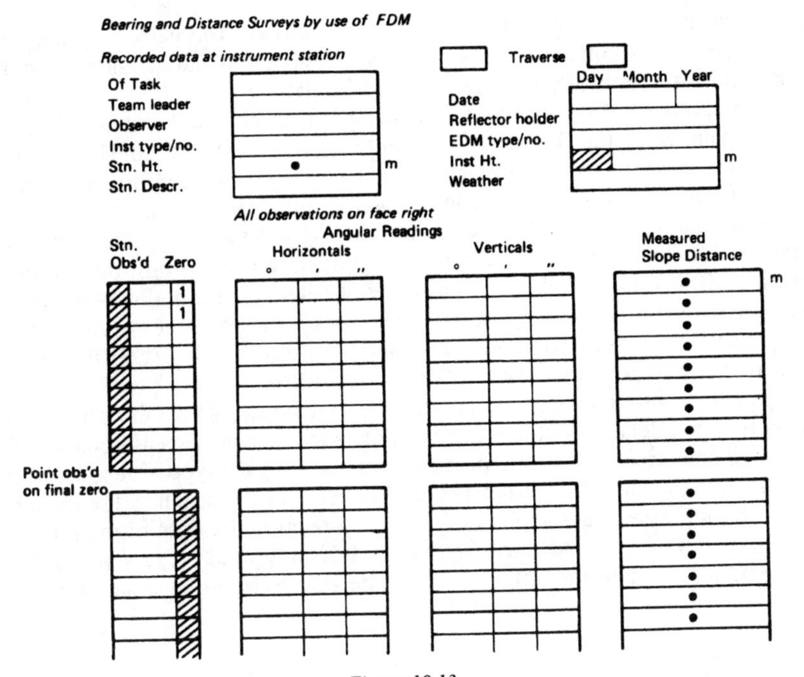

Bearing and Distance Surveys by use of FDM

Recorded data at instrument station ☐ Traverse ☐

Of Task
Team leader
Observer
Inst type/no.
Stn. Ht. • m
Stn. Descr.

Day Month Year
Date
Reflector holder
EDM type/no.
Inst Ht. m
Weather

All observations on face right

Angular Readings

Stn. Obs'd	Zero	Horizontals ° ′ ″	Verticals ° ′ ″	Measured Slope Distance m
	1			•
	1			•

Point obs'd on final zero

Figure 10.13

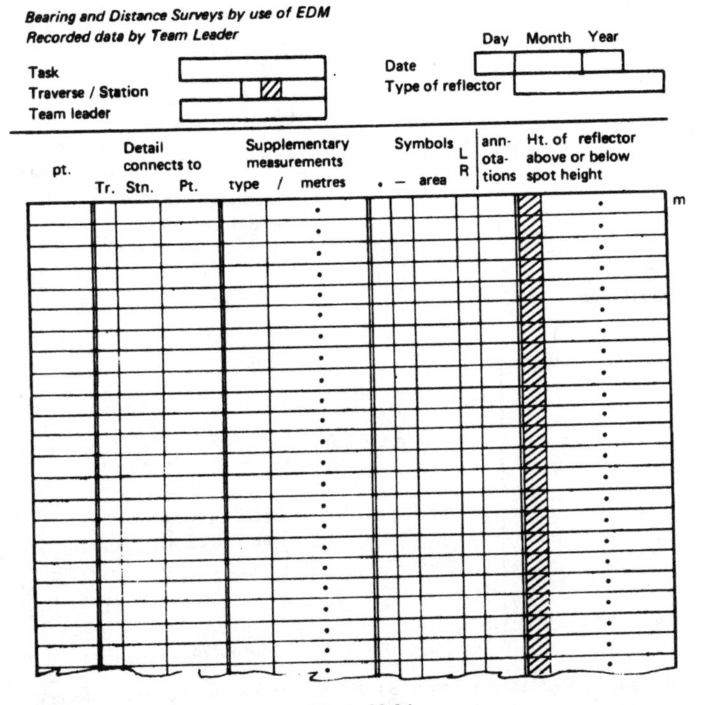

Bearing and Distance Surveys by use of EDM
Recorded data by Team Leader

Task
Traverse / Station
Team leader

Day Month Year
Date
Type of reflector

pt.	Detail connects to Tr.	Stn.	Pt.	Supplementary measurements type / metres	Symbols . — area	L R	annotations	Ht. of reflector above or below spot height m
				•				•

Figure 10.14

vary from the very simple to the very comprehensive. Too basic a system may inhibit proper booking, while too complex a system may unduly slow down the work and some sensible compromise is desirable.

The upper part of the form is for data common to all detail and height points. The remainder of the form is divided into five vertical blocks. The left-hand block contains the detail and height point numbers, then all information on the same line across the sheet refers to the point number in the left-hand block.

The second vertical block is used to indicate what detail, if any, the point in the first block connects to. If the detail is to be observed in strings, then in many cases it will join back to the previous point of detail and the annotation 'pp' (for previous point) may be adequate. In other cases it may be necessary to indicate the possible traverse, station and point number, for example 2/12/105 indicates that the point connects to point 105 of station 12 on traverse 2.

The central block is used for corrections to distance, where the prism is not actually at the detail point, together with the supplementary measurements. The left-hand column is for the two characters representing the type of supplementary measurement while the right-hand column is used for the distance. For example, ⊗ R 3.69 indicates that the point of detail is 3.69 m to the right of the prism, or alternatively ⊗ + 1.50 indicates the point is 1.50 m beyond the prism. □ L 5.17, rectangular feature to the left, of 5.17 m side length. If the information is to be

TABLE 10.2

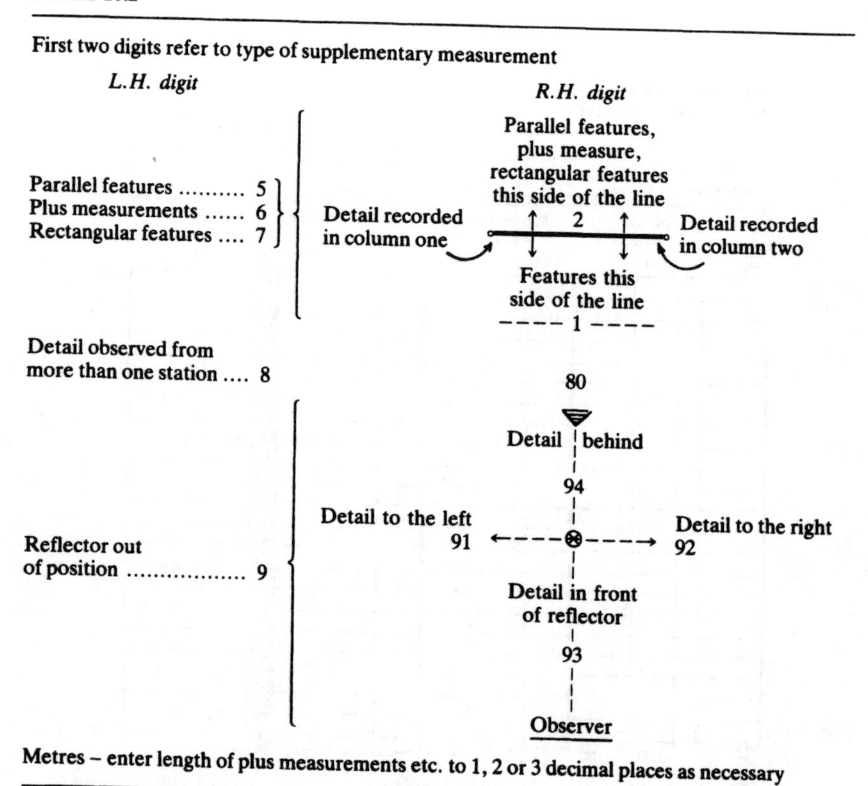

programmed to produce a set of rectangular co-ordinates for plotting, then an all-numeric system may be preferable, as shown in *Table 10.2*.

The fourth block allows first for a column for a point symbol to be used by the plotter to represent the detail or height point, then a column for a line type symbol, if any, and then a column for a relevant area symbol such as a roof, vegetation, slope, etc., followed by R or L if required to indicate right or left. Finally, the annotation column may be made use of to give additional information to clarify the data or point.

The reference may be very simple, such as just using the codes 1 to 9 for points and lines, with no area symbols or annotations at all. In this case, the area details and annotations would be added after the initial plotting stage.

Alternatively, a more comprehensive coding may be used as set out in *Table 10.3*. This allows for nine different point symbols, nine line symbols, up to 99 area symbols and also an indication whether these are to be portrayed left or right of the line symbol. Similarly up to 99 annotations may be used. The number of possible codings could, of course, be greatly increased by adding alpha codings or using alphanumeric codes. The sample set of codings in *Table 10.3* are taken from an actual task and it can be useful to have the set in use printed on the reverse of the field sheet.

TABLE 10.3

Point symbols		*Line symbols*		*Annotations*	
•	1	Solid line	1		
o	2	Pecked line	2	Boundary post	06
Single trees–coniferous	4	Elec. transmission line	3	Boundary stone	07
–others	5	Pipe line, at or close to GL	4	Cattlegrid	10
Culvert	7	–overhead	5	Chimney	11
Spot height only	8	Bank	6	Finger post	19
Bench mark	9	Drain	7	Flagstaff	23
		Direction of flow ←	8	Footbridge	24
		Line of constant slope	9	Footpath	25
				Fountain	26
				Gun	33
				Issues	37
				Letter box	42
				Mast	48
				Milepost	49
				Milestone	50
				Monument	51
				Platform	60
				Post } to prevent	61
				Posts } vehicular access	62
				Pump	63
				Spring	77
				Statue	78
				Tank	80
				Telephone call box	81
				Water tap	93
				Water trough	94
				Weighbridge	95
				Well	96

Area symbols
Archways 01, Pylons 02,
Roofs, glazed 03, others 04,
Sloping masonry-top 05, bottom 06,
Steps 07, 08,
Bog, marsh 10, Bracken 11, Bushes, shrubs 12,
Furze, gorse, 14, Heath, heather, bilberry 15,
Rough grassland 19, Undergrowth, underwood 22,
Orchard 16
Man-made slopes-top 80, bottom 81

Present recorded point of detail

Area symbol to the left enter '1' | Area symbol to the right enter '2'

A previously recorded point of detail

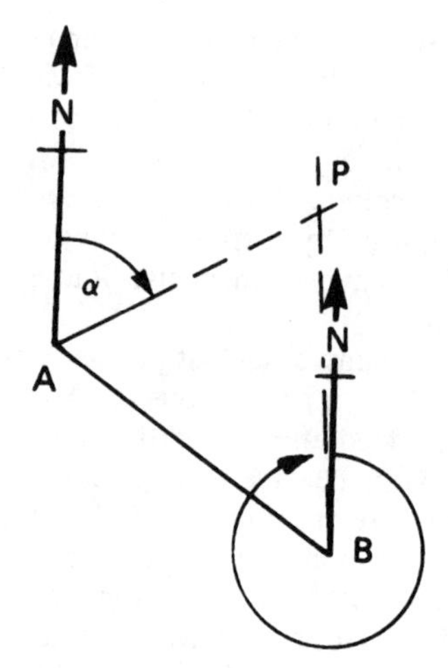

Figure 10.15

The final right-hand block of the form in *Figure 10.14* is used in heighting, to record the height of the reflector above or below the spot height being observed.

10.3.9 Office procedure

The computation of traverses, preliminary plotting considerations, the construction of grids, co-ordinate plotting, contour interpolation and plan completion after plotting are basically the same for both EDM and ODM radiation, they are dealt with elsewhere in the text and will not be repeated here. This section deals with aspects directly relevant to EDM radiation survey office work, including the computations which may be required before plotting and the methods of plotting and drawing detail.

Some or all of the calculations set out here may require to be carried out before or during the plotting of the survey, depending upon the actual equipment used. Early EDM equipment measured slope distance only, and it was necessary for the user to measure the vertical angle then calculate horizontal distance and height difference. Modern equipment may calculate the rectangular co-ordinates as well as the other values.

10.3.9.1 Horizontal distance

Horizontal distance D = slope distance $d \times$ cosine VA,
± a correction if the reflector is held in front of or behind the point
(VA represents vertical angle observed.)

10.3.9.2 Height difference

Height difference H = slope distance $d \times$ sine VA

10.3.9.3 Reduced level at reflector point

RL at point = station height ± H + instrument height i – reflector height h_r

This is simplified if the target is set to the same height as the instrument.

If the plotting is to be carried out by rectangular co-ordinates and not by polars, then the following may be required.

10.3.9.4 Orientation correction for conversion of horizontal angular readings to bearings

Orientation correction = bearing (instrument to RO, from co-ordinates) – horizontal angular reading to the RO.

10.3.9.5 Bearing to detail or spot point

Bearing to detail point = horizontal angular reading to detail point + orientation correction.

10.3.9.6 Partial co-ordinates of line to a point

ΔE and ΔN obtained using P → R key, or
ΔE = horizontal distance × sine bearing and
ΔN = horizontal distance × cosine bearing

10.3.9.7 Point co-ordinates

E of point = E of station + ΔE, and
N of point = N of station + ΔN

10.3.9.8 Reflector offset correction

If the reflector is held to the left or right of the point (supplementary code 91 or 92 in *Table 10.2*) then the point co-ordinates may be calculated from a two-legged traverse. Bearing of the supplementary correction is bearing from station to reflector ±90°. If detail is to the right, code 92, add 90, if to the left, code 91, subtract 90.

Alternatively, the station to reflector and reflector to point distances may be entered with the R → P key to deduce the station to point distance and the bearing correction.

10.3.9.9 Co-ordinate corrections for plus measurements, rectangular and parallel features

These may be carried out in a similar manner to the above.

10.3.9.10 Detail fix by intersection

Use the sine formula to calculate the unmeasured side lengths, then calculate co-ordinates by bearing and distance.

An alternative is to use the intersection formula, reference *Figure 10.15.*

$$N_P = (N_A \tan\alpha - N_B \tan\beta - E_A + E_B)/(\tan\alpha - \tan\beta)$$

and

$$E_P = E_A + \tan\alpha\,(N_P - N_A)$$

or

$$E_P = E_B + \tan\beta\,((N_P - N_B)$$

(Preferable to use bearings closest to 0° – 180°.)

10.3.10 Plotting detail and spot height points

10.3.10.1 Plotting by polars

Plotting equipment may be a 360° protractor or a polar co-ordinatograph. The protractor radius should be at least as great as the maximum measured length to be plotted at the plotting scale. (It is always bad practice to extrapolate beyond the circumference of a protractor.) The protractor should be placed centrally over the station point, then oriented by reference to the observed readings to the RO and forward station. The protractor may be kept in place by drafting tape or paperweights. The polar co-ordinatograph will remain in place by its own weight.

Using the co-ordinatograph, each observed horizontal angular reading and its distance are plotted with the instrument's pricker. The point is circled in pencil, then annotated with its point number or, preferably, with its reduced level if the plan is to be contoured.

Using the protractor, plot all the bearings first, indicating each by a light tick in pencil and annotating their appropriate reference numbers. Remove the protractor and, with the aid of a scale rule, plot the required distance along each of the bearings. The plotted points should be circled and annotated as for the co-ordinatograph. Connecting all the bearings from the station mark to the plotted ticks by a pencil line should be avoided, since the station mark will be obliterated.

10.3.10.2 Plotting by rectangular co-ordinatograph

Due to the variation in models, the manufacturer's instructions should be referred to for guidance.

10.3.10.3 Drawing detail and contours

Using the scale rule and set square, plot all those points where the detail was to the left or right of the reflector, supplementary codes 91 and 92. Connect together all the points of detail as indicated in columns one and two of the team leader's form. Plot the remaining supplementary points of detail, codes 51 to 80. Pen in the detail in accordance with the symbols block, *Figure 10.14*. Interpolate the contours along all lines of constant slope, line symbol, code 9, *Table 10.3* and *Figure 10.14*. Interpolate the contours between all adjacent spot heights except between spot heights either side of a line of constant slope. Note that it may be more convenient to interpolate the contours on a separate trace of the area. See Chapter 8 for further information.

Plan completion is carried out as detailed in Section 10.2.3.4.

10.4 ODM radiation detail survey

A variety of optical distance measurement methods have been devised over the last 200 years, but the development of electromagnetic distance measurement techniques has rendered most optical methods obsolete. The one form of ODM which is still in use is the oldest method, vertical staff tacheometry or stadia measurement, which is known to have been used by James Watt in Scotland in 1771.

The principles of stadia measurement with a vertical staff have been described in Section 7.2 and will not be repeated here. The general field procedure in stadia detail survey is largely similar to that of EDM detail survey, although obviously the equipment and computation methods differ.

10.4.1 Basis of the method

The basis of vertical staff tacheometry is essentially the same as for EDM – detail and spot heights are observed from an instrument station to all required points in the area. However, the instrument used is a normal theodolite equipped with stadia lines on the diaphragm, generally with a multiplying factor of 100, and instead of observing on a prism reflector observations are taken to a graduated staff held vertically on the detail points.

In the past special tacheometer staves with one centimetre graduation blocks were available, some being fitted with an extending foot. Today normal levelling staves are used. Note that a competent observer can keep several staff men occupied.

The control observations, instrument orientation and plotting methods are essentially the same as for EDM radiation.

10.4.2 Advantages and disadvantages

The advantages of stadia radiation are again that only one instrument is needed and all required measurements in an area may be obtained from the one set-up at the instrument station. In addition, no expensive specialist equipment is needed, only a normal theodolite and one or more levelling staves. The usual station marking materials are required.

The disadvantages of stadia measurement are the relatively low accuracy obtainable, generally about 1:500 in distance and ±50 mm for heights, and also the sighting distance limit of about 100 m.

10.4.3 Preliminary considerations

These are basically the same as for EDM radiation, but the possible sighting distances are much less than can be obtained with EDM and this will affect the layout of the survey and stations.

10.4.4 Field procedure

The reconnaissance of the area, the selection and marking of instrument stations, the preparation of station point sketches and the selection of the LRO are all as described in Section 10.3.4.

The theodolite is set up and centred over the station mark in the usual way, then the height from the top of the station mark to the transit axis of the theodolite is noted. In order to determine heights, the reduced level of the station mark must be known.

10.4.5 Stadia radiation survey observations

10.4.5.1 Control

Any control observations at a station should be carried out before the detail observations are made. If control consists of a tacheometer traverse, then the traverse angles should be measured by simple reversal, and the leg lengths measured in both directions.

10.4.5.2 Detail and spot heights

After completion of any control observations at a station and orienting on the LRO, the detail points should be observed. The observation procedure is:

(i) Direct the theodolite telescope on the level staff, bisect the staff with the vertical hair, apply the horizontal clamp, read and book the horizontal angle.
(ii) Set the middle hair to the same reading on the staff as the instrument axis height and clamp the vertical circle, read and book the vertical circle.
(iii) Read and book the middle hair, then the upper hair, then the lower hair.

Taking these readings is slow and laborious and in the past a variety of modified observing techniques were developed to speed-up the work and simplify the calculation problems.

The level staff should be held vertically at each point of detail in turn, with its face normal to the sight line from the instrument, preferably using a staff bubble to maintain verticality. A non-vertical staff is a significant source of error in stadia work.

Alternative measurement techniques may be required, as covered in Section 10.3.6, and the communication needs are basically as set out in Section 10.3.7. If a telescopic levelling staff is in use the observer should be able to signal the staff man to extend the staff, and he should also be able to signal him to slowly swing the staff backwards and forwards, if necessary to help decipher an obscured reading.

10.4.6 Recording observed data

The method of booking stadia observations is a matter of individual preference – every surveyor appears to use a variety of his own, either a hard cover book or separate sheets with a clipboard, suitably ruled. Large organizations of course lay down standard methods to achieve uniformity. The method illustrated in *Table 10.4* assumes the use of a modern optical theodolite with the vertical circle giving *zenith distances* (angle measured from the zenith down to the sight line) on face left, the normal observing position.

The layout is very similar to that shown for basic EDM, the top row of the table giving the information as to the instrument station, the job, the equipment, and the surveyor, but with i representing the instrument's axis height above the mark. Again each line of the bookings refers to one observed point, columns 1 to 3 being the same as for EDM, columns 4 to 6, the staff readings and column 7, $100 \times$ staff intercept s. Column 1 gives the identity of the point observed, column 2 the bearing to the point, column 3 the vertical angle as the zenith distance to the collimation line pointed on the mid-hair reading, and columns 4, 5 and 6 the staff readings. All the other columns except 12 are completed in the office after the fieldwork is complete.

It is generally simpler in practice to make the middle hair reading the same as the instrument's axis height, since this simplifies the later arithmetic. Note that in the example when observing on Point 3 it was impractical to sight on 1.530 so the sight was made on 2.530 instead, again to help simplify the arithmetic and avoid mistakes. The vertical circle readings have been taken to the nearest minute, this is generally adequate.

10.4.7 Reduction of observations

Referring to the sample part booking page in *Table 10.4* columns 7 to 11 are calculated in the office. The basic calculations for stadia observations have been covered in Section 7.2 and will be summarized here.

10.4.7.1 Staff intercept, s

The staff intercept, s, is obtained by subtracting the lower hair reading from the upper hair reading. Column 7 shows the value of s multiplied by 100, the normal stadia multiplication constant. Note that the readings for the three hairs can be checked by comparing *upper hair minus mid-hair* and *mid-hair minus lower hair*, which should be the same.

10.4.7.2 Horizontal distance, D

The horizontal distance, D, shown in column 8, is calculated as $100\, s \sin^2 z$, since the zenith distance or zenith angle has been measured and not the angle from the horizontal. If the vertical angle of elevation or depression had been measured then the value would be $100\, s \cos^2 VA$.

TABLE 10.4 Sample vertical staff tacheometry readings and reductions

	Inst. Station B.					*Date:*	*20/7/94*			*Job:*	*Shek Kip Mei*
	Red. Level	*135.243*	*Inst.Ht*	*1.53 m*							*Woh Chai St*
Point	*H.Angle*	*Zen. Angle*	*Vert.*	*Staff*	*Readgs*		*D*	*H*	Δh	*Reduced*	*Remarks*
Obs'd	° ' "	° ' "	*middle*	*upper*	*lower*	$100S$	$100S \sin^2 z$	$50S \sin^2 z$	$i \pm H - m$	*level*	*or symbol*
Stn.A	00.00.00	(LRO)									
Stn.C	145.25.20										
Point.1	20.30.00	79.00.00	1.530	1.757	1.302	45.5	43.84	+8.52	+8.52	143.77	*Crnr Bdg*
Point.2	43.20.00	81.20.00	1.530	1.779	1.281	49.8	48.63	+7.53	+7.53	142.77	*Top Bank*
Point.3	80.30.00	93.00.00	2.530	3.104	1.956	114.8	114.5	–6.00	–7.00	128.24	*Ed. Rd*
1	2	3	4	5	6	7	8	9	10	11	12

10.4.7.3 Height from instrument axis to mid-hair, H

The vertical height between the instrument axis and the mid-hair point on the observed staff is entered as H in column 9. Again, since z is used, H is calculated as $50\ s \sin 2z$.

10.4.7.4 Height difference between station point and detail point, Δh

The value of Δh, shown in column 10, is calculated as $i \pm H - m$, where i is the instrument axis height above the station marker, H is the height difference between instrument axis and mid-hair reading on the staff, and m is the middle hair reading. If m is kept equal to i then $\Delta h = H$.

10.4.7.5 Reduced level of detail point

The reduced level of the detail point, shown in column 11, is calculated as the station point reduced level (135.23) $\pm\ \Delta h$.

10.4.7.6 Bearings and co-ordinates

Sections 10.3.9.4 to 10.3.9.9 as described for EDM radiation generally apply also to stadia radiation.

10.4.8 Plotting

The plotting of a stadia radiation detail survey is the same as for EDM radiation, described in Section 10.3.10.

10.5 Theodolite and tape radiation detail survey

Where a stadia radiation survey is being carried out and there is much detail lying close to, and at approximately the same level as, the instrument station, then the detail may be picked up economically by this method. Bearings are taken to the detail with the theodolite and the distances measured horizontally with a steel or synthetic tape. This will reduce the observation and reduction labour if there is much close detail, but it is limited by the length of the tape used, and of course levels must be assumed to be the same as that of the instrument station.

10.6 Digital terrain modelling

A digital terrain model (DTM) is essentially a mathematical model of part of the surface of the Earth, the data being stored in digital form in a computer database, such that both planimetric detail and height information is available, i.e. the E and N co-ordinates and the reduced levels (RL) of the points held. This model may be used to produce a graphical screen display or a printout, and with contours as required, and the model may be viewed in perspective, from different angles, and may be used to calculate volumes, plan route lines, etc. For a large area the data is best acquired by air photo survey and photogrammetry, but for limited areas the ideal method is EDM radiation using a total station and an automatic data recorder.

Following the acquisition and storage of the data, the DTM may be produced using a rectangular grid or it may be based on triangulation. In the former, a rectangular grid is superimposed over the area and the reduced levels of the grid nodes interpolated from the field data, but this tends to smooth out the surface and the contours produced may be somewhat deformed.

In the triangulation method, the field survey detail points are taken as the node points of the DTM, the resulting model forming a network of triangles, each vertex being a data point. This provides a much more accurate representation of the ground surface.

Sections of levels (longitudinal or cross-sections) may be produced by interrogating a DTM, and this is particularly useful in highway and similar design.

Chapter 11

Building surveys

11.1 Introduction

Building surveys are usually for the purpose of preparing plans of an existing building which is to be altered – record plans are seldom available and when they do exist they are generally inaccurate. It may be necessary to prepare a plan of a very old building for conservation records. A survey should provide all information necessary to prepare not merely a ground plan but also a plan of each floor in the building, together with measured *elevations* (external views) and *sections* ('cut-away' views) of the buildings. While in land survey the surveyor may have to describe boundary features, vegetation and street furniture (post boxes, etc.), in the case of buildings he may have to describe the constructional details and the building services.

The essential difference between these surveys and the tape and offset surveys described in Chapter 4 lies in the plotting scales used. Tape and offset survey scales are typically 1:5000 to 1:500 and building survey scales 1:100 or 1:50. The larger scales are needed in order to communicate the information adequately to the client. This may require measurements accurate to the nearest 10 mm, and occasionally 5 mm, and offsets may become unacceptable unless over very short distances of a metre or less. It is not difficult to keep lines straight generally, since the lengths of the lines to be measured are often under one tape length. However, the network of lines required for the framework and the techniques used are very similar to linear survey control with tape and offset detail survey, except in the booking. A small task could be sketched in detail on one sheet of paper, then all the measurements taken and noted on that same sheet. It is not always essential to letter each station, since they are all on the same sheet and confusion is unlikely to arise. A larger survey might require two or more sheets of sketches and in this case it may be helpful to identify points by letters – it is also important to ensure that the several sketches are adequately connected by common points and lines of detail and do not become a series of unconnected parts.

The ground plan of the building is often required in its relationship to the site as a whole; this means that a *site plan* or *plot survey* will be necessary. This may be at a different scale from the building drawings. If the site plan is to be at a scale of 1:500 then tape and offset survey techniques may be used. At smaller scales a tracing from the National or Municipal large-scale maps of the site may be acceptable, but this may involve copyright approval and the payment of fees to the mapping agency.

At larger scales, tape and offset survey methods are unacceptable for plot surveys, due to the increased time and effort needed to achieve the desired accuracy. Alternative techniques include intersection or radiation (bearing and distance), as referred to in Chapters 1 and 10. Conversion of the field observations to rectangular co-ordinates may simplify the plotting. Spot levels or contours may be required over a site, but these are considered in Chapters 5 and 8.

On occasion, surveyors may be requested to measure the movement in buildings and other structures. Techniques for this are described at the end of this chapter.

11.2 The plot survey

11.2.1 Typical equipment

The typical equipment is as for tape and offset survey, but additional equipment includes

2 m or 3 m retractable steel tape rule,
2 m folding rod,
builder's line and plumb-bob,
manhole key,

and possibly

hand lamp or torch,
theodolite,
electromagnetic distance measuring equipment.

As regards the last item, the Leica DISTO is a hand-held laser distance measurer which can be used on surveys of buildings and small sites. It needs no reflector, the visible laser beam is simply pointed at a surface and the press of a button displays the distance to single millimetres up to 30 m. Values can be added and subtracted, and areas and volumes calculated. More sophisticated models of the Disto may be linked to an electronic theodolite or RS232-linked to a PC.

11.2.2 Field procedure

11.2.2.1 Preliminaries

Check the exact purpose of the survey, e.g. simple site plan, site and building plan with drainage layout, or a full building survey, and the scale expected. The survey provides information for the plan, but the client may require details as to the services available, future development in the area, planning restrictions, building lines, etc. It may be necessary to contact the local authority for the area, gas and electricity companies, telecommunication services and other interested parties.

Arrange access to the site and buildings and all permissions.
Check equipment, standardize tapes as needed.
Obtain a sketch map of the site, or an outdated plan, or even a plan at a smaller scale, to provide a rough guide in reconnaissance and planning line layout.

11.2.2.2 Site reconnaissance

A reconnaissance must be carried out, as described in Chapter 4, to fix the positions of lines. Typically the lines should break the area into triangles, each triangle having a check and the triangles tied to one long base line through the site if at all possible. The base line is of particular importance on long, narrow sites and it may be advisable to consider the use of a theodolite to establish the line, especially if the ground is covered with vegetation. Offsets will generally be few since it is often possible to run lines along straight boundary features. On occasion lines may have to be run through the building(s) and in such cases it is helpful if the lines are either parallel or perpendicular to the building faces or walls.

Where the plotting scale is to be 1:200 or greater, or if the site is long and narrow, it may be more effective to use radiation techniques as described in Chapter 10. Where the site is small and no assistant is available, intersection techniques may be suitable.

11.2.2.3 Linear measurement (and possibly angular measurement)

The methods to be used for direct linear measurement, angular measurement and indirect distance measurement have been covered in the preceding chapters.

11.2.2.4 Booking the measurements

When the site is small and there is only a small number of offsets, the booking may be carried out on a single sheet of paper, with single (skeleton) lines representing the survey lines. If the site is large, perhaps with many offsets, it may be better to use the conventional tape and offset survey booking methods. Where relevant the guidelines of Chapters 4, 9 and 10 should be followed, whichever style is used, particularly as regards clarity and accuracy.

Figure 11.1 illustrates a 'single page' survey booking. Note that the techniques of chainage, offsets and straights are similar, including the use of running

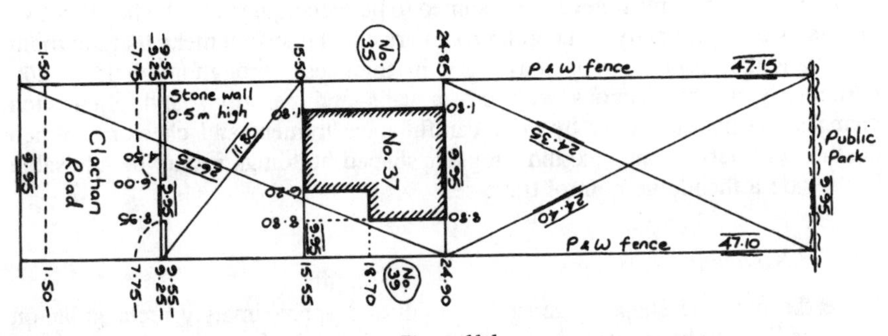

Figure 11.1

measurements and overall line lengths to maintain accuracy and avoid ambiguity. Site plans may also include the outlines of neighbouring buildings, footways, roads, sewers, drains, lamp-posts, telegraph poles, etc.

11.3 The building survey

11.3.1 Equipment

The equipment is generally as identified in Section 11.2.1, but in addition the following may be needed

sectional ladder,
binoculars,
surveyor's level and levelling staff.

11.3.2 Field procedure

11.3.2.1 Preliminaries

Check exactly what information is to be provided by the survey and obtain any existing plans, sections, elevations. Obtain keys for all unoccupied buildings and any necessary entry permissions. Check that all equipment needed is available and in good order.

11.3.2.2 Reconnaissance

A thorough 'recce' of the building must be carried out in order to assess the size and nature of the task and decide how best to tackle it. A building is usually a series of blocks, of different shapes and sizes at different levels and the inexperienced surveyor must avoid the temptation to measure them separately and then fit them together like a '3-D' jigsaw puzzle. The survey principles of Section 1.7 must be remembered, particularly 'working from the whole to the part' and the need for the independent check. However, it is always difficult to avoid the fragmentation of the measurements and when this occurs reference to the control framework must be made.

It is often difficult to build-up satisfactory triangles, but every possible diagonal measurement should be taken within rooms and every wall must be measured and, taken together, these can form the required triangular frameworks. Buildings or rooms must never be assumed to be rectangular and the fact that two diagonals of a notionally rectangular room are equal does not mean that the room is rectangular. Again, walls may vary in thickness throughout their length, particularly at intersections with cross walls and partitions and often such changes are detected only by very careful measurement and checking. When dealing with large, complex and irregular shaped buildings it may be advisable to provide a theodolite control traverse.

11.3.2.3 Sketching

Since the building shape is generally small and approximately rectangular on plan, the lines to be measured are usually run along the features to be surveyed, the sketches being drawn first, the bookings (measurement notes) being added later. (This is the opposite of the methods used in tape and offset survey.) These freehand sketches should be as large as possible, preferably on A4 plain or 5 mm square ruled paper and roughly true to shape, but no attempt should be made to sketch to scale – the most important consideration is that all details are clear and the required measurements can be shown without crowding or ambiguity. Sketches are required as follows:

(i) Plans of each floor of the building, including any basement. A plan within the roof space and/or a plan of the roof may also be needed.

Floor plans are drawn in the same way as external sites, that is to say looking down on the floor being drawn, but as if the building had been sliced through horizontally at about 1 m above the level of the floor concerned. It may be possible to trace the ground floor plan sketch as a basis of the detailed sketches for other floors.

Details to be shown include wall openings (doors, windows, hatches, etc.), changes in wall thickness, changes in floor levels (steps, stairs, ramps, etc.) and similar construction detail, with information as to materials, span directions, etc. Services installations and fixtures and fittings (gas, heating, water, electricity, ventilation, power, lighting, communication, cooking, sanitation, washing, etc.) and drainage layouts are all required.

Roof plans include gutters, direction of falls, covering materials, parapets, skylights, stacks, rainwater heads, etc.

(ii) Elevations of the exterior of the building. Elevation sketches need not be completed in their entirety where detail is repetitive. Detail to be shown is as for plans, but also includes sills and lintels, external piping and any detail which is visible on elevation but cannot be shown on plans.

(iii) Sections drawn to show what would be visible if the building was cut through vertically. A section need not be in one straight line across the building in plan (or vertically), it may be stepped if it is more appropriate for the job. In some such cases the section actually becomes a combined section and elevation in effect.

Details shown on sections will be similar to those shown on plans and elevations. The section lines should be drawn on the plan sketches.

As in tape and offset survey, the representation of detail by point, line or area symbols and by annotation by abbreviations on the sketches should be similar to that required on the plotted plans, elevations and sections. Some common abbreviations are listed here. Where relevant, the published recommendations of professional institutions and other bodies, such as the British Standards Institution, should be followed.

Common abbreviations are

agg	aggregate	m	metre
asp	asphalt	MH	manhole
bit	bitumen	mm	millimetre
BM	bench mark	MS	mild steel
bwk	brickwork	PI	petrol interceptor
CI	cast iron	RC	reinforced concrete
conc	concrete	RE	rodding eye

FFL	finished floor level	RWP	rainwater pipe
G	gulley	T&G	tongued and grooved
GL	ground level	TBM	temporary bench mark
GT	grease trap	VP	vent pipe
LP	lamp-post	WI	wrought iron

Figures 11.2 to *11.4* show examples of the representation of detail. Where detail is too small to show up clearly on the main sketch, larger detail sketches should be made on a separate sheet of paper, suitably referenced so that they may be referred to later.

11.3.2.4 System of measurement

External linear measurement may be carried out at ground level as described for taping lines in Chapter 4, but taping is often carried out with the tape held at chest height and making use of the hook attached to the zero end of the tape.

It is conventional to take all measurements from left to right, i.e. anti-clockwise around the outside of the building and clockwise around the interior of rooms,

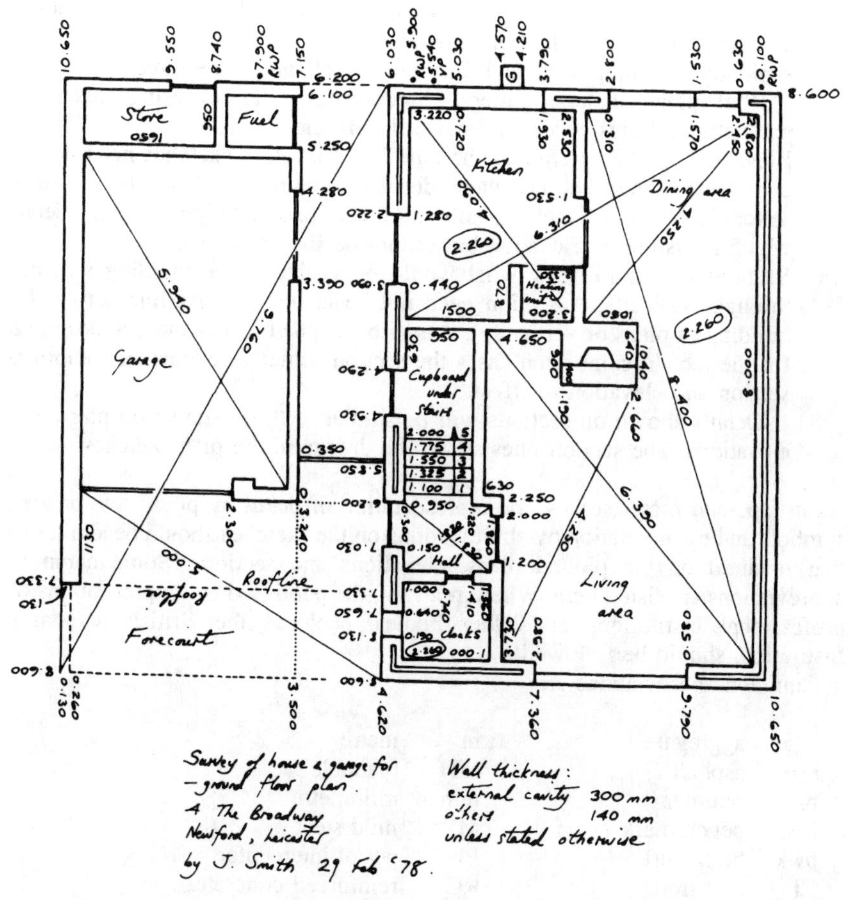

Figure 11.2

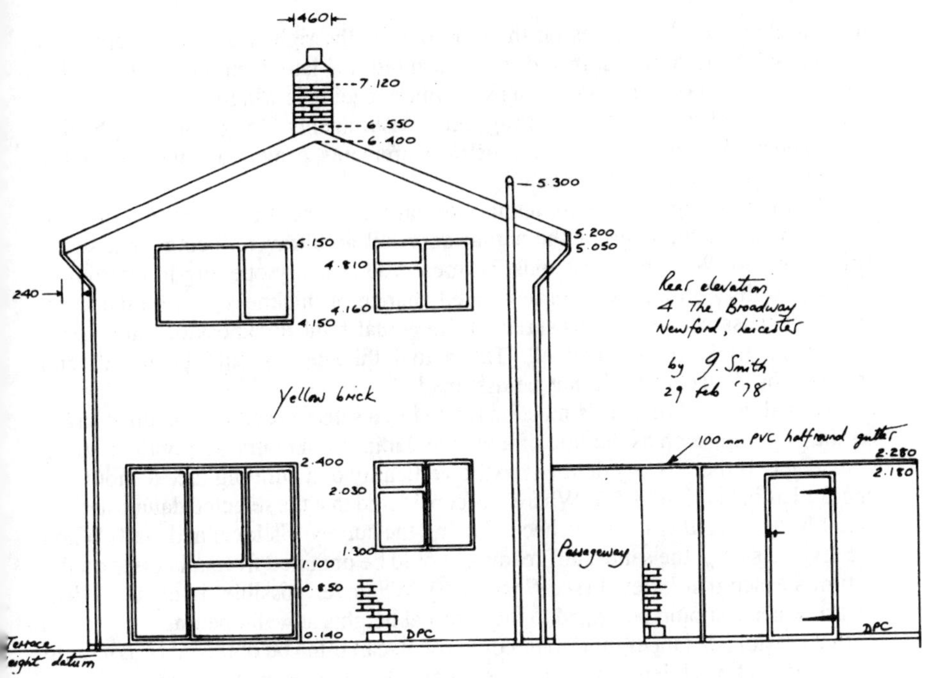

Figure 11.3

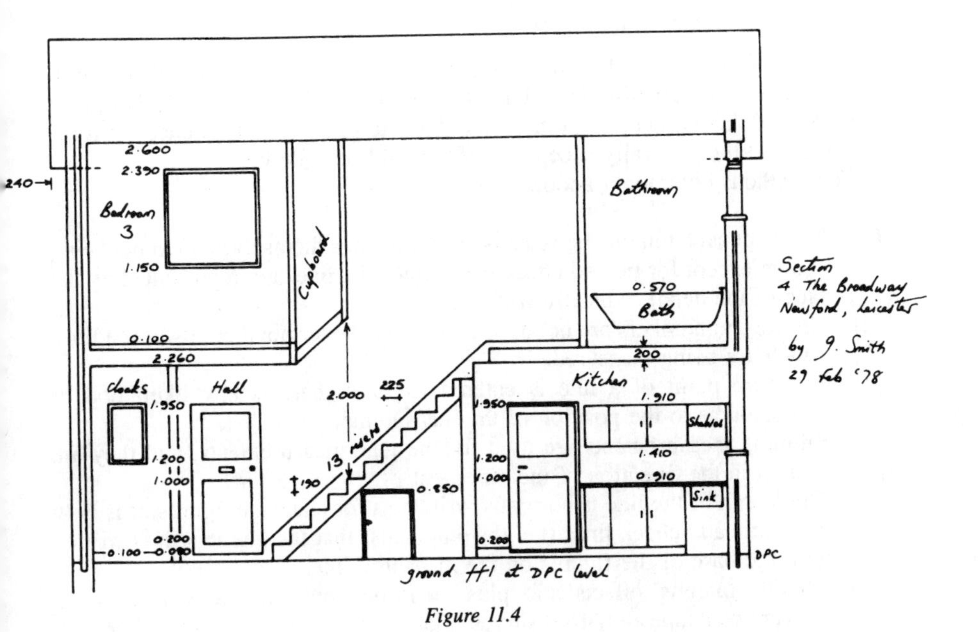

Figure 11.4

since in this way the figures on the tape will be the right way up and errors in reading will be minimized. In addition, when plotting later from the bookings this method reduces possible confusion as to which reading is which.

For ease and accuracy in plotting, all measurements along one straight line should be booked as one set of running measurements, provided that the accuracy of measurement can be maintained.

When measuring door openings in internal walls, the measurement along the wall is made to the edge of the actual door, all architraves, facings and so on being ignored. Wall thickness must be measured at every opening in internal or external walls – in this way an unnoticed change of thickness may be detected. Wall thickness should be measured as the actual total thickness of materials – rendering, brick, plaster and all. The actual thickness should be noted, and standard brick sizes should not be assumed.

External vertical distances must be related to a selected suitable datum surface on the building, such as the line of a visible damp-proof course, a plinth or string course, etc. If there is any doubt as to the verticality of a building face it should be checked as in Section 11.5.1. Where it is considered that the selected datum surface is not horizontal, this may be checked using the surveyor's level and staff. When the level is set up, the same staff reading should be obtained from all points on the datum surface if it is level (see Chapter 5). Wherever possible, internal vertical measurements should be linked to the external heights at wall openings.

Inaccessible distances, horizontal or vertical, can often be obtained by counting the number of brick lengths or courses involved then directly measuring a similar length of accessible brickwork. It must be emphasized that this practice should only be used where it is not possible to gain access to the feature to be measured.

The 2 m rod is used to measure heights, but the tape may be dropped for very long or overall heights.

11.3.2.5 Booking the measurements

Considerable licence is allowed in the recording of measurements in this type of work, since it is usually plotted by the surveyor himself, but it should be remembered that on occasion others may have to plot the work, hence clarity is essential and the generally accepted rules should be followed.

The methods commonly adopted are as follows

(i) Single or skeleton line booking is used, the lines themselves often not being shown except for tie and check lines, since the lines are represented by the face of the detail, typically wall faces.
(ii) Letters and numbers are not used to identify the terminal points of the lines which are being measured.
(iii) The zero point of a line is entered as $\ominus$ if there is any likelihood of confusion as to the position of the measurement zero.
(iv) Running measurements are used for the tape measurements, and they are written in the direction of measurement.
(v) The length of the line is normally written as for chainage figures, it is only on the tie and check lines (i.e. the diagonals) that the line length is written with the base of the figures on the measured line.
(vi) Offsets, running offsets and plus measurements may be written as in conventional tape and offset survey booking and may be entered either right

or left of the line as space permits, provided there is no ambiguity. Occasionally running offsets are also written as short survey lines, but not tied out. To help minimize ambiguity, survey line figures may be written as metres and decimals of a metre, while offsets and plus measurements can be written simply as the number of millimetres, i.e. no decimal point shown.

(vii) Floor-to-ceiling heights should be written in the centre of the floor plan of the area concerned, the figures being encircled to indicate that they are not horizontal dimensions.

Figures 11.2, *11.3* and *11.4* show example bookings for a floor plan, an elevation and a section, using these methods.

11.4 Office procedure – plotting

Plotting consists of several separate plots, including site plan, floor plans, elevations and sections and, depending upon the finished size of these, more than one sheet of drawing material may be needed. The site plan will often be at 1:500, while the remainder will be at 1:100 or 1:50 and occasionally large details at 1:10 or 1:5 will be needed.

11.4.1 Draughting equipment and materials

This is generally as described for tape and offset survey, with the addition of

drawing board plus T-square,
set squares, 45° and 60°,
(adjustable set square is useful in addition, but is no substitute for the two simple squares),
protractor,
150 mm compasses/dividers,
springbow compasses/dividers,
steel pricker.

For the alteration or conversion of an existing building, a heavy-grade tracing paper may be suitable. Some individuals, however, prefer to carry out all original plotting on cartridge paper. If a strong, durable and dimensionally stable material is necessary, a modern polyester-based translucent film may be used.

A measured survey of a small building and its site should generally plot on a single sheet of one of the recommended A-size drawing sheets, possibly A2 or A1, without being cramped.

11.4.2 Layout of the survey on the drawing material

The ground floor plan is generally placed at the bottom left-hand corner of the sheet, with the front of the building towards the bottom of the sheet. The first floor plan is then placed alongside the ground floor plan with the same aspect or orientation and the other floor plans similarly.

The front elevation of the building should be drawn immediately above the ground floor plan, again the other elevations are then drawn across the sheet in a row and level with the front elevation. Finally, the sections and site plan should be placed wherever they will conveniently fit into the general arrangement, although the sections should, for preference, be placed alongside the elevations.

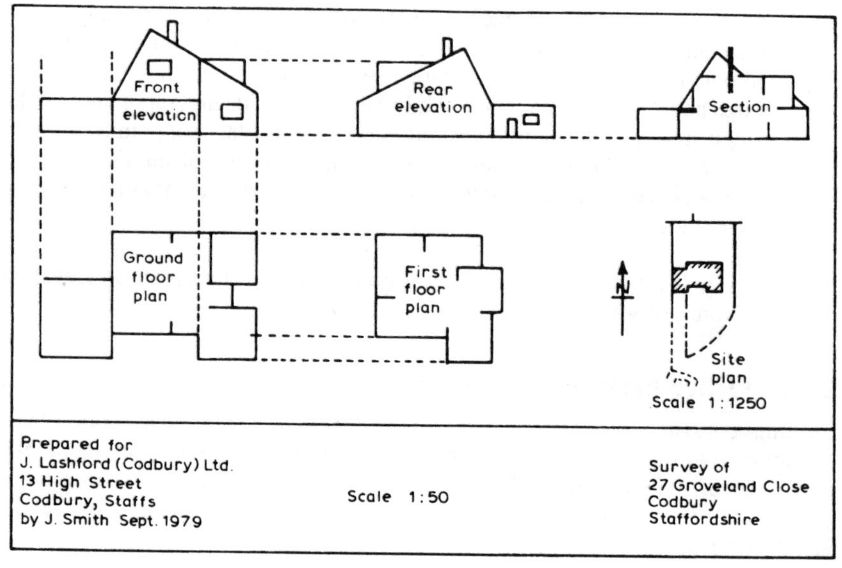

Figure 11.5

Wherever possible, it is important to arrange that common lines on plans, sections and elevations should lie on extensions of their respective lines on the drawing, see *Figure 11.5*. This presentation conforms closely to the 'first angle (or European)' projection, and it is the projection recommended in BS 1192:1969, *Recommendations for Building Drawing Practice*.

Variations are inevitable, thus, for example, a small, single-storey extension at the rear of a building might not require a front elevation to be shown and in each case the surveyor must decide the layout applicable in the circumstances. In all cases the site plan should be drawn with north towards the top of the sheet. BS 1192 also contains recommendations regarding marginal and border information.

The final consideration regarding the layout is that the whole presentation should be well balanced and pleasing to the eye, as with all drawings.

11.4.3 Plotting the plans

The site plan is plotted as covered in Chapter 10. The building itself is plotted in a similar manner, although the process is rather more complex. The procedure is as follows.

(1) Draw the longest external wall to scale, in the preferred location on the drawing material.
(2) Using the measured wall thickness, plot the alignment (but not the detail) of the interior face of the same wall. Remember the possibility of different thicknesses along the length of the wall, and make allowance as needed.
(3) On the external wall face, plot the position of all wall openings and raise right angles from these to cut the alignment of the inner face of the wall.

Note that there may be rebated or splayed jambs and the construction on either side of an opening may differ; due allowance must be made for such details.

(4) Using the internal wall face measurements, scale off and mark the position of all walls joining into the external wall.
(5) Using the conventional tape and offset survey technique of plotting triangles by swinging arcs, plot the lines representing the faces of all the internal walls of the rooms adjoining the external wall already plotted, using wall lengths and room diagonals to build up the triangles.
(6) Repeat the process until all the ground floor walls have been plotted, applying all possible checks (matching internal and external measurements at openings, checking overall external wall lengths, etc.) since errors can build up rapidly when a building is plotted in this manner.
(7) When the ground floor plan framework is complete and considered to be correct, plot all the detail involved.
(8) Plot the first floor plan by using the ground floor external dimensions (adjusted where necessary) and projecting across from the ground floor plan by the use of T-square and set square. Alternatively, trace the outline of the ground floor plan carefully, then locate the tracing paper where the first floor plan is to be plotted and 'prick' through the corners of the building on to the drawing material. The pricker marks may be joined in pencil to form the outline of the floor plan.
(9) Plot the interior of the first floor plan, in a manner similar to that used for the ground floor plan.
(10) Plot the remaining floor plans in the same way, adapting as necessary if a roof plan is to be drawn.

11.4.4 Plotting the elevations

The procedure to plot the front elevation is as follows.

(1) Project the lines of the external walls and openings of the front of the building upwards from the ground floor plan, as in *Figure 11.5*.
(2) Select an appropriate location and draw a line to represent the chosen datum line for heights, cutting across the lines projected up from the ground floor plan.
(3) Plot all the measured heights on the elevation, above or below the datum line as required, then complete the elevation drawing to show all the field sketch detail.
(4) Check the elevation, correct as necessary.

The remaining elevations should be plotted by a similar combination of projecting, tracing and scaling, as practicable.

11.4.5 Plotting the sections

For preference, the sections should be plotted alongside the elevations, allowing heights to be projected from elevations to sections and reducing the amount of scaling required. The sections are basically plotted in a similar manner to the elevations.

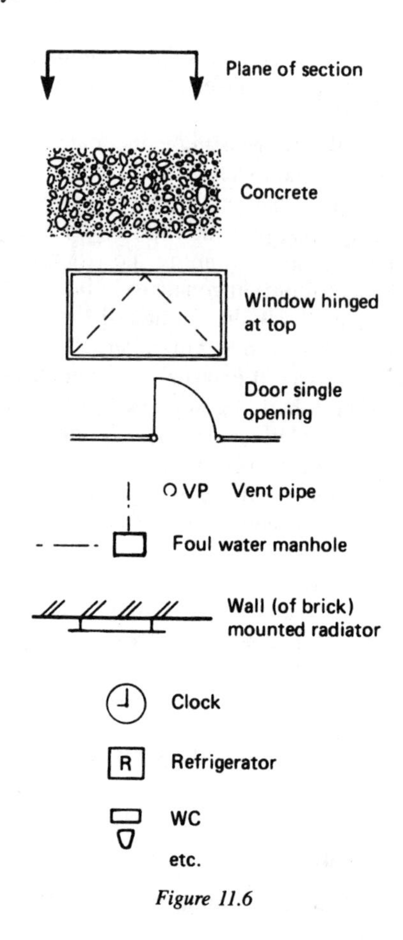

Figure 11.6

11.4.6 Completing the drawings

The section on the completion of survey plots in Chapter 10 is equally relevant to building surveys. Significant differences, however, include a probably greater use of graphical symbols, some of which are illustrated in *Figure 11.6*. There will also be a greater use of descriptive names, abbreviations, annotations and numbers, and it may be necessary to show certain horizontal dimensions and heights.

Examples

wall thicknesses,
timber sizes,
materials,
unusual structural details,
floor levels and changes of level,
ceiling heights,
direction of stair rise, number of risers, going,
window and door types,

services information – rising main, meters, etc.,
principal dimensions,
north point or points,
scale(s) of the drawing as representative fraction,
title and address of property,
information as to surveyor, when surveyed, etc.

With regard to stairs, each tread in a flight should be numbered in succession, 1, 2, 3, . . ., commencing from the lowest tread. The up direction of the stairs should be shown by an arrow and the word 'up'.

The north point should show the approximate direction of north, in order to indicate the aspect of the building. It may be placed on the site plan, or alongside the ground floor plan.

The principal dimensions, one length and one breadth, should be shown in each room of each floor plan. A simple arrowed line with a figured dimension is sufficient. The typical floor-to-ceiling height may be shown on one of the sections by a dimension line.

11.5 Building movement

Buildings and other man-made structures are subject to continual movement and when such movement becomes excessive and causes distortion of the fabric (e.g. leaning, bulging and cracking walls), a surveyor may be called in to monitor and record the movement. Movement in structures is recorded as angular or linear changes of position of two or more reference points on the structure with respect to one another and to external points or an external baseline. Measurements for this purpose are sometimes termed 'deformation measurements' and the movements determined may be mathematically converted to measurements on a three-dimensional x, y and z co-ordinate system. The following sections describe possible survey techniques.

11.5.1 Leaning walls and buildings

11.5.1.1 Methods of measuring the amount a wall is 'out of plumb', or its departure from the vertical

(i) By plumb-bob and string with a scale rule. A plumb-bob may be suspended from a projecting wall face, then, when the bob has stabilized, the horizontal distance from the string to the wall face may be measured. This is most suited to short vertical distances and is unreliable in windy conditions. On very short lengths it may be possible to use a bricklayer's spirit level, but this is difficult.

(ii) By optical plumbing techniques. This involves marking a point at high level on the wall, then marking another point on the ground, vertically below the first point. Again, the horizontal distance from the ground point to the face of the wall may be measured. The ground point is located by the vertical alignment control techniques described in Chapter 12.

(iii) By triangulation from a measured base. This entails marking survey points on the structure and establishing a survey base line near the building in such a position that the base line and the pairs of directions from its ends to each survey point form well-conditioned triangles in plan. The base length must be carefully measured, the difference in height of its two end-points determined by levelling (Chapter 5) and the base length reduced to the horizontal. If arbitrary co-ordinates are allotted to one end of the base line, and an arbitrary bearing to its direction, then three-dimensional co-ordinates may be calculated for both ends of the base line.

If, for each survey mark on the building, the horizontal angles to the mark from each end of the base line are measured, then the triangles so formed may be solved and the plan co-ordinates of the marks calculated. Measurement of the vertical angles to each mark from the base line ends, together with the calculated horizontal distances from base line ends to the marks, will allow calculation of the heights of the marks.

The horizontal and vertical angles must be measured by theodolite and angles correct to one second of arc may be required. With a one-second optical-reading theodolite this will entail perhaps ten measures of each horizontal angle, that is face left and face right on five zeros. Some of the latest electronic theodolites are eminently suited to deformation measurements.

(iv) By photogrammetric techniques. Photogrammetry is a technique for determining measurements from photographs. The equipment is expensive, but in some very old buildings or large and complex structures, it is the only practicable method. A more advanced text should be consulted for this.

11.5.1.2 Methods of monitoring movement (if any) in leaning walls and buildings

In monitoring movements in structures, it is necessary to take measurements of the types described immediately above, and to repeat these at intervals of time and compare the successive sets of results. This will usually entail setting up semi-permanent instrument stations well clear of the structure in such positions that they will not, themselves, be affected by movement of the structure.

(i) By the methods in sections 11.5.1.1.

(ii) By setting a theodolite on a fixed line of sight and observing the movement of scales fixed to the building. If horizontal and vertical graduated scales are fixed to the building and the theodolite can be directed in exactly the same line on each visit, variation in the scale readings against the theodolite cross-hairs will demonstrate movement of the structure.

(iii) By electromagnetic distance measurements. If EDM reflectors are placed on survey marks on the structure and a base line established, as in (iii) above, then a suitable total station instrument may be used to obtain high accuracy measurements of the distances from the base line terminals to the survey marks and the three-dimensional co-ordinates computed by trilateration methods. Dam deformation measurements are often made in this way, although they may also be made by the triangulation methods described above.

11.5.2 Bulging walls

Most of the methods of Section 11.5.1 are appropriate, but for short distances (e.g. measurements between wall faces) catenary taping is a possibility. Catenary taping may be useful inside buildings, trenches, tunnels and vaults.

11.5.3 Cracks in walls

To check on the movement at cracks, reference 'tell tale' marks may be placed on either side of the crack and changes in their relative positions over a period of time can indicate linear or linear and angular movement. The marks may be centre punched rivets or screws, with distances between them measured by a rule or vernier callipers. (At least three marks are needed to check on angular movements.)

The 'Avongard Tell-tale', a proprietary device, consists of two overlapping rectangular plastic plates with suitable scale markings, and when fixed over the crack each plate moves independently of the other.

11.5.4 Settlement and subsidence

Small vertical movements may be monitored by some of the methods detailed above. Large vertical movements, as in mining subsidence, may be measured by levelling from an area known to be stable, at suitable intervals of time. Large movements such as the movement of cliff faces may be monitored using a network of traverse stations at accessible points at the top, on the face and at the bottom of the cliff. Such traverses are best measured using EDM equipment, the measurements and computing of the three-dimensional co-ordinates of the stations being repeated at intervals.

Chapter 12

Setting out

12.1 Introduction

Setting out is the name used for all the operations required for the correct positioning of works on or adjacent to the ground, together with their three-dimensional control during construction. Setting out operations are the reverse of normal survey, since the dimensions are known but the points have to be located on the ground or on the structure.

Setting out for traditional building work does not require high accuracy, but with the development of precisely dimensioned frames and components it is frequently found that insufficient attention has been paid to the initial site dimensioning and the consequent poor fit of beams, panels, etc., may be extremely expensive.

Although the setting out of every job is different, certain jobs are standard and standard solutions are available for many tasks. The various requirements in setting out are considered here under four main headings – plan control, height control, vertical alignment control and excavation control. The headings are self-explanatory, but excavation control combines all the others in a specialized aspect.

Setting out operations concerned with *curved lines* are considered separately under the heading *Curve ranging* in Chapter 13.

12.1.1 Setting out documents and data

The documents and data available to the setting out surveyor or engineer may vary from a simple verbal communication to a most complex array of plans and specifications. Setting out plans or drawings may include *block plans*, *site plans*, *location drawings* and *detail drawings* and between them they should identify the site, relate new work to any existing and include the fundamental setting out data, i.e. the positions to be occupied by the various spaces and elements of the buildings or works and their dimensions.

In addition there may be a specification, schedules, survey data, possibly computer printout and mathematical tables. A *specification* describes material and workmanship, but it may contain information relevant to the setting out, such as the depth of bed for drain lines. A *schedule* tabulates information on numerous and repetitive items and again there may be relevant data such as the heights and invert levels of manholes. The *survey data* relates to the original site survey and may include information on traverse stations, bench marks, etc. *Computer printouts* may be included with the survey data and are often used where the position of detail for setting out is defined by two- or three-dimensional rectangular coordinates. Computer printouts are tending to replace the use of specialist mathematical tables for the setting out of high-speed rail and motor roads. See the Bibliography for more advanced texts on these matters.

12.1.2 The essentials of efficient setting out

These may be considered to be *accuracy, timeliness* and *clarity.* Lack of *accuracy* can be costly in time and money, since time must be wasted in rectifying mistakes. All measuring equipment should be carefully checked before the start of the particular task.

Timeliness involves not merely setting out to programme, but having the setting out completed early and in advance. If there is time in hand, then it will be possible to cope with emergencies such as vandalized marks, unforeseen changes in the construction programme, etc. If the setting out is early, then the surveyor need not work in haste or under undue pressure and both of these tend to lead to undetected mistakes.

Clarity means that all the personnel involved should be able to understand the surveyor's setting out and the meaning of all pegs and other marks defining the location and height of the new works. Colour coding of pegs and marks will assist in this aim. Again, although setting out is the reverse of surveying, it requires a skilled application of survey techniques, adherence to the principles of survey covered in Chapter 1, and a sound grasp of geometry.

12.1.3 Stages in setting out

12.1.3.1 Document inspection

The relevant documents as identified in Section 12.1.1 must be examined and checked for any information relating to the setting out, including any existing site survey data and all must be checked for adequacy and accuracy. If no setting out drawings are provided the surveyor must make up his own from the detailed drawings of the works. The accuracies required must be specified, since excessively high standards of accuracy are expensive and wasteful.

12.1.3.2 Site inspection

The nature of the site must be examined and the existence of obstacles to measurement and ranging noted, bearing in mind the methods and equipment to be used. The nature of the ground will affect the type of markers to be used and also the reliability of existing survey marks. Existing control must be verified, including bench marks, and any changes noted.

12.1.3.3 Proving the site drawings

The size and shape of existing detail on the site must be checked against the drawings and it must be verified that the new spaces and elements will fit on to the site in the manner shown on the drawings. As examples, the sum of individual measurements along a line must be compared with the overall length of the line, gradients must be checked against the actual levels shown, temporary bench marks should be checked by re-levelling from the OS bench marks, etc.

12.1.3.4 Preparing setting out data

Any calculations such as bearings and distances, co-ordinate transformations, curve computations, etc., should, as far as possible, be carried out before commencing the task on site.

12.1.3.5 Site setting out procedure

Assuming the use of traditional methods, with distances set out by tape and angles by theodolite, the detailed *plan* of the new works is set out first. When this is complete, suitable *check measurements* must be made to *prove the setting out*, according to the circumstances. For example, a rectangular plan may be checked by calculating the lengths of diagonals then measuring them and comparing the actual and theoretical distances. When the plan is correct *heights* may be fixed by ordinary levelling from the verified temporary bench marks.

EDM equipment is increasingly being used today in setting out by polar radiation, with each point being fixed in all three dimensions, using the angular bearing, horizontal distance and vertical height difference. This must be checked, but gross errors in line and level may be checked by eye, short check measurements may be taped and heights can be checked by levelling. Another form of check is to set out the same point twice, from two different instrument positions.

12.1.3.6 Setting out profiles, etc.

If pegs or other markers are set out to define the lines of the faces of new buildings, or on the centre-lines of roads, etc., such marks will be destroyed when the construction work commences. Accordingly, it is customary to establish a *reference frame* of accurately located lines outside the actual construction and clear of all obstructions, then the new works are located by measurement from the reference frame as required.

Typically, in a small building, this is arranged by first placing pegs to accurately mark the corners and lines of the walls (pegs A, B, C and D in *Figure 12.1*), then setting up *profile boards* opposite the ends of each wall. The exact

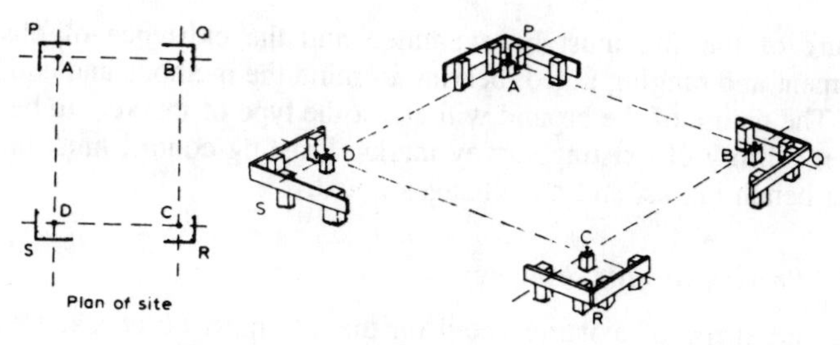

Figure 12.1

alignments of the walls are marked on the top of the profile boards, either by saw cuts or fine nails, preferably by theodolite projection from the fine marks on pegs A, B, C and D. When the pegs have been removed during the later construction work the wall lines and intersections may be relocated by running builder's string lines between the fine marks on the profiles. The tops of profile boards are usually set about 300–450 mm above ground level, and related to a specified height such as finished floor level.

On large buildings and similar structures, the reference frame may consist of two lines at right angles to one another, set out by theodolite. A better form, however, is a frame of four lines located by theodolite, outside and parallel to the proposed structure. Pegs may be placed on these lines to mark, for example, the *centre-lines* or *faces* of rows or lines of stanchions, columns, or walls to be erected.

Offset pegs are used to indicate the lines and heights of kerbs, drains, sewers and so on, and may be used to reference road centre-line pegs. On occasion offset points may be set out directly by the surveyor or engineer, but sometimes they may be set out by site operatives, from the data supplied by the surveyor.

Marker pegs in building works are generally 50 mm square wooden pegs, length according to the purpose of the marker. If a fine point is to be set out in plan, such as an instrument station or the intersection of lines, a small nail can be placed to project 10 mm vertically from the top of the peg. This acts as a fine alignment mark and also as a fine mark for setting the theodolite or EDM instrument over the point. For more permanent marks a 13 mm steel reinforcing rod may be driven into the ground and surrounded by concrete, the mark identifier being scratched in the 'green' concrete, and the exact plan point centre-punched into the steel. For very long-term markers, a 1 m steel angle iron or similar may be driven into the ground

12.2 Plan control

Plan control covers the setting out of detail in two dimensions in plan. Any pair of points may be related then either by a bearing and a distance (polar co-ordinates or radiation) or by a difference in eastings and a difference in northings (rectangular co-ordinate methods). Plan control may be supplied at any height, either at ground level, at the bottom of an excavation, or at intermediate floor levels, as required.

12.2.1 Methods available

New detail may be located from

existing detail,
existing control,
new control tied to the existing control,
a grid superimposed over the site, or
a combination of any of the above.

Where a *grid* is used, it may be the grid of the existing control and survey, i.e. a *survey grid*, or it may be a special site grid. A *site grid* may exist on the site drawings, that is pre-contract planned and this may or may not be the survey grid. Alternatively the site grid may be positioned by the setting out surveyor or engineer on the drawings to coincide with the orientation of the major plan units shown.

Whatever the new site detail is set out from, the setting out will be the basic survey methods of supplying detail, which are rectangular offsets, radiation, intersection, or a combination of these. The method to be used should be that which will be the most economical in time and money, provided that the required accuracy can be achieved.

The traditional approach has been by *rectangular offsets* since this is most commonly the form in which the information is presented. On small sites the offsets are typically from a framework of lines tied to survey control (if any) or to existing detail.

On larger sites, the framework may be a *grid superimposed over the site*. In its simplest form this can be just the four corners of a rectangle, but it may be a series of grid points around the perimeter of the site from which other points and ordinates may be located. Detail is located either by intersection of ordinates, or by short offsets from them, or by bearing and distance (polars) using short-range EDM equipment.

The methods to be used will depend upon the equipment and labour available, the skills of the operators, the survey data provided, the size and nature of the job, the topography of the area, the method of presentation of the information and, of course, the accuracy of the site plan.

12.2.2 Equipment

The equipment used for setting out may include all the measuring and note-taking equipment mentioned earlier in the text. Ground marking materials are needed, particularly timber pegs, stakes and battens, nails, steel reinforcing rods, cement and sand, chalk, crayons, paint and tools such as hammers, hand saws, spades, cold chisels, centre punches, etc. A builder's line (propylene twine), plumb-bobs and profiles or profile boards (see *Figure 12.1*) may also be needed.

12.2.3 Common setting out tasks

From the surveying point of view, the actual operations involved in plan setting out reduce to

setting out a horizontal angle,
lining-in a point between two existing marks,
prolonging a straight line,
setting out an exact horizontal distance, and
setting out a curved line in plan.

12.2.3.1 Setting out a horizontal angle

To set out an angle using an optical theodolite, the method is essentially an application of the *simple reversal* technique used to measure horizontal angles. The procedure is as follows

(i) Set the theodolite over the point on the line at which the angle is to be raised, centring and levelling up as usual.
(ii) *On face left*, sight on to a target placed on the far end of the line (the *reference object*) and read and book the horizontal circle reading. Any specified zero may be set on the circle before sighting the RO, but generally a zero of 00° 00' 00" is used.
(iii) Release the upper or horizontal plate and telescope clamps and turn the instrument through the exact desired angle, directing an assistant to place an arrow or other mark at the required distance on the alignment defined by the telescope central vertical cross-hair.
(iv) Repeat the whole operation on *face right* using a new zero in a different quadrant of the horizontal circle, say 180° 00' 00".
(v) If the two marks do not coincide, the mean position between the two indicates the required alignment, provided that both have been placed with equal care. A large difference indicates a mistake, of course.

Using these methods, BS5606: 1990 *Guide to Accuracy in Building*, Table 14.1 quotes a deviation range of ±5 mm in a distance of 50 m using a 20" glass arc theodolite centred by optical plummet, and ±2 mm in 80 m using a 1" theodolite.

Right angles may be set out using a steel tape to construct a 3:4:5 triangle and a large wooden right-angled triangle frame is often used on small building jobs. The optical square may be used for rough preliminary work, but it is not suitable for setting out an accurate right angle.

12.2.3.2 Lining-in between two existing marks

If high accuracy is not required, perhaps on rough preliminary location work, then the methods described under linear survey may be used with points lined-in by eye. For normal accuracy setting out, a theodolite is used for lining-in. Depending upon the intervisibility of the end points and also on whether it is possible to set the theodolite up over an end point, three distinct cases may arise.

Case 1 is where the end points are intervisible (visible one from another) and the theodolite can be set up over one end.
Case 2 is where the end points are not intervisible but both are visible from some intermediate point, or alternatively where the end points are intervisible but the theodolite cannot be set up over an end point.
Case 3 arises where there is no intermediate point at which both of the end points are visible.

Case 1 solution: Set up the theodolite over one end point, sight on the other end point, direct an assistant to place a mark at the required intermediate position, and when the mark has been placed then check again.

Case 2 solution: The procedure is as follows, shown in *Figure 12.2*:

(i) Use the linear survey techniques of 'lifting a line' to locate the approximate position of an intermediate point on the line such that both ends of the line are visible from the intermediate point.

(ii) Set up the theodolite over the approximate point, levelling and centring as usual.

(iii) On *face left*, sight on to the more distant of the end points and bisect the target carefully, then *transit* the telescope and direct an assistant to place an arrow exactly on the line of collimation and in the vicinity of the nearer end point.

(iv) Keeping the instrument on *face right*, turn the telescope, sight on the more distant point again, bisect the target carefully, then transit the telescope and place another arrow on the collimation line and close to the nearer end point.

(v) The mean position of the two arrows now defines the alignment from the far end point through the instrument station. Measure the offset from this alignment to the near end point.

(vi) Calculate the proportional distance the instrument must be moved in order to bring it on to the correct alignment joining the end points.

(vii) Move the instrument the required distance, repeat the observations, and repeat as necessary until the instrument is on the correct alignment. In the early stages it may be necessary to move the tripod, but later it will be possible to simply slide the theodolite over the tripod head into its correct position.

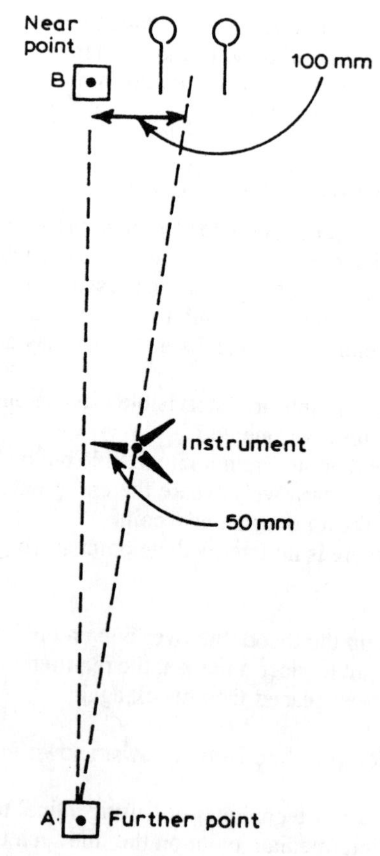

Figure 12.2

Case 3 solution: If the end points are not intervisible and there is no intermediate point from which both ends can be viewed, as for example in a wooded area, then a possible solution may be to run a traverse between the end points but as close to the alignment as possible. If the traverse is co-ordinated, then the co-ordinates of the required intermediate point may be computed and the point set out by bearing and distance from the nearest traverse station.

12.2.3.3 Prolonging a straight line

In site setting out it is normally necessary to use a theodolite for this task. Assuming a line AB is to be extended to the point C, two cases arise.

Case 1 arises where the theodolite can be set up on A and the position of C is visible. All that is necessary is to set up the instrument on A, point towards B, bisect the target carefully, finishing with the horizontal clamps applied, then re-focus and raise or lower the telescope and direct an assistant to place a mark on the collimation line at C.

Case 2 is where the theodolite cannot be set over A, or the position of C will not be visible from A. In this case, set up the instrument over B, sight the target at A on *face left*, transit the telescope and direct the assistant to place a fine mark exactly on the collimation line in the vicinity of C. Repeat the process on *face right*, then in the absence of gross error the mean of the two marks is on the required line.

12.2.3.4 Setting out an exact horizontal distance

In setting out an exact horizontal distance, the alignment is fixed by theodolite and the required distance is then fixed by steel taping with a spring balance to control tension, or by EDM. When using a steel tape, two cases arise, *Case 1* where the distance is less than one tape length and *Case 2* where the distance exceeds one tape length.

Case 1 solution: Set the theodolite over the start point, place a peg on the required alignment using the methods described above, then measure the desired distance along the line to the peg and move the peg as necessary. The peg alignment position and the distance must be checked alternately, as needed, until the peg is at the correct distance and on the true alignment. The most suitable marker for this would be the wooden peg and nail as described earlier.

Case 2 solution: Set the theodolite over the start point, place a temporary peg accurately on the desired alignment and within 3 or 4 m of the required distance (approximate taping only). Now measure the distance to the temporary peg carefully and calculate the additional distance required. Move the theodolite and set it up over the temporary peg, then set out the additional distance to the final peg in the same way as shown for Case 1 above. Finally carry out a check measurement of the distance from the start point to the final peg to ensure that the peg is at the required position.

For steel tape work, BS5606 quotes the following possible deviations:

Ordinary work, carbon steel 30 m tape, sag and slope corrected:
± 10 mm at distances up to 25 m;

± 15 mm at distances exceeding 25 m.
Precision tape, 30 m carbon steel, tension/slope/sag/temperature corrected:
± 3 mm up to 10 m, ± 6 mm up to 30 m.

Where distances are to be set out by simple EDM equipment, a similar approach may be used. Thus the reflector is placed on line at the approximate distance, the distance to the reflector determined accurately and the added distance to be set out is calculated and fixed by a short taped measurement from the reflector position. A peg may be located at this new position, then the reflector placed at the peg and the distance determined as a check.

Some EDM instruments have a *stake out mode*, where the reflector is placed on the required line and a specified distance is entered into the instrument. When set in this mode, the instrument will find the difference between the prism position and the specified distance. The prism may be moved after each of successive measurements until the difference is zero and the prism is at the specified distance.

12.2.3.5 Setting out a curved line in plan

The most common form of horizontal curve in practice is an *arc of a circle*, termed a *circular curve*. A *simple circular curve* is an arc of constant radius, joining two tangents. A *compound circular curve* consists of two simple curves, of different radii, meeting on a common tangent and having their centres lying on the same side of the common tangent. A *reverse circular curve* is similar to a compound curve, but the two simple curves have their centres on opposite sides of the common tangent.

Other curve forms may be encountered occasionally, such as the *ellipse* or the *parabola* and these, together with short radius simple curves, may be marked on the ground using chalk, crayon, scratch marks, etc., the curves being defined by templates or by locating centres or focii and swinging arcs of a length of tape or cord. The simplest example is a simple circular arc which may be fixed by locating the centre of the circle, placing a peg there, then swinging the desired radius from the centre. Similar techniques may be used for ellipses, etc., and arcs may also be marked by driving pegs at intervals along their length.

Roads, railways and canals have plan alignments consisting generally of straight lines connected by very large radius curves and the methods mentioned above are impractical. Setting out such curves is termed *curve ranging*, and the methods used are treated separately in Chapter 13.

12.2.4 Use of rectangular and polar co-ordinates

Rectangular co-ordinates and *polar co-ordinates*, both introduced in Chapter 1, are very widely used in plan setting out works today. All the methods previously mentioned are relevant, including rectangular co-ordinates from bearing and distance (P → R), bearing and distance from rectangular co-ordinates (R → P), co-ordinate grid transformation and all the control link methods such as resection, etc.

Figure 12.3 shows the data for setting out three co-ordinated points (*TBM35*, *A12* and *A16*) by polars from an EDM instrument station *BaseA1*. The rectangular co-ordinates of all the points are listed, together with an RO, point

Setting out Station and Point Detail

Station 1 (BaseA1)		Co-ordinates	Bearings
	Stn.	458001.387 303599.786	
	R.O.	459000.000 304500.000	47 57 59 (A1RO1)
	Check	460000.000 305000.000	54 59 07 (A1RO2)

Points set out from Station 1

	E N	H.Dist.	Horizontal Angles from R.O.	Horizontal Angles from check R.O.
Point 1 (TBM 35)	458060.000 303603.445	58.727	38 27 41	31 26 34
Point 2 (A12)	458103.546 303650.662	114.126	15 33 35	8 32 28
Point 3 (A16)	458177.568 303667.234	188.650	21 05 06	14 03 59

Figure 12.3 Co-ordinate data for setting out

A1RO1, and a check RO, *A1RO2*. The horizontal angles and horizontal distances to set out each of these three points from *BaseA1* are shown, together with the angles from the check RO. The method of calculating polars and vice versa are covered in Section 1.11.

The use of these methods will be assumed where relevant and they will not be treated further here.

12.3 Height control

The control of *heights* or *levels* in construction work is generally a simple matter. Accurate temporary bench marks should be fixed on site at points where they are unlikely to be disturbed, fenced off if necessary. The TBMs should be levelled from national or municipal bench marks such as OBMs and a careful record kept of their reduced levels. Thereafter the height of any part of the site or the construction may be determined by reference to the site TBMs, using the surveyor's level and staff, or EDM equipment with trigonometric heighting, or a rotating laser giving a reference plane, or specialist site operative's equipment such as the water level or the site operator's automatic level.

When using the surveyor's level it may occasionally be necessary to fix heights above the level of the collimation line and in this case the staff may be used *inverted*, the readings then being *negative* values. This is detailed in Chapter 8.

Where the point to be heighted is more than a staff length above the collimation line, then a white steel tape may be pulled taut downwards from the point, the tape being read in the same way as a staff. An alternative method is *trigonometric heighting*, but this is not generally as accurate as levelling, although it may be adequate for some tasks.

12.3.1 Accuracy of levelling in construction works

If permissible deviations or accuracies are not specified for a task then BS 5606: 1990 *Accuracy in building* or BS 7334 : Part 3 : 1990 *Methods for determining accuracy in use: Optical levelling instruments*, may provide guidance. As an example, these indicate that when check levelling between a given TBM and an OBM they should agree within 10 mm, while in checking points on steel structures the height differences should agree within 2 mm.

12.3.2 Equipment

All the heighting equipment previously referred to may be used in site levelling or height control, including surveyor's levels, EDM equipment, lasers and all their ancillary equipment, together with pegs, profiles, etc.

Lasers, not previously described, have become established as efficient setting out instruments for both height and alignment control. Laser is an acronym for light amplification by stimulated emission of radiation and a laser projects a narrow beam of light, providing an alternative to the visible builder's line or the line of sight through an optical instrument. The line of the beam, depending upon the application, may be horizontal, vertical, or at any desired gradient. In some instruments, the beam of light may be rotated at speed, thus establishing a visible horizontal or vertical reference plane.

Figure 12.4 illustrates such an instrument, the Leica Wild LNA20 Automatic Laser Level. The beam of the built-in laser diode is passed through a rotating deflector prism, thus sweeping out an invisible plane which can be located with the hand-held LPD20 laser detector. A pendulum compensator automatically levels up the laser beam plane, and a vertical adapter allows the beam to sweep in a vertical plane for checking verticals. The detector can be set for two different degrees of sensitivity, ±0.8 mm for indoor and short-range work and ±2.4 mm for external and long-range work.

Figure 12.4

Other laser levels can be used to control mechanical plant in excavating over an area to a required level, with the detector fixed to the blade of the bulldozer or other plant. A bright display tells the driver if he is operating at the correct height.

Laser eyepieces may be fitted to theodolites and levels to project a visible collimation line, used for levelling and for alignment generally, such as in tunnel work. Pipe laying lasers provide a visible red beam which can be used to control the horizontal and vertical alignment of drainage pipes.

Like any other bright light, lasers can cause permanent damage to the eyes and the light source and any reflection or refraction of it should not be stared at or viewed through a theodolite or level. Wherever possible, lasers should be set up well above or below eye height. The precautions to be observed depend upon the class of laser instrument in use.

12.3.3 Common setting out tasks

The tasks involved in height control may be reduced to establishing a point or mark at some given height, establishing two or more such points to define a horizontal or inclined line or plane and establishing a series of points in a curve in a vertical plane, this latter then being known as a *vertical curve*. Vertical curves are considered separately in Chapter 13.

12.3.3.1 Establishing a point at a specified height

The plan location of the required peg or mark must be fixed first and in this case, it will be assumed that a peg is to be placed as a height marker. A peg must be driven at the required position, ensuring that the top of the peg is *above* the required height, then the level of the top of the peg determined by careful levelling from a TBM. When the peg top level is known, the difference between peg level and required level may be scaled down the side of the peg and a horizontal mark drawn across the peg to indicate the desired height reference. As a check, the levelling staff zero should be placed against the drawn line and the height of the mark determined again by levelling and calculating its reduced level.

12.3.3.2 Establishing horizontal or sloping lines or planes

These tasks are carried out by placing appropriate pegs as described in the previous Section 12.3.3.1, according to the demands of the specific task.

12.4 Vertical alignment control

The control of vertical alignment in construction works has two distinct aspects. These are first, the vertical transfer of control points or lines to higher or lower levels and second, the provision of vertical control lines and the checking of verticality of construction elements.

12.4.1 Vertical transfers of control points or lines

The methods for establishing a reference frame at ground floor level have been outlined already. For the plan control of higher floors, or basements, etc., it is necessary either to reconstruct the original reference frame at the new levels or to transfer fixed points of the ground frame to the new levels and use these to construct new reference frames.

As usual, the accuracy required for vertical position transfer depends upon the particular job specification. For tall buildings, a typical requirement is that the error in position of the reference lines at tenth floor level should not exceed ±2.5 mm.

The plumb-bob and string is widely used on small buildings and within individual stories of large buildings, but precision is difficult for large height differences. BS5606 suggests a possible deviation of ± 5 mm in 10 m for a 3 kg plumb-bob immersed in oil, in still conditions.

Vertical transfer is often carried out by *two-theodolite intersection*. In this method, a ground point is observed from two directions in plan (preferably at right angles) then the point may be re-located at high level by elevating the telescopes and sight lines of the two theodolites and locating the intersection of the collimation lines. This is similar in principle to the method of prolonging a straight line. If only one theodolite is available it will be considerably slower and more difficult.

The principal method of vertical transfer for the control of large buildings is *optical plumbing*, where an instrument sight line is directed vertically upwards from ground control points, either using a specialist instrument or a theodolite with special attachments. Variations of this method make use of the laser to provide a visible vertical line.

12.4.2 Vertical transfer by optical plumbing

Although there are certain tasks for which the suspended wire plummet is the best method of vertical position transfer, optical plumbing is generally the most effective method in the construction of tall buildings.

The normal technique is to locate control points at ground level then set an instrument exactly over these and plumb upwards to locate new reference points on each floor in turn. On each floor, the new points are used to set out reference lines for the positioning of construction elements on that floor. The points are located at high level by using suitable aiming targets according to the particular circumstances. For plumbing in open-frame structures, or outside buildings, aiming targets on offset brackets may be used. When plumbing inside buildings which have the floors laid or constructed as the building rises, four or more (as necessary) holes, about 200 mm square, are left open in each floor in suitable positions near the corners of the floor plan. When a floor is formed, aiming targets or graduated scales are placed over each hole and accurately positioned for the floor reference frame setting out. When work is complete on the floor, the targets are removed to allow clear sighting from the instrument at ground level up to the targets on the next floor.

Although some instruments have ranges of 100 or 200 m, the practical limit for optical plumbing is often taken as about 10 stories or about 30 m. All floors up to the tenth are, therefore, fixed by sighting from ground level, then the instrument is set up again at the tenth floor level and sights are made up to the twentieth floor and so on.

Targets for sighting are often best designed for the individual job, depending upon the instrument positions, the structure, the sight length and the instrument to be used. A target may be as simple as a board with the plumb-point pencilled on the underside, then a nail driven through the mark to provide an instrument station on the upper surface, or alternatively graduated transparent sheets or scale rules.

With all optical plumbing instruments, the vertical observation should be made from four distinct orientations of the instrument in plan, at 90° to one another, then the mean position of the four target points defined should be free from instrument error.

12.4.3 Optical plumbing instruments

Optical plumbing instruments all rely on either a bubble tube or an automatic levelling compensator, then the horizontal sight line is deviated through 90° by a prism to provide a vertical sight line.

12.4.3.1 Theodolite telescope roof plummet

This is a small horizontal telescope with a 90° prism built in. It is attached to the top of a theodolite telescope and when the main telescope is exactly horizontal it provides a vertical line of sight. The particular application of this attachment is in locating a theodolite under an overhead mark, or in fixing an overhead mark above a theodolite station. The Wild version, used on T1 and T2 theodolites, has a range of 10 m, the accuracy claimed is 1:5000.

12.4.3.2 Small independent optical plummets

These consist of a small horizontal telescope fitted with a pentaprism to provide a vertical sight line, supported by a theodolite tribrach and tripod. Levelling up is by reference to bubble tubes. Relatively short range, accuracy possibly 1:10 000.

12.4.3.3 Theodolite with diagonal eyepieces

Most instrument manufacturers produce *long diagonal eyepieces* for their theodolites and when these are fitted it is possible to observe objects at large angles of elevation (up to the zenith) and also to read the circles. Provided the theodolite is in good adjustment, then when the vertical circle is set to 0° (or 90°, as appropriate) the sight line is directed to the zenith tracing a vertical line through the instrument centre. The theodolite's own optical plummet should be used for centring over the ground mark, in the usual way.

BS5606 suggests that an optical theodolite used in this way (and remembering to take four sets of observations at 90° in plan) will give a possible deviation of ±5 mm at a height of 30 m.

12.4.3.4 Objective pentaprism

This is a prism which is attached to the objective end of an automatic level's telescope, deflecting the line of sight through 90°. Wild claim that it converts their NA2 automatic level into a high precision optical plummet for upward or downward plumbing, with an accuracy of 1:100 000. The pentaprism may also be used with a Wild theodolite, when 1:70 000 is claimed.

Note that with this instrument, it must again be used in four directions in plan and opposite pairs of indicated points must be connected so that the intersection of the lines defines the plumb point. (When a pentaprism is fitted on a theodolite telescope, the telescope must be sighted horizontally for plumbing and thus with both instruments the prism provides a vertical line which is offset from the plan centre of the instrument.)

12.4.3.5 Independent precision plummets

These are similar to the instrument described in Section 12.4.3.2, but with a high power telescope and a built-in pentaprism. Verticality may be controlled either by bubble tubes or an automatic compensator unit like an automatic level. Some versions have two telescopes, sighting, respectively, up and down, others a single telescope sighting one way only. Manufacturers claim accuracies between 1:50 000 and 1:200 000, with ranges of 100 to 200 m. For an automatic precision plummet BS5606 suggests a possible deviation of ±5 mm at a range of 100 m.

12.4.4 Vertical transfer by laser

Alignment lasers which project a *vertical reference line* may be used for plumbing, accuracy being checked by rotating the laser instrument about its vertical axis in the same way as for optical plummets.

A laser beam projected through a precise optical plummet was used to control the verticality in setting out Australia's tallest structure, the 300 m high Centrepoint Tower in Sydney, primarily a 250 m high 7 m diameter steel shaft.

Laser attachments may be obtained for theodolites, these using a beam splitter so that the laser beam is projected along the collimation line while the observer can sight along the collimation line through the eyepiece in the usual way. Such devices may be used to sight at any inclination.

12.4.5 Vertical control lines and checking verticals

12.4.5.1 Vertical control lines

These lines are often required so that walls, columns, lift guide rails, etc. may be erected properly vertical. Traditionally the best example is the *suspended plumb-bob*, providing a visible, physical line from which craftsmen may measure offset distances themselves as required. When the plumb-bob has settled and stopped swinging and the line is correctly positioned, the lower end of the wire or string may be made fast and tightened by turnbuckle. Today the plummet is being replaced by the laser, either an *alignment laser* projecting a vertical line or a *rotating laser* in a *vertical plane* being used to identify the position of objects such as internal walls.

12.4.5.2 Checking verticals

The checking of verticality of columns, walls, stanchions, etc., may be carried out with the plumb-bob and string as in the previous section.

The theodolite or a rotating laser set in a vertical plane may be used in steel erection to check stanchions for both plumb and alignment. The method is to set out a line parallel to the centre-line of the stanchions and offset some distance from it and set up the instrument on the line but well back from the nearest stanchion. If the instrument is aligned accurately along the offset line, any point on any stanchion may be checked by an offset measurement from the collimation plane or the rotating beam plane. To check plumb, the procedure must be repeated with other offset lines at 90° in plan.

Optical plummets and alignment lasers may also be used for checking verticals.

12.5 Excavation control

The typical tasks in excavation or earthworks are (1) excavation or fill over an area, to a new level, (2) excavation for cutting and filling embankments for roads, etc., and (3) excavation for trenches and pipelaying. The general control for heights is provided by the usual site bench marks, located well away from areas of activity and plant movement.

12.5.1 Area excavation or fill

Traditionally, height control pegs should be placed over the area concerned, their tops carefully levelled from site benchmarks. To *fill* over an area, pegs are best driven until their heads are at the required new level. Where this is impracticable, pegs should be marked with the new level, or the distance to be measured vertically to reach the new level. (See *Figure 12.5* and note that in *(d)* the top of the peg is 0.5 m above the new level required – ambiguity must be avoided.)

For *excavating* over an area, pegs may be placed similarly and marked to show how high the peg top is above the required level, see *Figure 12.5(e)*.

To provide site operative control of earthworks, *sight rails* or *profiles* may be set up as in *Figure 12.5(c)* or *(f)*. The sight rails should be constructed at a convenient viewing height of 1 to 1.5 m above existing ground level, such that a line of sight between the tops of any pair is parallel to the required level and preferably at some multiple of 0.1 m.

Site operatives use these rails in conjunction with a *traveller* or *boning rod* to control excavation or fill by sighting from one sight rail to the next. When the traveller rests on the surface of the earth and its top is in line with the line of sight between the rails, the ground is at the correct level, see *Figure 12.5(g)*.

Gradients may be established in the same way, with the sight rails set so as to be parallel to the plane of the slope and at some convenient height above it, *Figure 12.5(h)*. *Double sight rails* may be required on steep slopes, *Figure 12.5(j)*.

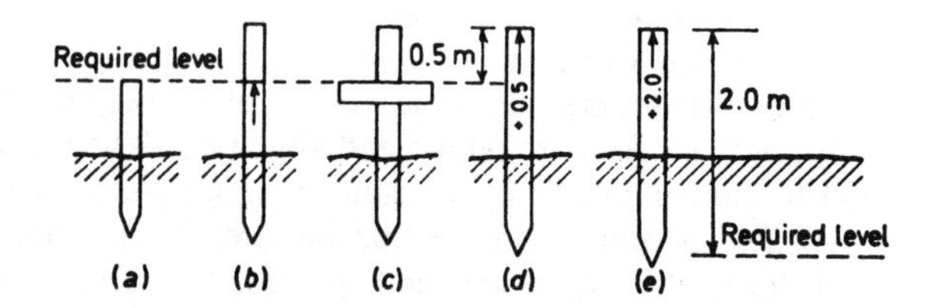

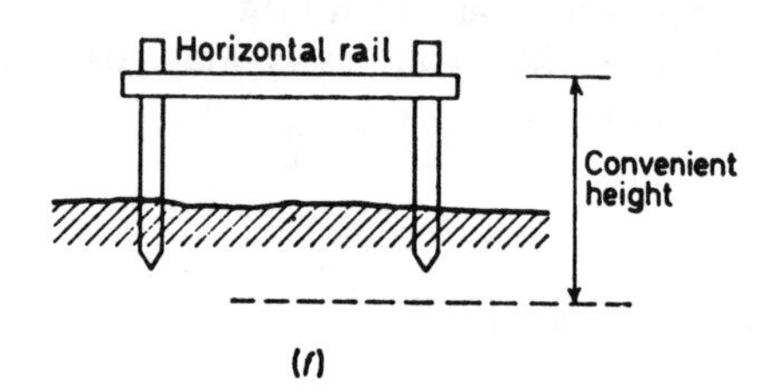

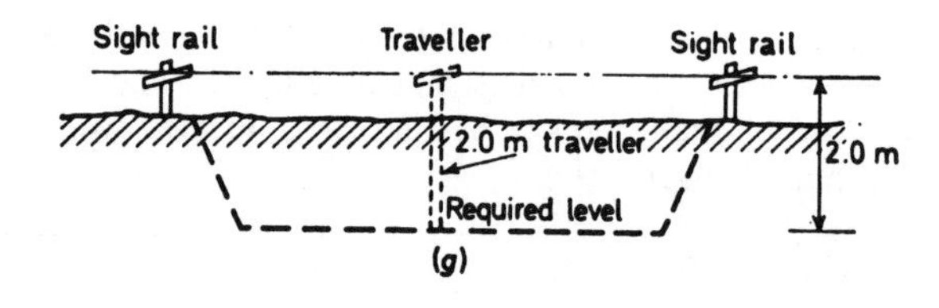

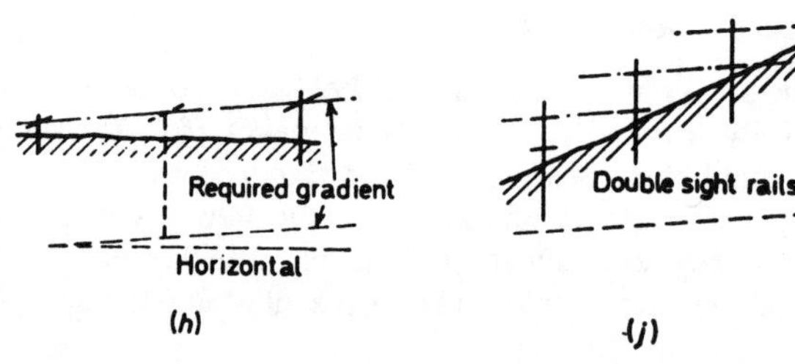

Figure 12.5

Earthworks are carried out to the levels required between lines of pegs or sight rails, leaving these standing on mounds and finally the mounds are cut away to level up the remainder.

An alternative approach is to use a rotating laser and to dispense with sight rails and travellers, although markers are necessary to indicate the extent of the area to be regraded. The laser is set up clear of the working area, at such a position and height that the beam will not be at the eye level of the site operatives and others adjacent to the site and such that the rotating beam is at some convenient constant height above the finished plane surface. *Sensor units* may be attached to the earth moving vehicles at the appropriate height, with the detectors outside and the receivers inside the cab.

12.5.2 Cut and fill for roads, etc.

These are essentially controlled for height in the same way as general areas, the difference lying in the formation of the side slopes. The position of the limits of side slopes is indicated by *slope stakes* or *batter pegs*. These are often offset at a standard distance beyond the outer limit of the slope, for example one metre, since if placed on the exact alignment they may be disturbed by the site operatives. The position of these pegs is obtained either by scaling off from the working drawings or by a combination of scaling and calculation to obtain compatibility between the distance and the height difference from the centre-line to the outer limit of the side slope, maintaining the correct side slope gradient. Again, sight rails, also known as *batter boards*, are used to control the angle of slope and may be constructed at a convenient height above the slope, or on the slope, or often on a prolongation of the line of slope, see *Figure 12.6*.

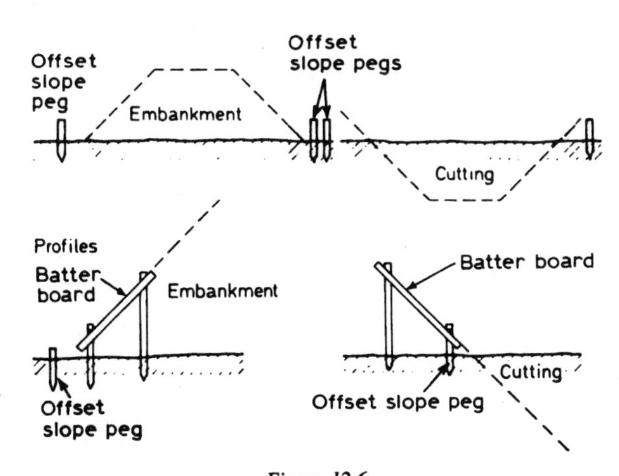

Figure 12.6

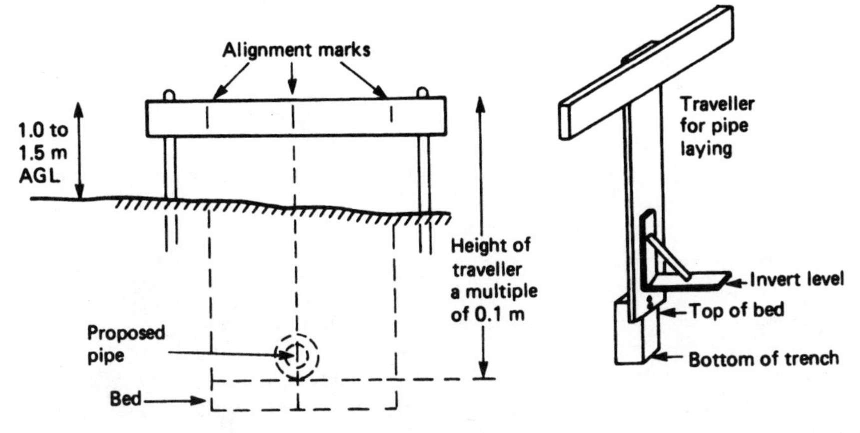

Figure 12.7

12.5.3 Trench excavation

The traditional method of control is by an application of *sight rails* and *traveller*. In this case, either a *two-post sight rail* is set across the trench line or, if mechanical excavation is to be used, a *T-shaped sight rail* is set to one side of the trench. The alignment of the trench and the pipe centre is controlled either by marks on the two-post sight rails or by offset pegs.

The *traveller* may be used to set out and check the following – the depth of trench, the thickness of the bed upon which the pipe is to rest and the level of the inside bottom surface of the pipe, known as the invert level. This is achieved by having the traveller made to be correct at the top of the bed, an extension piece being added for the initial excavation, and an angle bracket fitted for the invert level, see *Figure 12.7*.

An alternative approach is to use an alignment laser to control both the formation of the trench and the laying of the pipe. Initially the sensor may be in the mechanical excavator and, later, the beam may be directed down the centre of the pipe as it is laid.

Chapter 13

Curve ranging

13.1 Introduction

Curve ranging will be considered here on the assumption that it is aimed at setting out roads, but the same principles are, of course, applied in railways and the setting out of large radius curves generally. Although the traditional method for constructing roads is to lay out straight lines in plan then connect them together with *circular* or *transition/circular/transition curves*, it should be noted that the road alignment may be treated as a continuous curve (a *cubic spline*) which may be passed through specified points and rectangular co-ordinates of points at appropriate intervals along the road line may be computed. The mathematics of the cubic spline are somewhat more advanced and will not be covered here.

Roads are also constructed in straight vertical (sloping) alignments connected together by vertical parabolic curves.

13.2 Horizontal circular curves

Figure 13.1 shows two straight lines, ab and bc, representing the centre-lines of two successive road straights intersecting at b, and a circular arc connecting the lines. The circular arc is part of the circumference of a circle, the lines ab and bc are tangential to the circle and they are tangents which touch or meet the circle at a and c.

The following list defines the various parts of the figure, together with the name and symbol used for each in practice. Note that there is some lack of agreement in terminology between English-speaking practitioners.

Distance ab = bc = tangent length = T
Distance da = dc = de = radius of the curve = R
Arc length aec = length of the curve = L
Distance eb = external distance (or apex length) = E
Distance ac = long chord
Distance ef = mid ordinate = M

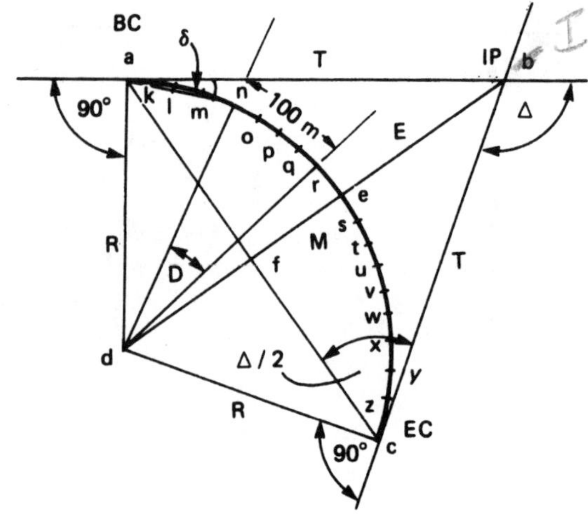

Figure 13.1

Angle Δ = deflection angle of the curve
= ∠adc at centre of circle

Angle D = degree of the curve,
= angle subtended at the centre by an arc of 100 m

∠abc = supplement of Δ,
= intersection angle of the curve

∠bac = ∠bca = Δ/2 = total tangent deflection angle

∠ban = tangent deflection angle of the arc an, from point a
= angle at a between the tangent and the chord an = δ

∠bad = ∠bcd = 90°

(angle between tangent and radius drawn to the point of contact)

Assuming a right-hand curve, moving from a towards c,

Point a = tangent point 1 = TP1
= beginning of curve = BC
= point of curvature = PC
} alternatives

Point c = tangent point 2 = TP2
= end of curve = EC
= point of tangency = PT
} alternatives

Point b = point of intersection = IP
Point d = centre of the circle
Point e = midpoint or crown point of the curve

Note that in dealing with large radius curves it is not practical to scale the details of a highway curve from a drawing – the required tangent lines (road centre-lines) must be separately pegged out on the ground, their intersection accurately located and pegged, then a theodolite set up over the intersection point peg and the deflection angle Δ measured by theodolite.

Road centre-lines are usually marked with pegs at uniform intervals and when the pegs are identified by their distance from the start of the job it is known as *through chainage*. The *uniform interval* is a sub-multiple of 100 m, commonly 20 m. A common convention for expressing the chainage of a peg is to state the number of hundreds of metres followed by a plus sign and then the remaining distance, e.g. a peg at 17 120 m from the start has a chainage of 171 + 20.0. The *tangent points* are unlikely to be at exact multiples of 20 and could have values such as 171 + 22.5 and 173 + 13.1, i.e. 17 122.5 m and 17 313.1 m from the start. The lower value is of course at the beginning of curve BC, while the greater is at the end of curve EC.

Figure 13.1 shows how the curve may be marked by pegs at these *standard intervals*, forming a set of *standard length chords*. The first and last chords will actually be *sub-chords* of less than the standard length.

13.2.1 General circular curve calculations

13.2.1.1 To calculate curve radius, given the degree of the curve (Figure 13.2(a))

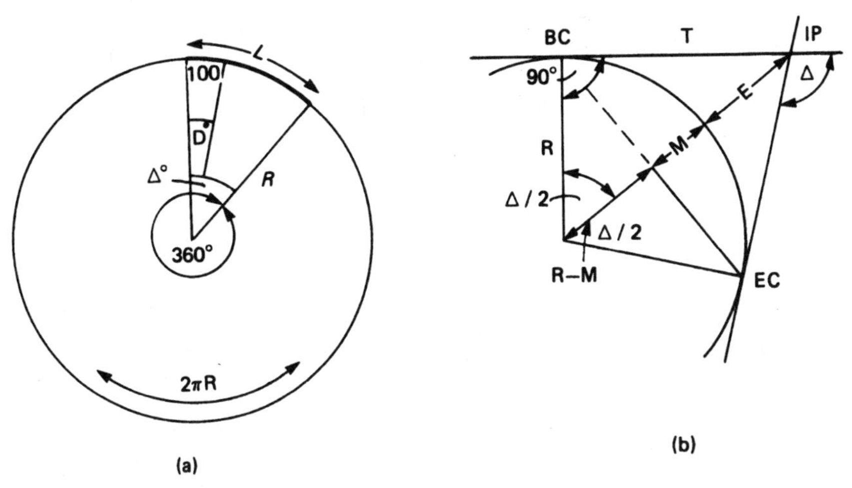

Figure 13.2

$$2\pi R/100 = 360°/D°,$$

therefore

$$R = 36\,000/2\pi D = 18\,000/\pi D \tag{13.1}$$

13.2.1.2 To calculate degree of curve from given radius

$$D = 18\,000/\pi R \tag{13.2}$$

13.2.1.3 To calculate tangent length (Figure 13.2(b))

$$T = R\tan(\Delta 2) \quad (13.3)$$

13.2.1.4 To calculate the curve length (Figure 13.2(a))

$$L/2\pi R = \Delta°/360°$$

therefore

$$L = 2\pi R\Delta/360 = \pi R\Delta/180 \quad (13.4)$$

or

$$L/100 = \Delta°/D°$$

therefore

$$L = 100/\Delta D \quad (13.5)$$

13.2.1.5 To calculate external distance (Figure 13.2(b))

$$\cos(\Delta/2) = R/(R + E)$$

therefore

$$R + E = R/\cos(\Delta/2)$$

and

$$E = (R/\cos(\Delta/2)) - R$$

therefore

$$E = R((1/\cos(\Delta/2)) - 1) \quad (13.6)$$

13.2.1.6 To calculate the mid-ordinate

$$\cos(\Delta/2) = (R - M)/R$$

therefore

$$R\cos(\Delta/2) = R - M$$

and

$$M = R - R\cos(\Delta/2) = R(1 - \cos(\Delta/2)) \quad (13.7)$$

13.2.1.7 To calculate the long chord length (Figure 13.2(b))

$$\sin(\Delta/2) = \text{half length of long chord}/R$$

therefore

$$\text{long chord length} = 2R\sin(\Delta/2) \tag{13.8}$$

13.2.1.8 To calculate tangential deflection angles for arcs

In *Figure 13.3*, a chord of length C is shown, starting at a tangent point (the beginning of curve in this case). If the *tangential deflection angle* (the angle between the tangent and the chord) of the arc is δ, then the angle subtended by the arc/chord at the centre of the circle is 2δ. (A geometrical theorem states that the angle between a tangent and a chord to the tangent point is equal to half the angle subtended by the chord at the centre of the circle.)

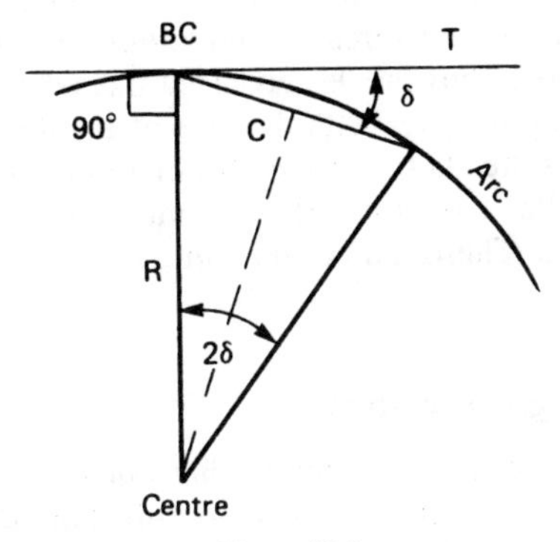

Figure 13.3

From *Figures 13.2* and *13.3*,

$$2\delta/\text{arc length} = \Delta/L = D/100 = 360/2\pi R$$

therefore

$$\delta\ (\text{degrees}) = \text{arc length} \times \Delta/2L \tag{13.9}$$

$$= \text{arc length} \times D/200 \tag{13.10}$$

$$= \text{arc length} \times 90/\pi R \tag{13.11}$$

Alternatively,

$$\delta\ (\text{minutes}) = \text{arc length} \times 1718.87/R \text{ minutes} \tag{13.12}$$

This latter form is the most useful in setting out. Note that the tangential deflection angle varies directly with the *arc* length, thus if a *standard arc length* is used a *standard angle may be calculated* and for a multiple of the arc length the same multiple of the angle can be used. In practice, if a standard chord is $\leq R/20$, then for all practical purposes the chord and its arc may be considered to be of the same length.

Referring again to *Figure 13.3*,

$$\sin \delta = C/2R \quad (13.13)$$

thus

$$C = 2R \sin \delta \quad (13.14)$$

and this may be useful in various applications.

13.2.1.9 *Curve specification*

The greater the speed of the traffic on a road, the larger radius curves will be required if traffic is to negotiate the road safely. National highway authorities specify design speeds for new roads, and minimum circular curve radii for particular design speeds. Thus the design speed fixes the minimum curve radius, and then the curve length varies with the size of the deflection angle.

If the circumstances are such that the minimum circular arc radius cannot be achieved, then a smaller radius circular arc must be used in conjunction with spiral transition curves. These are outlined in Section 13.3.

13.2.2 Linear setting out methods

Three main methods are available for setting out large radius circular curves using linear measuring equipment only – *deflection distances*, *offsets from tangent* and *offsets from the long chord*. In each case one or both of the tangent lines and tangent points must be located on the ground before the curve can be pegged and the radius must be known. Setting out commences from a known tangent point.

13.2.2.1 *Deflection distances method (Figure 13.4)*

A standard chord length of C must be calculated, $\leqslant R/20$, of suitable length. In the figure, a standard chord runs from the tangent point a to the first point on the curve b, the angle between the chord and tangent is δ and the angle at the centre of the circle is 2δ. The perpendicular distance from the point b to the tangent line is termed the *offset distance*. Given R and C, the offset distance must be calculated.

Now,

$$\sin \delta = \text{offset distance}/C$$

therefore

$$\textit{offset distance} = C \sin \delta$$

but from Equation (13.13),

$$\sin \delta = C/2R,$$

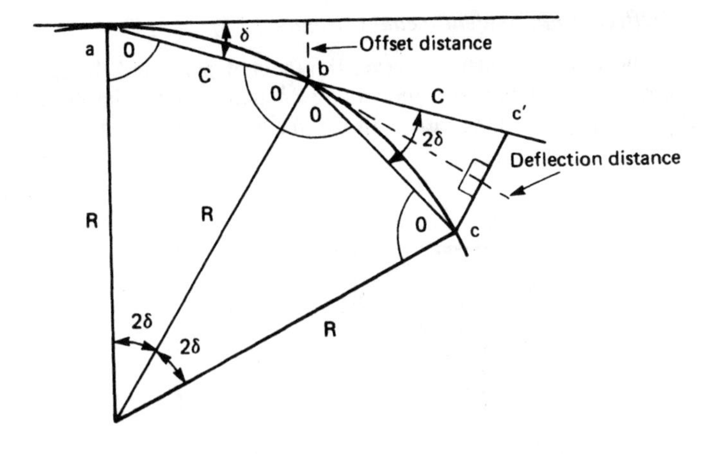

Figure 13.4

therefore

$$\textit{offset distance} = C^2/2R \tag{13.15}$$

In practice, C and the offset distance are calculated, then the first curve point is fixed by swinging a tape of length of C from the point a, until the required offset distance from the tangent is achieved.

To set out the next curve point, the first chord must be extended (on the same alignment) by a further distance C and a peg inserted at c′, then the tape swung towards the curve by a distance of twice the offset distance, locating the new point c on the curve. The distance c′c is termed the *deflection distance,* calculated from

$$\sin \delta = \text{deflection distance}/2C$$

so that

$$\textit{deflection distance} = 2C \sin \delta$$

but

$$\sin \delta = C/2R$$

therefore

$$\begin{aligned}\textit{deflection distance} &= 2C^2/2R = C^2/R \qquad (13.16)\\ &= 2 \times \text{offset distance}\end{aligned}$$

All subsequent points on the curve may be located in the same way as point c.

13.2.2.2 Offsets from tangent method (Figure 13.5)

The figure shows a tangent to a curve, the tangent point and the radius at the tangent point. x = a distance measured along the tangent from the TP, and y = the offset from the tangent to the curve. In the shaded triangle,

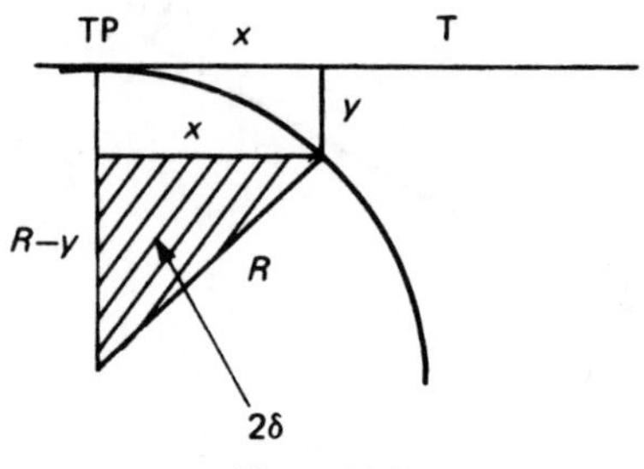

Figure 13.5

$$(R - y)^2 = R^2 - x^2$$

therefore

$$R - y = \sqrt{(R^2 - x^2)}$$

and

$$y = R - \sqrt{(R^2 - x^2)} \tag{13.17}$$

Thus, for any distance, x, along the tangent, an offset distance y, may be calculated.

Where it is required to set out pegs at specified distances along the curve, using the central angle 2δ of the triangle,

$$\sin 2\delta = x/R,$$

or

$$x = R \sin 2\delta \tag{13.18}$$

and

$$\cos 2\delta = (R - y)/R$$

so that

$$R \cos 2\delta = R - y$$

and

$$y = R - R\cos 2\delta$$
$$= R(1 - \cos 2\delta) \quad (13.19)$$

The value to use for 2δ is determined from the specified arc/chord length, as shown earlier, i.e. $2\delta = (\text{arc} \times 180)/\pi R$. Alternatively, it may be required to set out the curve by dividing its length into a number of equal chords, and 2δ may be deduced from the deflection angle of the curve and the number of chords.

13.2.2.3 Offsets from long chord method (Figure 13.6)

In the figure, l is an ordinate from a point on the long chord to a point on the curve, while k is the distance along the long chord from a TP to the ordinate. In the shaded triangle,

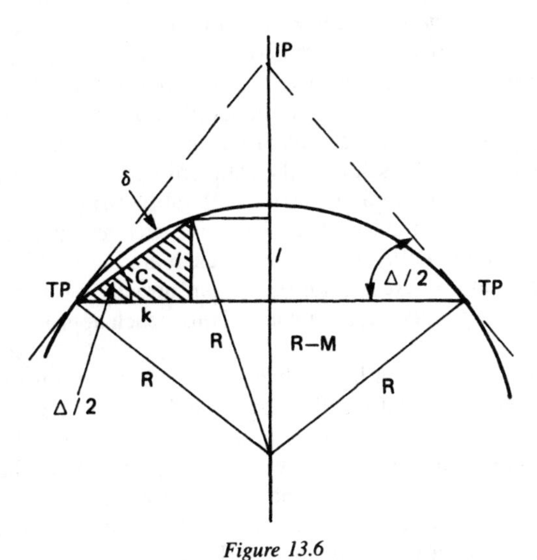

Figure 13.6

$$\cos(\Delta/2 - \delta) = k/C \text{ and } \sin(\Delta/2 - \delta) = l/C$$

but by Equation (13.14), $C = 2R\sin\delta$, therefore

$$k = 2R\sin\delta\cos(\Delta/2 - \delta) \quad (13.20)$$

and

$$l = 2R\sin\delta\sin(\Delta/2 - \delta) \quad (13.21)$$

Again, δ may be deduced as required as in the preceding method.

If arbitrary values of k are selected, then Pythagoras' theorem may be used to calculate the appropriate l values.

$$(l + (R - M))^2 = R^2 - ((\text{long chord}/2) - k)^2$$

or

$$l = \sqrt{\{R^2 - ((\text{long chord}/2) - k)^2\}} - (R - M) \tag{13.22}$$

If the long chord and mid-ordinate values have not been calculated, then from Equations (13.7) and (13.8)

$$l = \sqrt{\{R^2 - (R \sin(\Delta/2) - k)^2\}} - R\cos(\Delta/2) \tag{13.23}$$

13.2.3 Tangential deflection angles method

This method, the most common for large radius curves, requires the use of a *theodolite* to set out *deflection angles* from the *tangents*, together with a steel tape to set out *short standard chord lengths* along the curve. Generally the standard peg interval of 20 m is used for these chords and the even chainage points are carried through so that there are sub-chords at the beginning and end of the curve. As stated earlier, if the standard chord is not greater than one-twentieth of the radius, then the standard chord and its arc may, for practical purposes, be considered the same. If the standard chord must be less than 20 m, then it should be a sub-multiple of 20 m, preferably 10 m.

The essence of the method is that successive arcs are measured out from a tangent point, while the deflection angle for each arc is set off from the tangent line. Strictly speaking an arc cannot be set out directly so instead the arcs are built up by setting out successive standard chords, each approximating to a short arc.

The curve should be set out in two halves – one half set out from the BC, the other half from the EC – and ideally they should meet at the *crown point,* CP.

The CP should, therefore, be located and pegged before setting out the two halves and the inevitable small errors will show up by the two halves failing to close on to the CP. A small discrepancy is to be expected and, provided it is reasonable, a few pegs in each half are adjusted, the adjustment diminishing with increase in distance of a peg from the CP. If the misclosure is large, mistakes should be looked for, such as a missed deflection angle. The acceptable misclosure cannot be defined, being dependent upon the specification, the equipment used and experience.

(*Note*: Tangential deflection angles are sometimes referred to as 'tangential angles', and sometimes as 'deflection angles'. In the remainder of this chapter, for brevity, tangential deflection angles will simply be referred to as 'deflection angles'.)

13.2.3.1 Example circular curve calculation

Curve No. 5R (meaning fifth curve along the road and curving to right-hand side).

Required radius = 699.80 m
Centre-line pegs at 20 m. Chainage of IP from the start, 72 + 54.5. Δ measured, found to be 10° 15' 00".

Calculation sheet (distances to 0.05 m)

IP = 72 + 54.50 Δ = 10° 15' 00"
R = 699.80 m $\Delta/2$ = 05° 07' 30"
Chords = 20 m $\Delta/4$ = 02° 33' 45"
$T = R\tan\Delta/2$ = 62.75 m
$L = \pi R\Delta/180$ = 125.20 m
$L/2$ = 62.6 m
E (external distance) = $R((1/\cos(\Delta/2)) - 1)$ = 2.80 m

Chainages

IP = 72 + 54.50
$-T$ = 0 + 62.75
BC = 71 + 91.75
$+L/2$ = 0 + 62.60
CP = 72 + 54.35   Check BC = 71 + 91.75
$+L/2$ 0 + 62.60 $+L$ 1 + 25.20
EC = 73 + 16 .95   EC = 73 + 16.95√

Centre-line pegs required at	*chord lengths*
BC 71 + 91.75	0
72 + 00.0	8.25 sub-chord
72 + 20.0	20.0
72 + 40.0	20.0
CP 72 + 54.35	14.35 sub-chord
72 + 60.0	5.65 sub-chord
72 + 80.0	20.0
73 + 00.0	20.0
EC 73 + 16.95	16.95 sub-chord
Total	125.20 = L, check √

Deflection angles required, from $\delta = (1718.87/R) \times l$ minutes

for l = 8.25, δ = (1718.87/699.8) × 8.25
= 2.456 × 8.25 = 20.26' = 20' 15.6"
for l = 20.0, δ = 2.456 × 20.0 = 49.12' = 49' 7.2"
for 14.35, δ = 2.456 × 14.35 = 35.25' = 35' 15.0"
for 5.65, δ = 2.456 × 5.65 = 13.88' = 13' 52.8"
for 16.95, δ = 2.456 × 16.95 = 41.63' = 41' 37.8"

Setting-out table for curve

First half, theodolite set over BC

Peg sighted	*Chord*	*Circle reading*
IP	0	00° 00' 00"
		+ 20 15.6
72 + 00.0	8.25	00 20 15.6
		+ 49 07.2

72 + 20.0	20.0	01 09 22.8
		+ 49 07.2
72 + 40.0	20.0	01 58 30.0
		+ 35 15.0
CP → 72 + 54.35	14.35	02 33 45.0 = Δ/4 check√

Second half, theodolite set over EC

Peg sighted	*Chord*	*Circle reading*
IP	0	360 00 00
		− 41 37.8
73 + 00.0	16.95	359 18 22.2
		− 49 07.2
72 + 80.0	20.0	358 29 15.0
		− 49 07.2
72 + 60.0	20.0	357 40 07.8
		− 13 52.8
CP → 72 + 54.35	5.65	357 26 15.0
		− 13 52.8
	Check	360 00 00
		− 357 26 15.0
		02 33 45.0 = Δ/4

13.2.3.2 Field operations

The field operations for setting out the typical curve above fall into three stages, as shown in *Figure 13.7*.

Stage 1: Fix the IP by projecting the tangent straights to intersect, then measure angle Δ and find the chainage to the IP. The radius or degree of curve being specified, calculate all the items needed for setting out, as shown. Measure distance *T* from the IP and fix the BC and EC points on the tangents. Fix the CP by bisecting the angle BC-IP-EC and measuring out the distance *E*.

Stage 2: Set the theodolite over the BC point, orient it on the line to the IP (reading zero when sighted on the IP). As a check, set out the angles Δ/4 and Δ/2 from the tangent in turn, they should bisect the CP and EC pegs, respectively. Set on the first deflection angle of 00° 20' 16" and direct an assistant to place a peg on the sight line at distance 8.25 m from the BC. Mark the peg with the chainage, 72 + 00.0. (Note that although the angles have been shown to decimals of a second in the calculation, the rounded values to be used will depend upon the actual instrument in use.)

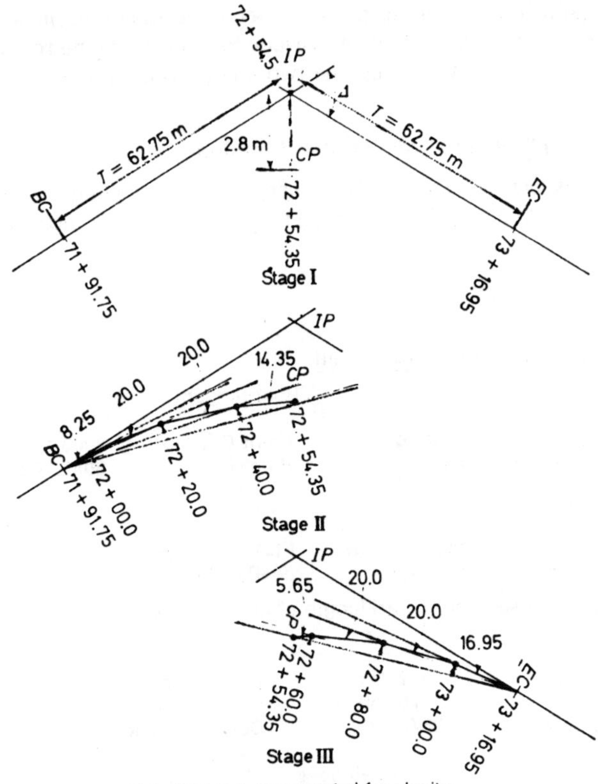

Figure 13.7

Set on the next total deflection angle 01° 09' 23", direct the assistant to pull the tape on and insert the next peg on the sight line at a distance of 20 m from peg 72 + 00.0, mark the new peg 72 + 20.0.

The procedure is repeated until the sub-chord to the CP is set out.

Stage 3: Move the theodolite to the EC point, orient it on the IP. Repeat the previous procedure, but swinging the theodolite *anti-clockwise* and *deducting the deflection angles from 360°*, until the CP is reached again.

13.2.4 Two theodolites method

This method uses two theodolites, one placed over each tangent point, then peg locations are fixed by the intersection of their deflection angles from both tangents. The calculation is carried out in the same way as in Section 13.2.3.1, but the Setting Out Table for the first half is extended to cover all points from BC round to the EC and the second half of the table is extended to cover all points from the EC round to the BC.

To fix a point on the curve, the two theodolite operators must collaborate carefully, taking it in turn to sight an assistant holding a ranging rod, until the rod is precisely on both sight lines for the two angles to the point.

13.2.5 Rectangular/polar co-ordinates method

This method is particularly suitable where the job is being carried out on a rectangular co-ordinate grid and the road tangent lines are defined by either three

Horizontal Circular Curve

Curve Ex. 5.10 (right)

Tangent alignment co-ordinates specified

Entry tangent	E	N	Exit tangent	E	N
Point A	2675.454	3748.621	Point C	2965.131	5087.163
Point B	2845.736	4972.814	Point D	3047.162	3796.734

Curve specification

Radius	Degree	BC Chainage	Pt.A Chainage
400.000	14 19 26	29+14.034	50+00.000

Through chainage – standard chord length of 20

Computed data

Tangent length	3952.951	Long chord length	795.935
Curve length	1175.960	Rise of long chord	359.730
Apex distance (ED)	3573.138	Deflection angle	168 26 38

	Chainages	E	N
Beginning of Curve	29+14.034	2388.068	1682.546
Crown Point of Curve	35+02.013	2799.196	2027.159
End of Curve	40+89.993	3183.448	1652.814
Intersection Point	68+66.985	2932.670	5597.803
Centre of Circle		2784.254	1627.438

First sub-chord 5.966 58 Standard chords Final Sub-chord 9.993

Tangential angle for standard chord – 1 25 57 5156.6 secs

Curve point details

Peg	Chainage	Chord	E	N	Tangential Angles At BC	Tangential Angles At EC
BC	29+14.034		2388.068	1682.546	00 00 00	84 13 19
		5.966				(275 46 41)
1	29+20.000		2388.935	1688.449	00 25 38	83 47 41
		20.000				(276 12 19)
2	29+40.000		2392.478	1708.131	1 51 35	82 21 44
		20.000				(277 38 16)
3	29+60.000		2397.000	1727.611	3 17 32	80 55 47
		20.000				(279 04 13)
4	29+80.000		2402.491	1746.840	4 43 28	79 29 51
		20.000				(280 30 09)
5	30+00.000		2408.936	1765.771	6 09 25	78 03 54
		20.000				
6	30+20.000					

Figure 13.8

pairs of co-ordinates (the IP and one point on each tangent) or four pairs of co-ordinates (two points on each tangent). The curve data for all points on the curve (IP, BC, EC, CP, etc.) and all the standard chainage peg positions must be calculated as for the normal tape and theodolite method, and then rectangular co-ordinates computed for all these points. See example printout, *Figure 13.8*, which shows both rectangular co-ordinates and deflection angle data for setting out part of a curve.

Given the rectangular co-ordinates of suitable instrument stations, polar co-ordinates of the curve points may then be computed from the rectangular co-ordinate differences and set out from the instrument stations using suitable EDM equipment.

This method avoids the need to set up the instrument on the actual curve, and allows the most suitable position to be chosen for the theodolite or total station.

13.2.6 Common problems

A variety of problems may crop up in practice, but some common ones are worth mentioning.

Where the IP of a curve is inaccessible, such as on tight curves on mountain roads, or where there are buildings on the line, a straight line may be laid between the two road straights and its length and the angles it makes with the straights measured. Thereafter, the inaccessible triangle at the apex may be solved and the distances from the IP to the cross-line calculated. After calculating the tangent length, the distance to BC and EC from the cross-line can be worked out and the points located as normal. See *Figure 13.9(a)*.

When setting out from one end of the curve by tape and theodolite, the sight line to one or more points may be obstructed. In this event, the theodolite should be moved to the last point sighted on the curve, set up there, then oriented so that when the reading on the circle is zero the telescope is sighted along the *tangent to the curve at that point*. Thereafter deflection angles may be set out again in the normal way. See *Figure 13.9(b)*.

If measurement along the curve line is obstructed, and it is not possible to measure the chord from one peg to the next peg position, then the new peg may be set out by direct measurement from the tangent point, the distance from the tangent point to the new peg being equal to $2 \times R \times$ sine of the deflection angle to the new peg.

If it is required to replace a missing peg, the chord bisection method shown in *Figure 13.10* may be used. The preceding and following pegs are located and joined by a chord, then the offset from the centre of the chord to the missing peg is equal to R versine 2δ. (*Note:* versine $\delta = 1 - \cos\delta$.)

13.3 Transition curves

When a vehicle of weight W travels around a level highway curve of radius R, at a constant velocity of v, then the radial force pushing the vehicle outwards is $F = Wv^2/gR$, where g is the acceleration due to gravity, and the resultant of W and F is inclined at an angle to the road surface, tending to cause the vehicle to skid towards the outside of the curve.

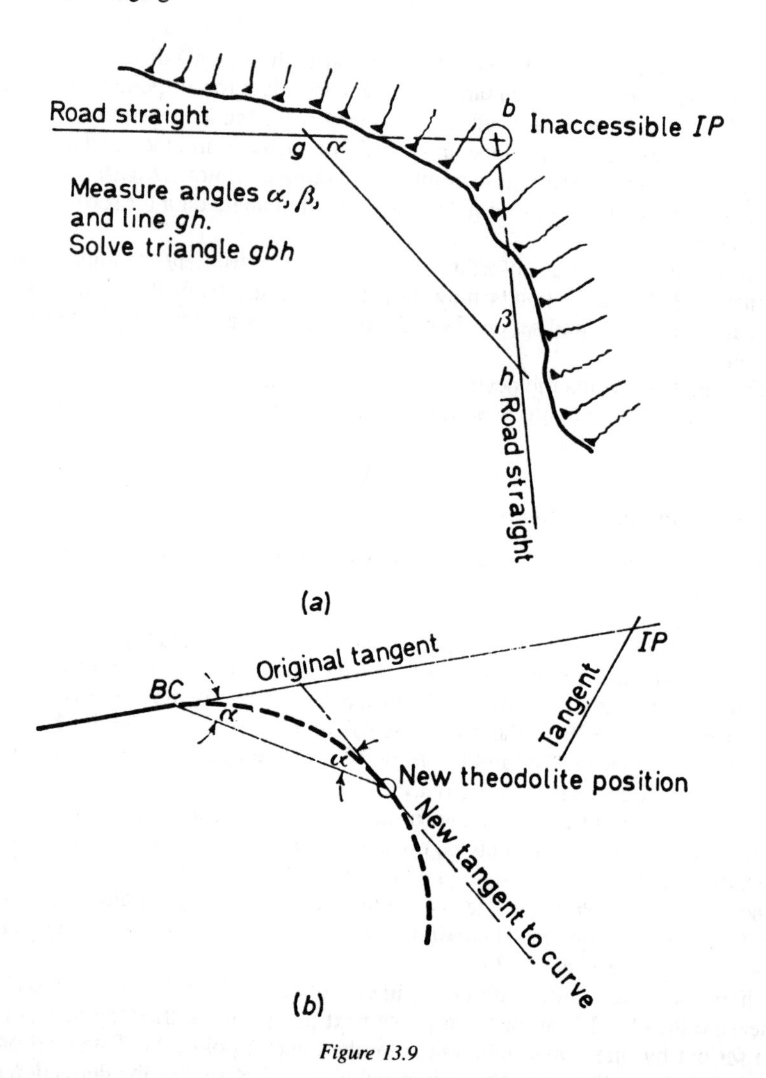

Figure 13.9

When the vehicle is travelling along the straight, the radius is infinite, hence the radial force is zero, but when the vehicle leaves the straight and enters a circular curve the radial force acting on the vehicle will increase immediately from zero to the full value of Wv^2/gR, assuming that there is no change in velocity. If the circular arc radius is very large then the value of F will be small and there will not be a serious problem.

However, if the circular arc radius cannot be made large enough to minimize the sudden lateral shock, a curve of gradually varying radius may be placed between the straight and the circular curve, so that the radial force builds up gradually rather than instantaneously. In addition, the cross-section of the carriageway can be tilted as shown in *Figure 13.11*, where the outer edge of the carriageway has been lifted by an amount *e*. This is the application of

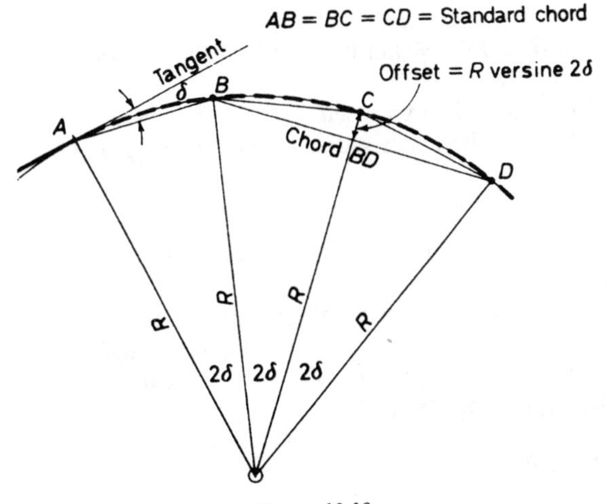

Figure 13.10

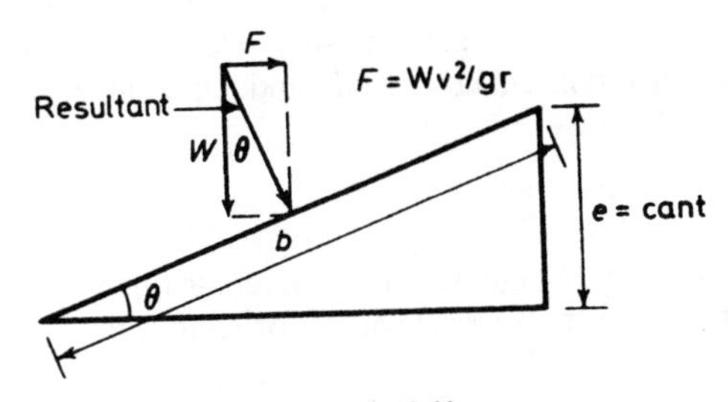

Figure 13.11

superelevation or *cant*, and it will be observed that in this case the amount of superelevation has been sufficient to cause the resultant of the forces F and W to act normal to the road surface, thus removing the outward push which would tend to cause the vehicle to skid off the road.

A *transition curve* has curvature which varies uniformly with distance along the curve. It is used to change the radius of a curve and also to introduce and gradually increase the amount of superelevation on the highway, so that the radial force increases gradually, minimizing both the discomfort to passengers and the risk of skidding. Transitions are typically used when it is not possible to set out a circular curve of large enough radius for the design speed of the highway, and then the radius must be changed gradually from infinity (the radius of a straight line, the road tangent) to the actual circular arc radius to be used, R, then back to the straight again, radius infinity. A transition curve may also be used between two circular curves of different radius.

13.3.1 Centrifugal ratio and circular arc radius

Figure 13.11 actually illustrates a cross-section through a carriageway on a curve, where the resultant of F and W is normal to the road surface under equilibrium conditions, the angle of inclination of the surface being θ, then

$$\textit{centrifugal ratio} = F/W = \tan\theta = Wv^2/WgR = v^2/gR \tag{13.24}$$

If the centrifugal ratio is expressed as V^2/gR, where V is the velocity in kilometres per hour, R the curve radius in metres and $g = 9.806\,\text{m/s}^2$, then

$$\textit{centrifugal ratio} = V^2/127\,R \tag{13.25}$$

{Obtained from $(V^2 \times 1000^2)/(9.806 \times 3600^2 \times R)$}

Typical values used for the centrifugal ratio are 0.21 to 0.25 for highways and 0.125 for railways. If a value is selected for the centrifugal ratio, say 0.22, then the minimum circular arc radius for a specified velocity may be obtained by transposing the expression, then

$$R = V^2/(127 \times 0.22) \tag{13.26}$$

It is recommended that generally *larger* radii be used than would be indicated by this expression. In United Kingdom practice, design is normally carried out on the basis of tables produced by the Transport Department (TD). These specify minimum radii for different design speeds and superelevations.

13.3.2 Superelevation

Referring to *Figure 13.11* again, with the centrifugal ratio equal to v^2/gr, carriageway width b and the superelevation or cant e, then

$$\textit{superelevation} = e = b\sin\theta \approx bv^2/gr \tag{13.27}$$

since $\sin\theta \approx \tan\theta$ when θ is small.

If the velocity v is constant then the superelevation varies directly with the inverse of the radius, that is to say $e \propto 1/r$, and the greater the radius the smaller the superelevation required, or the smaller the radius the greater the superelevation.

If superelevation is applied uniformly along the transition, then $e \propto l$, where l = distance along the transition, and $l \propto 1/r$, therefore lr is a constant, i.e. at any point on the transition curve the product of the distance along the transition and the radius at the point is a constant, say K.

The maximum superelevation will occur when r is a minimum, i.e. $r = R$, circular curve radius. When $\tan\theta = V^2/gR$, then theoretically

$$\text{maximum superelevation} = bV^2/gR$$

However, if the full amount of superelevation is used it may well be disconcerting to drivers approaching the curve, and vehicles travelling at less than the design speed may tend to slide down the road surface. Accordingly, the TD recommend that only 45% of the radial force should be balanced out by superelevation, leaving the remainder of the force to be absorbed by the friction of the road surfacing. Thus *maximum allowable superelevation* is given by

$$e = 0.45\ bV^2/gR \tag{13.28}$$

To obtain the superelevation as a percentage, then

$$e = 0.45 \times 100 \times V^2/gR$$

and taking V in kph and g as 9.806 m/s^2, then

$$e \text{ as percentage} = (0.45 \times 100 \times V^2 \times 1000^2)/(9.806 \times 3600^2 \times R)$$

$$= V^2/2.824\,R \tag{13.29}$$

The TD quote a value of $V^2/2.828\ R$, and specify a maximum permitted superelevation of 7%. The *minimum* cross-fall on any carriageway is 2.5% or 1 in 40.

13.3.3 Transition length

At any point in a transition curve the *radial acceleration* is equal to v^2/r, and the usual criterion for deciding the length of a transition curve is the *rate of change of radial acceleration, q*, expressed in metres per second cubed. The radial acceleration at the beginning of the transition is zero, while the radial acceleration at the junction of the transition and the circular arc is v^2/R, thus the change of radial acceleration over the transition length is v^2/R. If the vehicle velocity in metres per second is v, and the transition length is L metres, then the time taken over the transition is L/v seconds, and the rate of radial acceleration is

$$q = (v^2/R)/(L/v) = v^3/LR$$

and therefore

$$L = v^3/qR \text{ metres}$$

where v is in metres per second.

If the velocity is V kph, then

$$L = V^3/(3.6^3\ qR) = V^3/(46.7\ qR) \tag{13.30}$$

For reasons of passenger comfort, the customary value for q is 0.3 m/s^3 but 0.45 and 0.6 may be used in difficult circumstances.

13.3.4 Curves used for transitions

13.3.4.1 The clothoid

As shown in Section 13.3.2, the value of lr is required to be constant for a transition curve, so that the radius of curvature decreases in proportion to the length. The *clothoid* meets this requirement and is regarded as the ideal transition spiral. The equation of the clothoid or *Euler spiral* is

$$\phi = l^2/2RL \tag{13.31}$$

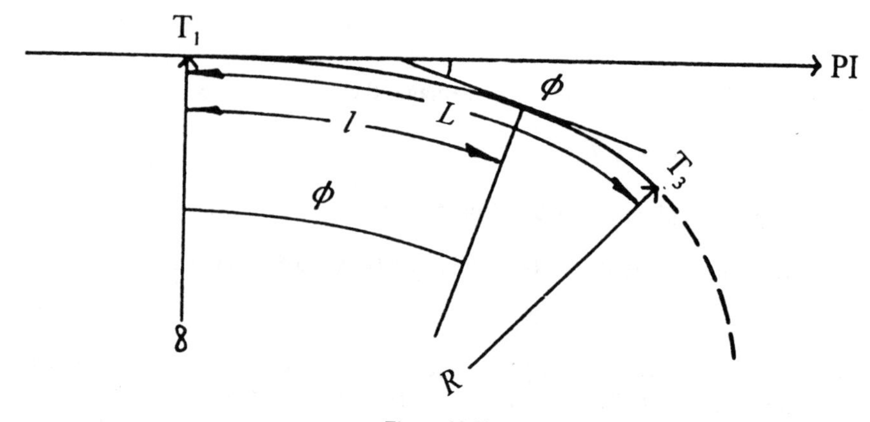

Figure 13.12

or

$$l = (2RL\phi)^{1/2} \tag{13.32}$$

where ϕ is the deviation angle between the straight and the tangent at a point distant l from the beginning of the transition curve, as in *Figure 13.12*.

In the figure the transition length is L, and it starts at T_1 and meets the circular curve at T_3, the radius changing from infinity at T_1 to R at T_3.

The clothoid gives rectilinear co-ordinates which are infinite series. In *Figure 13.13*, using x for distance along tangent to a point, and y for the perpendicular offset to the transition from the point, ϕ in radians and θ the deflection angle in degrees, then

$$x = l - l^5/10(2RL)^2 + l^9/216(2RL)^4 - l^{13}/9360(2RL)^6 + \ldots \tag{13.33}$$

$$y = l^3/6RL - l^7/42(2RL)^3 + l^{11}/1320(2RL)^5 - \ldots \tag{13.34}$$

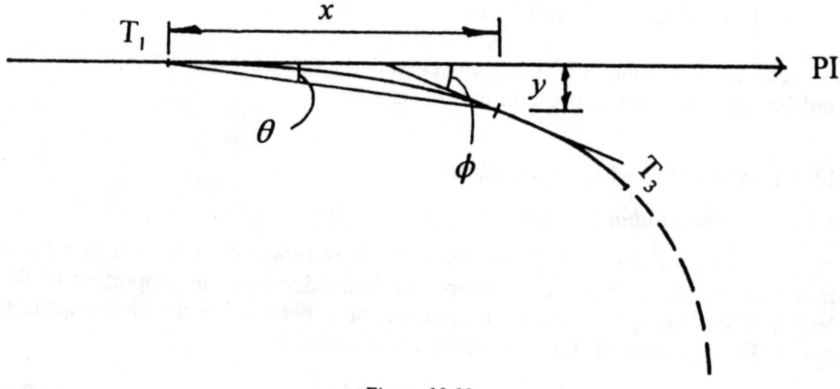

Figure 13.13

and

$$\tan\theta = \phi/3 + \phi^3/105 + \phi^5/5997 - \phi^7/198700 - \ldots \tag{13.35}$$

The third and subsequent terms in the series for x and y are very small and can be ignored in practice. The maximum value of ϕ will occur at the junction of the transition and circular arcs, where $\phi = L/2R$.

Clothoids are set out using the Highway Transition Tables prepared by the County Surveyors' Society. If these are not available and the range is limited, then the *cubic parabola* may be used, as it is relatively easily calculated.

13.3.4.2 The cubic parabola

If the first terms of the series for the clothoid are used, with $x = l$, the result is the cubic parabola with equation

$$y = x^3/6RL \tag{13.36}$$

and this may be used to set out the curve using offsets y at distances x along the tangent. This equation may also be used to calculate rectangular co-ordinates of points on the transition.

For the cubic parabola, the clothoid Equation (13.31) becomes

$$\phi = x^2/2RL \tag{13.37}$$

then for a deflection angle of θ, $\tan\theta = y/x = (x^3/6RL)/x = x^2/6RL$, but since the angles are small, $\tan\theta = \theta$ radians, therefore

$$\theta = x^2/6RL \text{ radians} \tag{13.38}$$

Taking Equations (13.37) and (13.38), then $\theta = \phi/3$, $\theta_{max} = \phi_{max}/3$. Further, the *angle consumed*, ϕ, is equal to θ plus the *back angle*, hence the back angle = $2\phi/3$.

The deflection angles are required in minutes, then

$$\begin{aligned}\theta = x^2/6RL &= (x^2 \times 180)/(\pi \times 6RL) \text{ degrees}\\ &= 1800\, x^2/\pi RL \text{ minutes}\end{aligned} \tag{13.39}$$

The transition may be set out by short chords and deflection angles from the tangent, the distance l along the transition curve being the sum of the short chords, in a similar manner to setting out a circular curve by theodolite with deflection angles and chords. The deflection angles are calculated using (13.39), with l substituted for x.

As a general rule the cubic parabola should not be used when the value of ϕ_{max} exceeds 12° (or θ_{max} exceeds 4°). Unless stated otherwise, the following sections refer to the cubic parabola transition curve.

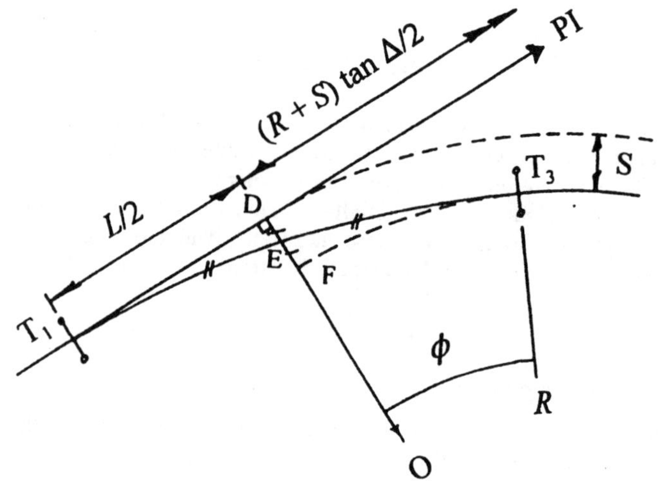

Figure 13.14

13.3.5 Shift of the circular arc

Figure 13.14 shows a transition curve meeting a circular arc, it is effectively an enlargement of part of *Figure 13.15*. The line from O to D is perpendicular to the tangent, cutting the transition at E. The circular arc is continued from T_3 as a dotted line, intersecting the line OD at F. It will be noted that the effect is as if a circular arc of radius R + DF had been moved inwards and its radius reduced to R. The distance DF is termed the *shift* of the circular arc, symbol S. For all practical purposes the radial line to D bisects the transition length and the transition bisects DF at E. Taking the distance T_1 to D as $L/2$, then $x = L/2$ and

$$\text{DE} = x^3/6RL = (L/2)^3/6RL = L^2/48 = S/2$$

therefore

$$\text{shift } S = L^2/24 \qquad (13.40)$$

13.3.6 Spiral and circular arc geometry summary

Figure 13.15 shows the basic geometry of a composite curve, consisting of a circular curve connected to two identical but handed transition curves. For simplicity here, the transition curves are cubic parabolas.

Angle Δ = the deviation angle of the whole curve, T_1 and T_2 are the tangent points of the whole curve, O the centre of the circular arc, as for a simple circular curve, and R = the radius of the circular arc. T_3 and T_4 are the junctions of the transitions and the circular arcs.

Transition length = AH = L.
Deviation angle for transition = ∠BKH = ϕ_m = angle consumed by transition.

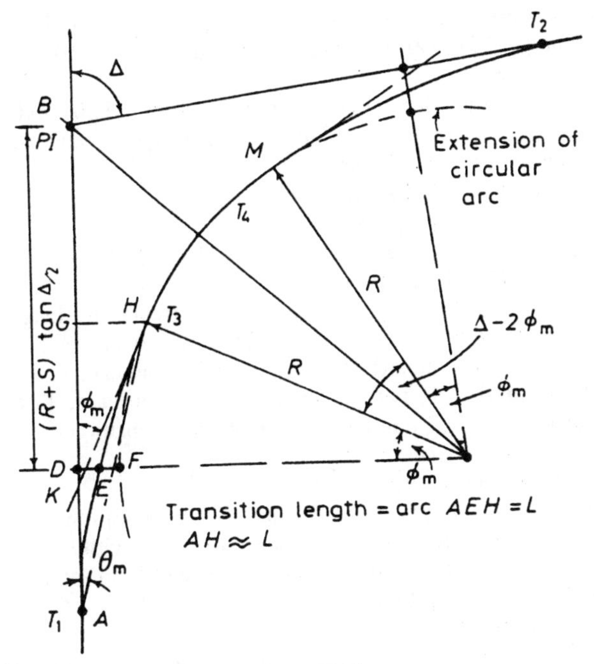

Figure 13.15

Deflection angle for transition = ∠BAH = $\theta_m = \phi_m/3$.
$\phi_m = L/2R$ radians = 3 × 1800 $L/\pi R$ minutes
$\theta_m = \phi_m/3 = L/6R$ radians = $1800L/\pi R$ minutes.

Angle consumed by the circular arc = $\Delta - 2\phi_m$.
Circular arc length equals = HM = $\pi R(\Delta - 2\phi_m)/180$ [angles in degrees]
Arc EH ≈ arc FH = $R\phi_m \approx L/2$.
AE = $L/2$.

Tangent length AB = $L/2 + (R + S)\tan(\Delta/2)$
Shift S = DF = $L^2/24R$.
DE = $S/2$.
Back angle = ∠KHA = $\phi_m - \theta_m = 2\theta_m$.

13.3.7 Setting out the spiral and circular arc

13.3.7.1 Setting out cubic parabola and circular arc

The given data normally is basically the design speed, carriageway width and value to use for q.

(i) Locate the PI, measure deflection angle Δ, and determine the chainage of the PI.

(ii) Find the minimum circular arc radius, R. If this is not specified, Equation (13.29) may be used, substituting 7 for e and the design speed for V, to find R.
(iii) Compute the transition length L from equation (13.30) and the shift S.
(iv) Calculate the total tangent length $T = L/2 + (R + S) \tan(\Delta/2)$.
(v) Calculate the through chainages:
chainage of T_1 = PI chainage $- T$
chainage of $T_3 = T_1$ chainage $+ L$
chainage of $T_4 = T_3$ chainage $+ \pi R(\Delta - 2\phi_m)/180$
chainage of $T_2 = T_4$ chainage $+ L$
Check: T_2 chainage $= T_1$ chainage $+ 2L +$ circular arc length.
(vi) Locate the tangent points T_1 and T_2 by taping from the PI.
(vii) Assuming 20 m standard chords, through chainage, calculate the sub-chords required at T_1 and T_3 for the first transition, at T_3 and T_4 for the circular arc, and at T_4 and T_2 for the second transition.
(viii) Set out the first transition, *either*
(a) by theodolite deflection angles:
With theodolite placed on T1, use (13.39) with l substituted for x, to compute deflection angles for chosen chord lengths l_1, $(l_1 + l_2)$, $(l_1 + l_2 + l_3)$,..., then set out the chords and angles in the same way as for a circular arc, up to T_3, or
(b) by tangent offsets:
Using (13.36), calculate appropriate offsets y at distances x along the tangent, until T_3 is reached.
(ix) Set out the circular arc. Move the theodolite to T_3, align on T_1, set off the back angle $2\theta_m$ to define the tangent to the circular arc, then set out the circular arc in the usual way.
(x) Repeat the procedure working back from T_2.

13.3.7.2 Setting out clothoid and circular arc

The procedure here is much the same, but appropriate quantities are obtained from the TD documents and the County Surveyors' Society *Highway Transition Curve Tables*.

13.4 Vertical curves

Vertical curves are required at the intersection of differing road/rail gradients. Curves may be convex (summit or crest) or concave (valley or sag). The requirements to be met by a vertical curve from the point of view of passenger comfort and safety are:

(i) constant rate of change of gradient;
(ii) uniform rate of increase of centrifugal force;
(iii) adequate sighting distances.

The simple parabola is normally used, owing to its simplicity and constant gradient change.

Gradient sign convention

Gradients rising to the right, positive, falling to the right negative.
Specified as a percentage, e.g. x % = 1 in $100/x$.
Left-hand gradient designated p%, right-hand gradient q%, thus the left (entry) tangent changes level by p metres in every 100 m horizontally, while the exit changes level by q metres in 100 m.

Desirable and *absolute maximium* gradients are recommended by the TD for all new roads. For motorways, dual carriageways and single carriageways, the desirable maxima are 3%, 4% and 6%, respectively. The absolute maxima are 4%, 8% and 8%, respectively.

Grade angle = deflection angle = difference in percentage grade = $q\% - p\%$. The correct signs of p and q must be inserted, according to the particular case, and in the case shown with p positive and q negative, i.e. respective signs +, –,

$$\text{grade difference} = (-q) - (+p) = -(p + q)$$

In effect, when the two signs are different the grade difference is the sum of the absolute values of p and q, and when the signs are the same it is the difference between the absolute values $(p - q)$.

13.4.1 The vertical curve parabola

13.4.1.1 Properties of the vertical curve parabola

Referring to *Figure 13.16*, the basic equation of the parabola is

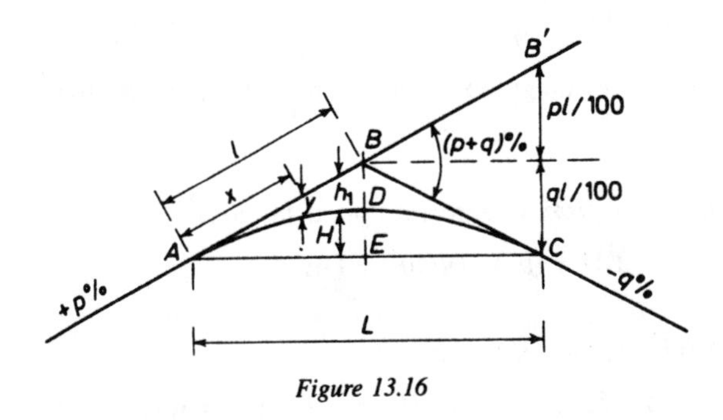

Figure 13.16

$$y = ax^2 \tag{13.41}$$

then

$$dy/dx = 2ax = \text{gradient of tangent to curve}$$

and

$$d^2y/dx^2 = 2a = \text{a constant, the rate of change of gradient}$$

Since the gradients involved are generally small, it is sufficiently accurate to assume that

(i) perpendicular offsets from the tangent are equal to vertical offsets,
(ii) the distance along the tangent is equal to the horizontal distance, which is equal in turn to the distance along the curve,
(iii) the vertical through the intersection B bisects AC,
(iv) BD = DE and D is the vertex of the parabola,
(v) a chord to a vertical curve has a rate of grade equal to that of the tangent at a point horizontally midway between the points of intercept of the chord.

13.4.1.2 *Offsets from the tangent to the curve*

The equation of the curve is $y = ax^2$, thus for a point on the tangent distant x from the beginning of the curve the offset $y = ax^2$. In *Figure 13.16*, considering the point B′ which is distant $2l$ from A, then the offset from B′ to the end point of the curve at C is B′C, and

$$B'C = pl/100 + ql/100 = (p + q)l/100 = ax^2 = a(2l)^2$$

hence

$$a = (p + q)/400l$$

and

$$y = (p + q)\, x^2/400l \qquad (13.42)$$

$(p + q)$, of course, is the grade difference here. If the grade signs were the same, it would be $(p - q)$.

13.4.1.3 *Distance from intersection point to curve*

When $x = l$, $y = BD = (p + q)l/400 = DE$.

13.4.1.4 *Highest point of curve*

Using A as datum for height, let H represent the height of the highest point of curve, and let this point be distant x from A; then

$$H = xp/100 - (p + q)x^2/400l$$

and for H a maximum, then $dH/dx = 0$, therefore

$$p/100 - 2(p + q)x/400l = 0$$

thus

$$x = 2pl/(p + q) \qquad (13.43)$$

13.4.2 Sight distances and curve length

Sight distance is the length of road over which a driver may safely see objects on the other side of the vertical curve, and it fixes the length of curve required and hence the design of the curve. In the past this varied with the road type and speed and a minimum curve length had to be calculated. Today constants have been introduced by the TD, known as *K-values*, and the minimum length of curve for a specific road is

$$\textit{minimum}\, L = K \times \text{grade difference} \tag{13.44}$$

K-values are listed in TD standards for *Full overtaking sight distance crests*, and if overtaking is not to be allowed for, K-values exceeding the listed *desirable minimum crest K-values* should be used. *Absolute minimum crest K-values* are also shown, for use in difficult conditions. For sag curves, *absolute minimum sag K-values* are listed. All of these K-values ensure adequate visibility and also comfort.

Examples

(i) For a single carriageway road, overtaking permitted, crest, 85 kph, K-value is 285, then minimum curve length = 285 × grade difference.

(ii) Same road, no overtaking permitted, the desirable minimum K-value is 55 and desirable minimum curve length is 55 × grade difference.

(iii) Same road, sag, minimum K-value is 20, minimum curve length is 20 × grade difference, suitable headlight distance.

13.4.3 Setting out vertical curves.

(i) Establish the required data – road type, crest/sag, single/dual carriageway, overtaking or not, entry and exit gradients, chainage and level of intersection.

(ii) Determine the required K-value and hence the curve length L and tangent lengths l.

(iii) Fix the chainages of tangent points (A and C in the figure) and the intersection B and their levels.

(iv) Compute grade levels at required distances along tangents.
Grade level = level of $A \pm px/100$, or level of $B \pm qx/100$.

(v) Compute offsets from tangent, $y = (p + q)x^2/400l$.

(vi) Compute curve levels, from level of A $\pm px/100 \pm (p + q)x^2/400l$, or similarly from point B.

Normally through chainage is used, as with other curves, 20 m chords, with sub-chords at the tangent points.

Chapter 14

Areas and volumes

14.1 Introduction

The surveyor is often required to determine areas of land and sometimes to subdivide areas and fix new boundaries. Volume calculations, e.g. in construction earthworks, disused quarries, opencast mining, etc., all involve a knowledge of area calculation methods first. Volume calculations in surveying are generally concerned with the volume of earthworks and may be subdivided into calculations based on cross-sections, used for roads, trenches, etc., and calculations based on contours or spot heights. Computer methods, including the use of digitisers, are not covered here.

The standard area units are the square kilometre (km), the hectare (ha) and the square metre (m). All may be used in surveying as appropriate.

$10\,000\,\text{m}^2 = 1\,\text{ha}$ (approx. 2.5 acres)
$100\,\text{ha} = 1\,\text{km}^2$ = the area of one OS National Grid 1:2500 map sheet

14.2 Areas of simple figures

14.2.1 Rectangle, square

$$\text{Area} = \text{length} \times \text{breadth} \qquad (14.1)$$

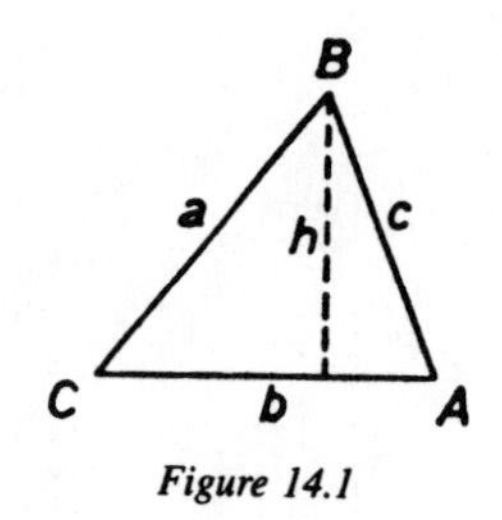

Figure 14.1

14.2.2 Triangle

Given base length b and perpendicular height h (*Figure 14.1*),

$$\text{area} = b \times h/2 \tag{14.2}$$

Given sides a, b and c and $s = (a + b + c)/2$,

$$\text{area} = \sqrt{\{s(s - a)(s - b)(s - c)\}} \tag{14.3}$$

Given two sides, a and b and included angle C,

$$\text{area} = (a\, b \sin C)/2 \tag{14.4}$$

Given side a and angles A, B and C,

$$\text{area} = (a^2 \sin B \sin C)/(2 \sin A) \tag{14.5}$$

14.2.3 Trapezium

Given parallel sides a and b and perpendicular distance between them h,

$$\text{area} = (a + b)\, h/2 \tag{14.6}$$

14.2.4 Circle

Given radius r and $\pi = 3.141592$,

$$\text{area} = \pi r^2 \tag{14.7}$$

14.2.5 Segment of a circle

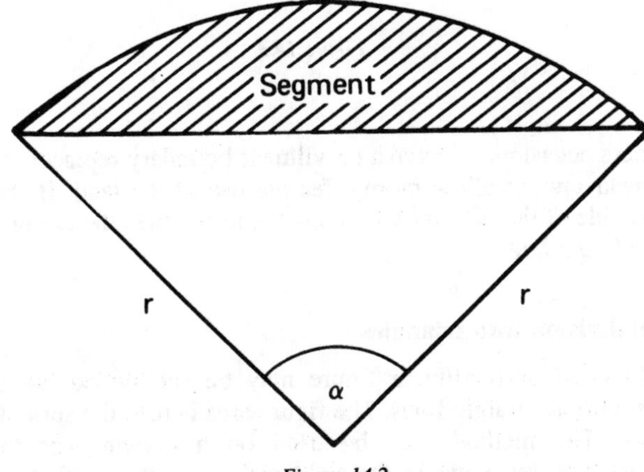

Figure 14.2

Area of the sector less the area of the triangle in *Figure 14.2*,

$$\text{area} = (\pi r^2 \alpha/360) - (r^2 \sin \alpha)/2 \qquad (14.8)$$

Note that $(r^2 \sin \alpha)/2$ comes from Equation (14.4).

14.3 Areas from drawings or plans

Areas may be obtained from drawings, from field measurements (with or without plotting) or from calculated data such as rectangular co-ordinates or partial co-ordinates. In some cases the methods may overlap, but the logical procedure is to determine the data available and then select an appropriate calculation method.

For calculation purposes, the boundaries of areas on plans and drawings may be made up of straight lines, a *rectilinear boundary*, or of irregular lines, known then as *curvilinear boundaries*. Where the boundary of an area is curvilinear, as in *Figure 14.3*, straight *give and take lines* may be drawn on the plan to replace the actual boundaries for calculation purposes. These lines should be placed so that the areas excluded by them are approximately equal (by eye) to the external areas taken in by them. When the boundaries have been 'averaged out' in this way the figure becomes rectilinear and the area can be calculated from triangles and trapeziums.

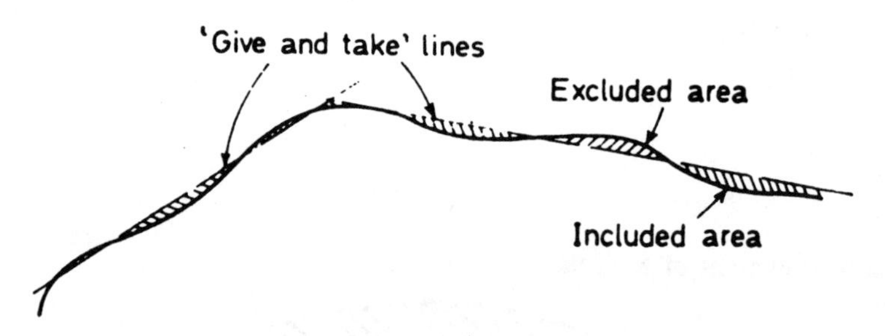

Figure 14.3

Landowners occasionally have a curvilinear boundary replaced on the ground by a rectilinear one, to allow more effective use of the land. If the area of the parcel either side of the boundary remains the same, then the boundary is known as a *give and take line*.

14.3.1 Subdivision into triangles

Any straight-sided or rectilinear figure may be subdivided into triangles by drawing appropriate straight lines. The figure area is then the sum of the areas of the triangles. The method may be used on a drawing, or the necessary measurements may be made in the field and the area calculated directly. On occasion it may be possible to use trapeziums, this slightly reducing the labour involved.

14.3.2 Counting squares or dots

The *squares method* uses a transparent or translucent overlay, with a grid drawn on the overlay, each grid square representing a unit of square measure. A typical type is 2 mm squared paper with each tenth line in a heavier gauge. At 1:500 scale each small square represents 1 m^2 and each large square 100 m^2. At 1:2500 scale the areas would be 25 m^2 and 0.25 ha, respectively.

The transparency is placed over the area to be measured so that as many large squares as possible fall within the boundary and a line of the grid is made to coincide with one or more of the rectilinear boundaries, if any. The squares are now counted, part squares at the boundary being counted as either in or outside the parcel as judged by eye, see *Figure 14.4*, the hatched area being counted.

The grid should then be shifted and rotated in plan and a fresh count made. Preferably this should be done still again and the results compared. An area correct to 0.5% should be expected.

An alternative method, considered by some to be a better method, is to replace

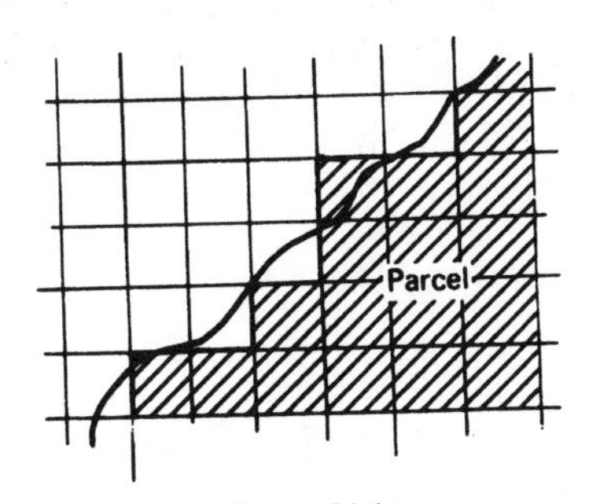

Figure 14.4

the grid of squares by a grid of dots at the same spacing as the grid lines, then count the dots included in the figure. Either method is suitable if there are many areas to measure, but it is uneconomical for a single parcel of land if the grid has to be specially produced by hand drawing.

14.3.3 Ordinates methods

14.3.3.1 Ordinates overlay

This method uses an overlay like the square counting method, but the overlay is ruled with equally spaced parallel ordinates, their distance apart being some convenient interval of, say, 8 or 10 mm. The ruled sheet is placed over the area to be measured and estimated give and take lines drawn in as in *Figure 14.5*. Knowing the distance between the ordinates, to scale, the distance between the vertical give and take lines may be scaled off for each block and the area of each

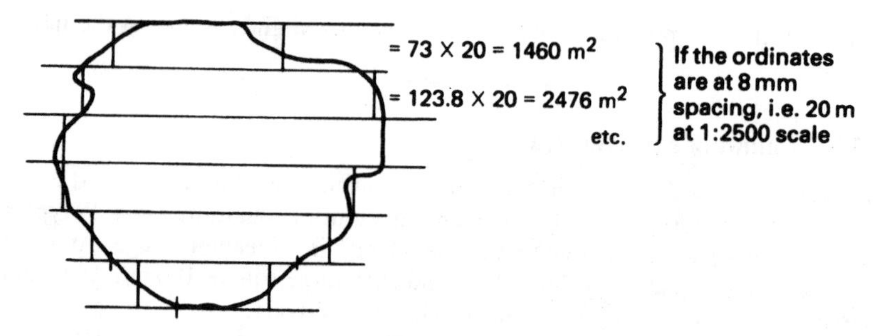

Figure 14.5

block calculated as a simple rectangle. The sum of the areas of the rectangles will give an estimate of the area of the whole figure.

14.3.3.2 Mean ordinate rule

In this method, a line is drawn through the centre of the area to be measured, as in *Figure 14.6*, the line length being *d*. The line is divided into equal intervals, of length *l* and ordinates drawn at right angles to the line, the length of the ordinates being scaled as $o_1, o_2, o_3, \ldots o_n$.

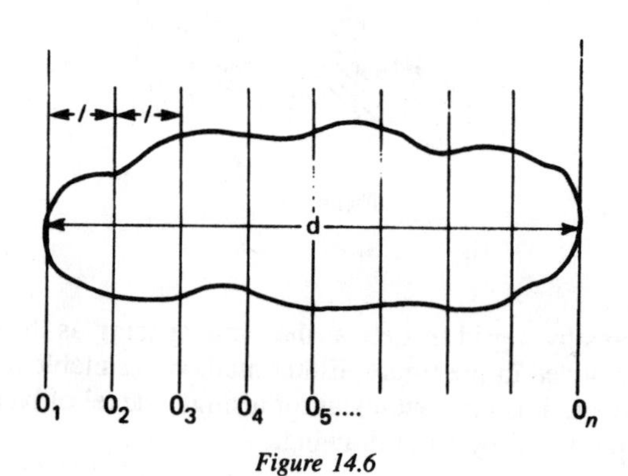

Figure 14.6

The *mean ordinate rule* states that the area is equal to the mean ordinate length by the total length of the line.

$$\text{area} = (d/n)(o_1 + o_2 + o_3 + \ldots + o_n) \tag{14.9}$$

where *d* is the total line length, *n* is the number of ordinates and the ordinates are $o_1, o_2, o_3, \ldots$. The method is rapid, but not very accurate.

14.3.3.3 Trapezoidal rule

In this method, the shape formed between each pair of ordinates is considered to be a trapezium, then summing the area of each trapezium gives

$$\text{area} = (l/2)(o_1 + 2o_2 + 2o_3 + 2o_4 + \dots 2o_{(n-1)} + o_n) \quad (14.10)$$

If the boundary is curvilinear then the area is a good approximation and the accuracy may be increased by increasing the number of ordinates.

14.3.3.4 Simpson's rule

This is very similar to the trapezoidal method, but assumes that the irregular boundary consists of a series of parabolic arcs between the ordinates rather than straight lines. In this case, however, the area must be divided into an *even* number of strips by an *odd* number of ordinates. If there are an odd number of strips, then the last strip area must be calculated separately and added to the area calculated for the even number of strips.

$$\text{area} = (l/3)(o_1 + 4o_2 + 2o_3 + 4o_4 + \dots + 2o_{(n-2)} + 4o_{(n-1)} + o_n) \quad (14.11)$$

The rule may be stated in words as:

> Where an area with curvilinear boundaries is divided into an even number of strips of equal width, the total area is equal to one-third the strip width multiplied by the sum of the first and last ordinates, twice the sum of the remaining odd ordinates and four times the sum of the even ordinates.

14.3.4 The planimeter

The *planimeter* is a mechanical device for integration, i.e. the calculation of the area under the curve of a function. Area measurement from drawings by planimeter is efficient and fast, particularly if the areas are small or very irregular in shape. The forms used in survey are known as *polar planimeters*, their accuracy typically being similar to that for counting squares, 0.5%. Recent electronic models are probably more accurate.

The elements of the instrument are illustrated in *Figure 14.7*.

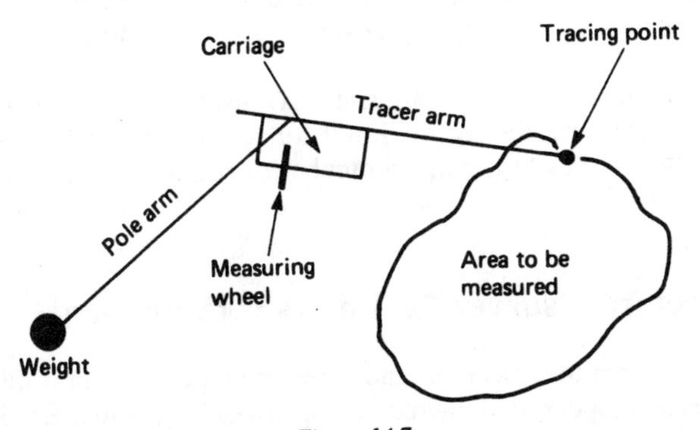

Figure 14.7

(1) The *pole arm*, with a needle pointed weight at one end, the *pole weight* sometimes being separate. The other end of the pole arm carries a pivot resting in a socket in the tracer arm.
(2) The *tracer arm*, fitted at one end with a *tracing point* with an adjustable support.
(3) The *carriage*, which may be fixed to, or may slide along, the tracer arm, has a *measuring wheel* and a *counting dial*.
(4) The *counting dial* records the number of revolutions made by the wheel, which may also be fitted with a vernier scale.

Planimeters with fixed carriages are known as *fixed arm polar planimeters*, and will record the area either in square inches or square centimetres only. *Sliding bar polar planimeters* (the carriage may slide along the tracer arm) have a scale on the tracer arm so that the carriage may be set at some particular map scale so that one revolution of the wheel will equal one specific unit of area. For example, at 1:1250 scale, one revolution of the measuring wheel is equal to one hectare, or at 1:100 scale, 100 m^2.

The simplest application of the planimeter is as follows. Set the carriage as necessary for the scale of the drawing, attach the pole arm to the tracer arm. Place the needle pointed weight outside the area to be measured, with the pole and tracer arms roughly at right angles to one another and the tracing point approximately in the centre of the area. Check, without moving the weight, that the tracing arm can reach any part of the figure boundary without the measuring wheel moving across the edge of the drawing material. If this cannot be done, try the pole weight in a different position or divide the area into several smaller areas. Locate and mark a starting point on the boundary, an ideal position being a 'dead' point where a slight movement of the tracing point causes no rotation of the wheel. Set the tracing point on the starting point, on older models note the reading, on newer models set the scale to zero. Carefully move the tracing point clockwise around the boundary, terminating exactly on the starting point, note the number of revolutions on the dial and the part revolutions on the measuring wheel drum and vernier as appropriate. Subtract the start reading, if any, then convert the result to an area in hectares or square metres.

The procedure should be repeated several times, obtaining several values for the area and when three consistent results have been achieved they may be meaned and the mean value accepted.

A checking rule or test bar should be carried in the instrument case, to test the instrument's accuracy. A check may also be made by circumscribing an area of known size.

Electronic models have a keyboard and LCD display. The survey scale may be entered at the keyboard, the required unit for results (m^2, km^2, acres, etc.) and whether the user wishes to average a number of results. An accuracy of ±0.2% or better is claimed.

14.4 Areas from survey field notes with no plan

On occasion the area of a piece of land is required, but no plan is available, e.g. for land or crop compensation payment, crop areas for subsidy, etc. Survey lines can usually be arranged so as to divide the survey area into triangles or trapeziums and then the formulae (14.2) to (14.6) may be used to calculate the survey area, direct from field bookings, without plotting a plan. If the field bookings are tied to a closed loop traverse it may be preferable to obtain the co-ordinates of the traverse then calculate the area as shown below.

Where there are narrow strips of land outside the survey triangles, offsets may be taken to act as ordinates and Simpson's or the Trapezoidal rule used to calculate the strip area. This is not generally practicable in a survey which is to be plotted, since in that case such offsets are usually taken at changes of direction of the detail or boundary.

14.5 Areas from co-ordinates

If a closed traverse has been co-ordinated, or the boundary of a property or a cross-section is defined by co-ordinated points, then it is possible to calculate the area of the figure concerned directly from the co-ordinates and a basic sketch of the traverse plan. *Figure 14.8* shows a sketch of a four-sided closed loop traverse. It will be evident that the area of the figure ABCD is equivalent to the addition and subtraction of the various trapeziums thus:

$$\text{area} = (N_B - N_A)(E_A + E_B)/2 + (N_C - N_B)(E_B + E_C)/2$$
$$- ((N_C - N_D)(E_C + E_D)/2 + (N_D - N_A)(E_D + E_A)/2 \qquad (14.12)$$

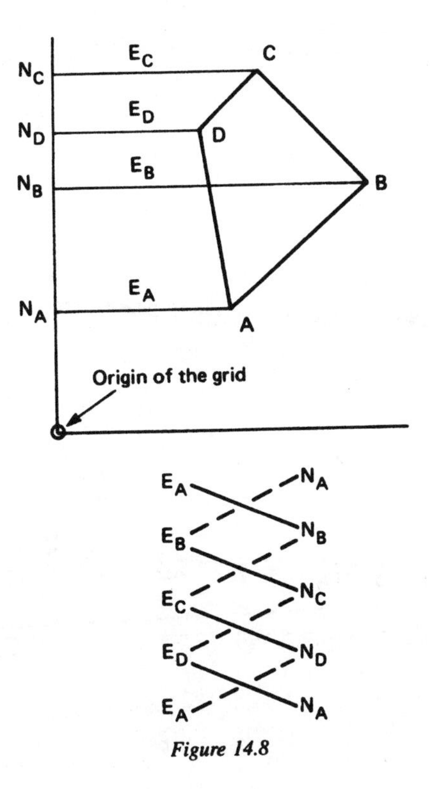

Figure 14.8

This can be developed into standard formulae, but the basic method is much easier to remember.

14.6 Alteration and subdivision of areas

Typical area management problems concern the replacement of one boundary by another (e.g. substituting a straight boundary for a curvilinear one) and the subdivision of areas of land. Existing boundaries may be described loosely as either being in the English system of general boundaries, a boundary being *mered* to a line on a map, or else being defined mathematically by a series of co-ordinated points, as in the fixed boundaries of the Australian Torrens system.

14.6.1 Methods for replacing a curvilinear boundary by a give and take line

14.6.1.1 Geometrical (graphic) methods

Figure 14.9 illustrates a boundary plotted on a drawing, the boundary consisting of three straight lines. It is required to replace these by a single straight line. The areas must, of course, be plotted to a convenient large scale.

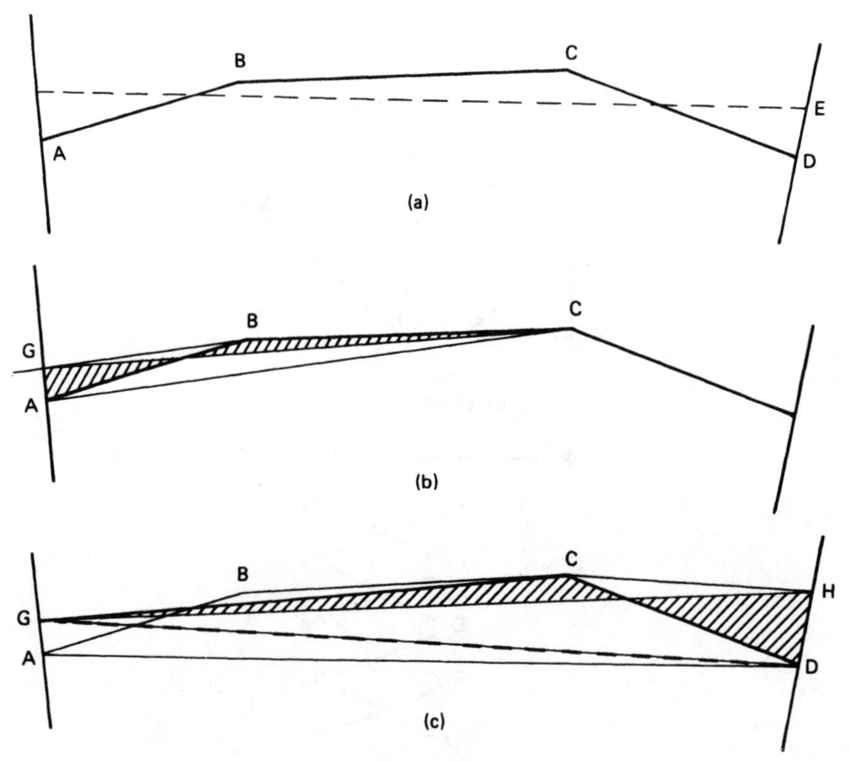

Figure 14.9

Method 1: Lightly draw an estimated give and take line, as in *Figure 14.9(a)*, then by counting squares, calculate the areas ABCDA and ADEFA. If the two areas are equal, the give and take line is in the correct position. If the areas are not equal, move the estimated line and repeat the area calculations, repeating the process as needed.
Method 2: Draw a trial and error give and take line, as in (1), but measure the area of the figure AFEDCBA, clockwise from A, passing through the points in that order, using a planimeter. If the give and take line is in the correct position then the total area of the figure will be zero, the included areas equalling the excluded area. If not equal to zero, move the give and take line and try again.
Method 3: In *Figure 14.9(b)*, construct the line BG parallel to CA, then the area of triangle ABC is equal to the area of triangle AGC. (The geometrical theorem that the areas of triangles on the same base and between the same parallels are equal.) Construct the line CH parallel to the line GD, then similarly the area of triangle GCD is equal to the area of triangle GHD. The area ABCDA in *Figure 14.9(c)* is now equal to the area AGHDA, thus the line GH is the required give and take line.

14.6.1.2 *Numerical methods*

Alternatives to graphical solution include direct field measurement and setting out or calculation from co-ordinates.

Method 1: Field method. *Figure 14.10* shows a field sketch of an existing fence to be replaced by a give and take line. Select a point A on the fence from which the new line will commence, then run a chain line from A to any point B on the opposite boundary. From chainage and offset measurements, calculate the area enclosed between the existing fences and the chain line AB.

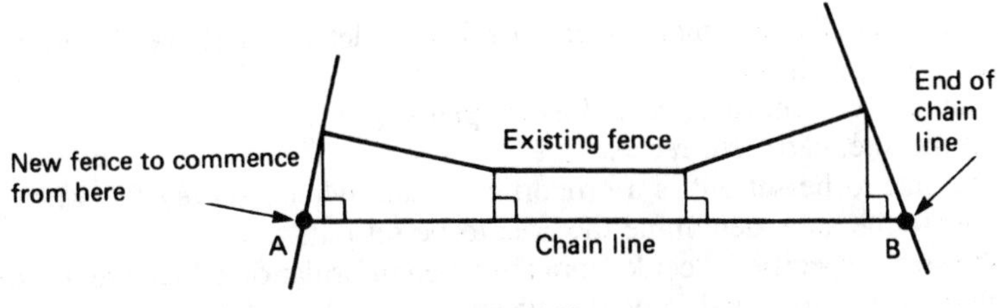

Figure 14.10

Calculate the height (h) of a triangle, of area equal to that enclosed area and with a base equal in length to AB. Set out a line parallel to the chain line AB and at a distance of h from the chain line. The point where this new parallel line cuts the right-hand boundary defines the other end of the required give and take line and this should be joined to point A.
Method 2: Co-ordinates. *Figure 14.11* illustrates existing boundaries defined by the co-ordinated points A, B, C, D, W, X, Y and Z. It is required to replace the boundary ABCD by a single give and take line.

Draw the line AD, and calculate the hatched area ABCD. Calculate the length DF, F being a point on the line DX, such that the area of the triangle ADF is equal

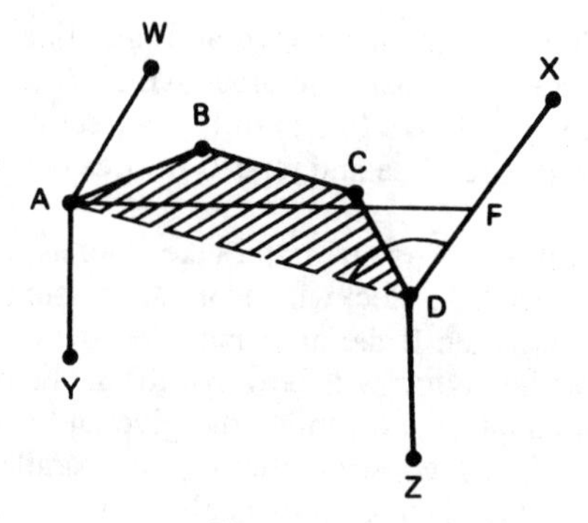

Figure 14.11

to the hatched area. The length AD and the angle ADF may be calculated from the known co-ordinates, then Equation (14.4) may be used to determine DF. The line AF is the required give and take line.

14.6.2 Subdivision of areas

Subdivision problems may involve dividing a plot of land into proportional parts, or parts of a specific size or shape. New boundaries may be required to be parallel to some feature, or to start and/or end on given points. Each task may be different, with different methods of solution and setting out and for some, graphical methods will suit, while others will need numerical methods. Setting out problems are covered in Chapter 12.

While no set methods can be laid down, the following general approach is recommended:

For each subdivision plot, ascertain relevant details of shape, location size, fixed points or limits.
Obtain all relevant data, co-ordinates, plans if any.
Visit the site, carry out recce.
If the size to be set out is a proportion of the whole, survey the whole site, calculate the area, determine the area to be set out.
If the size is specified, decide from the given information which are the known boundary positions and, roughly, where the new boundary will lie.
Insert pegs or other marks on your estimate of where the new rectilinear boundary or boundaries shall be located.
Survey the site as bounded by the existing features and the new pegs. Obtain the area of the new subdivision, this is unlikely to be correct but will provide a start basis.
Deduce how much the new peg or pegs must be moved to obtain the correct area.
Move the pegs, check the area, repeat as necessary.
If the area is to be a regular shape, it may be possible to set it out directly without an initial pegging of the site.
If the area is defined by rectangular co-ordinates, the new boundary points may be computed before visiting the site, using the formulae given earlier.

14.7 Volume calculations

Volume calculations in surveying are generally (but not exclusively) concerned with the volume of earthworks and may be subdivided into calculations based on cross-sections, as used for roads, trenches, etc., and calculations based on contours or spot height.

14.7.1 Volumes of simple solids

14.7.1.1 Cube, prism, cylinder

Given base area A and height h (in *Figure 14.12*),

$$\text{volume} = A \times h \tag{14.13}$$

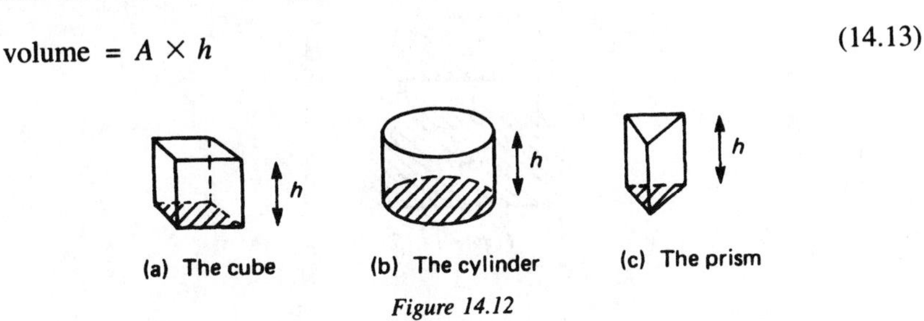

(a) The cube (b) The cylinder (c) The prism

Figure 14.12

14.7.1.2 Wedge

Given base area A and height h (in *Figure 14.13*),

$$\text{volume} = A \times h/2 \tag{14.12}$$

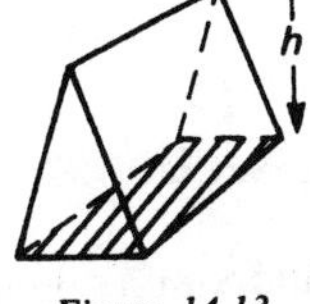

Figure 14.13

14.7.1.3 Pyramid, cone

Given base area A and height h (in *Figure 14.14*),

$$\text{volume} = A \times h/3 \tag{14.15}$$

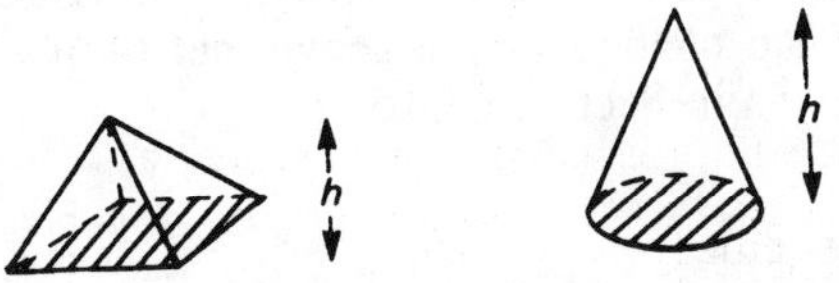

h h

(a) The pyramid (b) The cone

Figure 14.14

14.7.2 The prismoid

The prismoid is the commonest shape in earthworks calculations. A *prismoid* is a solid figure having two parallel end faces, not necessarily of the same shape, the sides of the solid being formed by continuous straight lines running from face to face. A prismoid may consist of any of the simple solids above, or it may be a truncated prism or wedge.

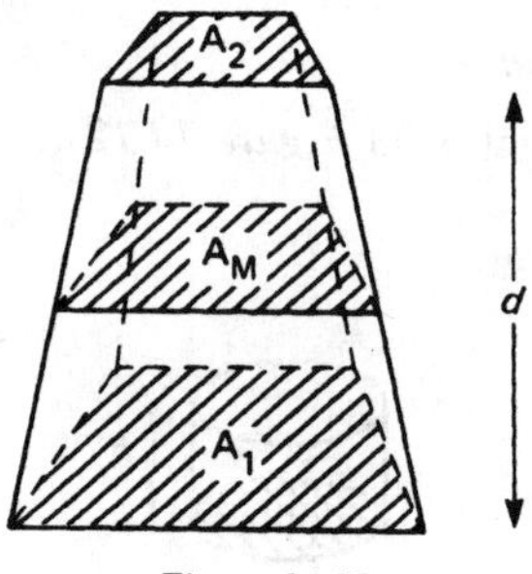

Figure 14.15

The common formula for all forms of prismoid is,

$$\text{volume} = (d/6)(A_1 + 4A_M + A_2) \tag{14.16}$$

where A_1 and A_2 are the end face areas, A_M is the area of the cross-section midway between the faces, and d is the distance between the end faces, as in *Figure 14.15*. A_M may be obtained from either the calculated or measured side lengths.

14.8 Volumes from cross-sections

In roadworks, railways, canals and similar long earthworks, cross-sections are taken as described in Chapter 8, at regular intervals, then the volumes of cut or fill are obtained from these together with the measured distance between cross-sections. The solid figure between successive cross-sections is typically a prismoid, except where the longitudinal section is along the line of a curve or in special circumstances such as an opening in an embankment.

Three basic calculation rules are used, these being the *mean areas rule*, the *trapezoidal rule* and the *prismoidal rule*, all very similar to the ordinate rules in area calculations. In all the rules the individual cross-section areas may be obtained by any of the methods given above, but further suggestions on area determination are given in Section 14.8.5.

14.8.1 Mean areas rule

Where successive cross-section areas are $A_1, A_2, A_3, \ldots A_n$, and distance from section A_1 to section A_n is L,

$$\text{volume} = (L/n)(A_1 + A_2 + A_3 + \ldots + A_n) \quad (14.17)$$

In effect, the mean or average of all the cross-section areas is obtained and multiplied by the overall length of the excavation. The method is easy to apply, but not very accurate, giving too large a volume. It is therefore rarely used.

14.8.2 Trapezoidal rule

This method is also known as the *method of end areas* or *averaging end areas.* Where two successive cross-sections are A_1 and A_2 and they are spaced a distance l apart, the volume contained between them is

$$\text{volume} = (l/2)(A_1 + A_2) \quad (14.18)$$

provided the cross-section area at the midway point between A_1 and A_2 is actually the mean of the areas of these two. If there is only a slight change between successive cross-sections, the accuracy is acceptable for many purposes and the form may be developed for a large number of consecutive cross-sections A_1, A_2, A_3, … ,A_n.

In this case,

$$\text{volume} = (l/2)(A_1 + 2A_2 + 2A_3 + 2A_4 + \ldots + 2A_{(n-1)} + A_n) \quad (14.19)$$

where l is the uniform distance between successive cross-sections. The similarity to the trapezoidal rule for areas will be evident. Again, the results are generally greater than the true volume.

14.8.3 Prismoidal rule

The best results are obtained if the volume of earth between successive cross-sections is a prismoid. The volume of the prismoid is obtained from Equation (14.16). Where a large number of cross-sections have been taken, every alternate section may be regarded as being an end face of a prismoid, the other cross-sections being taken as the respective A_M or middle cross-sections of the prismoids. Then,

$$\text{volume} = (l/3)(A_1 + 4A_2 + 2A_3 + 4A_4 + 2A_5 + \ldots + 2A_{(n-2)} + 4A_{(n-1)} + A_n) \quad (14.20)$$

where l is the uniform distance between successive cross-sections. Note that the value d in Equation (14.16) is equivalent to $2 \times l$ in Equation (14.20), hence $d/6$ becomes $l/3$. This is *Simpson's Rule* for *volumes,* similar to that for areas. Note that each cross-section appears once and their multipliers are respectively 1, 4, 2, 4, 2, 4, … 2, 4, 1, and there must be an odd number of cross-sections just as there is an odd number of offsets or ordinates in the area rule.

14.8.4 Roads and similar works curved in plan

Where roads, railways, etc., follow a curved line on plan, the volumes given by the standard formulae are incorrect, since the end faces are not parallel. The error is generally small and frequently disregarded for practical purposes. Details may be obtained in a specialist textbook on engineering survey.

14.8.5 Cross-section areas

Traditionally, cross-sections for volume calculations are plotted to scale, then the cut and fill areas deduced by planimeter, by counting squares, or by dividing the area into triangles and computing the triangle areas from scaled measurements. There are also formula methods and these may be more popular with increased use of calculators and computers. It should be noted, however, that where a cross-section is very complex, the data entry for a formula method by calculator may take longer than the traditional methods.

Formula methods rely, effectively, on making use of the reduced level at each change of slope of the cross-section, both of the existing ground surface and the final section shape, together with the distances of such points left or right of the centre-line. Most of these points may be determined, but the points where the new ground slopes intersect the existing ground slopes may have to be scaled or computed. In addition, the ground surface may vary irregularly across the width of the cross-section and the formation may not be horizontal. For roads, for example, the formation may be cambered, or superelevated. Further, many formula methods pose problems on cross-sections which are part in cut and part in fill.

14.8.6 Changes in soil volume

Excavated materials generally expand and their volume becomes greater than when measured in the ground. This affects haulage costs, of course and may be an important consideration. The change in volume is known as *swell* or *bulking* and may vary from 5% to 50% of the original volume. Loose material, when compacted in embankments or fill, may conversely *shrink*. Sand may have *8% shrinkage*, while a clay might be subject to 25% shrinkage. A soil survey may be needed to determine appropriate values for the area concerned.

14.9 Volumes from contours

Contour plans may be used to calculate volumes by treating the areas enclosed by successive contours as 'cross-sections' and the vertical interval between contours as the constant distance between cross-sections. The three methods – mean areas, trapezoidal and prismoidal, may be used as appropriate to calculate the volume of material contained between two specified closed contours. This may also be applied to the calculation of the amount of material in a stack such as dumped fill, sand-dunes and so on, but the commonest application is probably the determination of the volume of water which will be contained by a reservoir or dam. The method may also be used to calculate the cut or fill in constructing a horizontal or sloping base on an existing surface, as for the foundations of an industrial site or the playing areas of a sports complex.

For water volumes, each contour line on the side slopes may be regarded as one of successive water-lines and the successive plan areas of water at the various levels are used.

Plan areas enclosed by a contour line are best measured by planimeter, which is ideal for such irregular outlines. In calculating the volume of a mound or hollow it is unlikely that the highest or lowest point will coincide with the level of a contour line. It is therefore normal practice to calculate the volume of the solid between this spot height and its adjacent contour by the most relevant formula, i.e. $V = A \times h/3$, Equation (14.15).

14.10 Volumes from spot heights

The following method is most useful when an excavation is to be made with vertical sides, such as for a basement. Spot heights are taken over the whole area, on a uniform grid of squares, as outlined earlier for contouring. Each square of the grid may then be considered as the top end-face of a vertical prism running from formation level up to the original ground surface. The volume of each of these regular prisms is the product of the grid square area and the mean of the four corner heights. The corner heights are, of course, ground level minus formation level.

Figure 14.16 represents the plan of an area to be excavated, with A, B, C, ... J being points on the surface which have been levelled in the form of a square grid. The reduced levels are shown beside each point, the grid squares are 10 m on a side and the required formation level is 12.000 m above datum.

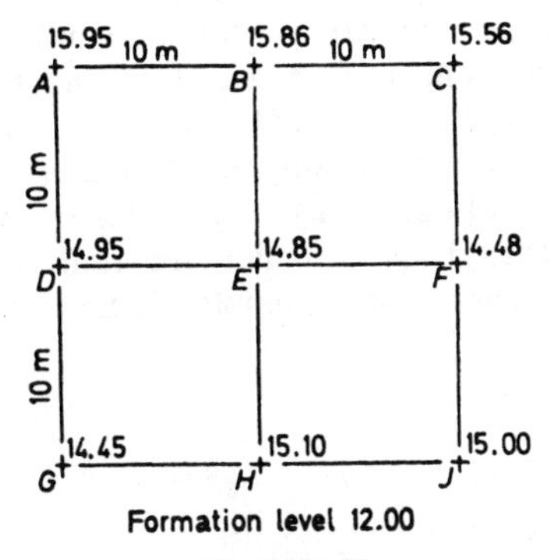

Figure 14.16

To calculate the excavation volume, each square prism could be treated separately and for each one the mean of the four heights taken and multiplied by the plan area of $10 \times 10\,m^2$. This would be a tedious procedure and unnecessary, since several of the corner heights appear in more than one prism. The calculation is normally made as follows, in tabular form:

Ground Point	*Prism corner height* (h_n)	*No. of squares occurs in* (n)	*Product* $h_n \times n$
A	3.95	1	3.95
B	3.86	2	7.72
C	3.56	1	3.56
D	2.95	2	5.90
E	2.85	4	11.40
F	2.48	2	4.96
G	2.45	1	2.45
H	3.10	2	6.20
J	3.00	1	3.00
			Total = 49.14 = $\Sigma(h_n \times n)$

$$\text{Volume} = 10 \times 10 \times 49.14/4$$
$$= 1228.5\,\text{m}^3$$

In this calculation, each corner height is multiplied by the number of times it appears as a corner height and the total of 49.14 is the final sum of the four heights for all the squares. When this figure is divided by four, the result is the sum of the mean heights and need only be multiplied by one square area.

The mean height of the site, or in this example, the mean height of the excavation, may be obtained from $\Sigma(h_n \times n)/\Sigma n$, which is 49.14/16 = 3.071 m (3.07125). The mean height × total area would also give the volume, in this example 3.07125 × 20 × 20 = 1228.5 m^3.

The grid could also be interpreted as a series of triangles, such as ABD, DBE, BCE, ECF and so on. The calculation would be similar, but each corner height would appear a different number of times and the quantity ($hn \times n$) would have to be divided by 3 instead of 4 before multiplying by the plan area of one triangle. There is no advantage, however, in using the triangle method, unless the diagonal of any square lies along a ridge or a valley. If this happens, it may be better to divide the square into triangles, thus better emulating the surface of the ground. The calculation is, of course, more complex if there is a mix of squares and triangles.

Bibliography

Allen, A.L., *Practical Surveying and Computations*, 2nd ed, Butterworth-Heinemann (1993).

Bannister, A. and Baker, R., *Solving Problems in Surveying*, Longman (1989).

Bannister, A., Raymond, S. and Baker, R., *Surveying*, 6th ed, Longman (1992).

Bowyer, J., *Guide to Domestic Building Surveys*, Architectural Press (1979).

Brighty, S.G., *Setting Out: a guide for site engineers*, Granada (1981).

British Standards Institution:

BS 5606: 1990 Guide to Accuracy in Building.

BS 5930: 1981 Code of Practice for Site Investigations.

BS 5964: 1990 Building Setting Out and Measurement.

BS 7307: 1990 Building Tolerances: Measurement of Buildings and Building Products.

BS 7308: 1990 Presentation of Accuracy Data in Building Construction.

BS 7334: 1990, 1992 Parts 1 to 10: Measuring Instruments for Building Construction.

Building Research Establishment:

BRE Digest 202: 1977 Site Use of the Theodolite and Surveyors Level.

BRE Digest 234: 1980 Accuracy in Setting Out.

Burnside, C.D., *Electromagnetic Distance Measurement*, 2nd ed, Collins (1982).

Cooper, M.A.R., *Control Surveys in Civil Engineering*, Collins (1987).

Cooper, M.A.R., *Fundamentals of Survey Analysis and Measurement*, Collins (1974).

County Surveyors Society, *Highway Transition Curve Tables (Metric)*, The Carrier Publishing Co. Ltd. (1969).

Department of Transport, *Departmental Standard TD 9/81, Road Layout and Geometry: Highway Link Design* (1985).

Irvine, W., *Surveying for Construction*, 4th ed, McGraw-Hill (1995).

Muskett, J., *Site Surveying*, BSP Professional (1988).

Ordnance Survey – various.

Petrie, G. and Kennie, T.J.M., *Terrain Modelling in Surveying and Civil Engineering*, Whittles Publishing in association with Thomas Telford (1990).

Schofield, W., *Engineering Surveying 1*, 3rd ed, Butterworths (1984).

Schofield, W., *Engineering Surveying 2*, Butterworths (1984).

Shepherd, F.A., *Advanced Engineering Surveying*, Edward Arnold (1981).

Shepherd, F.A., *Surveying Problems and Solutions*, Edward Arnold (1968).

Uren, J. and Price, W.F., *Surveying for Engineers*, 3rd ed, Macmillan (1994).

Index

Abney level, clinometer, 57
Abstract of traverse observations, 179
Accuracy, errors, precision:
  area calculations, 306–8
  areas from survey notes, 308
  building surveys, 241
  consistency, 9
  direct linear measurement, 64–9
  economy of, 9
  EDM surveys, 138–43
  general, 9, 10
  levelling, 102, 106, 265
  setting out, 257, 263, 265, 268, 270
  significant figures, 16
  stadia measurement, 136
  traverse survey, 172, 186, 190
  types of error, 11
Additive constant in stadia, 134
Adjustment of the level book, 103
Alidade, theodolite, 110
Alignment errors, 68
Alignment laser, 270
Altitude bubble, 117
Altitude correction (above MSL), 69
Angle of depression/elevation, 118
Angular measurement, 109
  Abney level, 57
  adjustments, 129–31
  bearings, 18
  booking, 126–8, 174–9
  measuring an angle, 124–8
  nadir, 117
  optical square, 56
  parallax error, 80
  prismatic compass, 168
  reference object (RO), 125
  repetition method, 128
  round of angles, 127
  setting on an angle, 122–4
  sighting targets, 124
  simple reversal technique, 125
  single horizontal angle, 125
  terms used, 125
  theodolite:
    electronic, 109, 120–2
    optical, 109, 111–20
  vertical angles, 118, 128
  zenith, zenith distance, 118, 134
  zero of measurement, 125
Applications of EDM:
  control, 144
  detail survey, 144
  levelling, 145
  remote elevation measurement, 145
  setting out, 144
Applications of levelling:
  contouring:
    direct, 149
    indirect, 150
  establishing a TBM, 146
  precise levelling, 156–8
  reciprocal levelling, 159
  sections, 153
  spot heights, 150
Areas, 302–12
  accuracy of calculations, 306, 308
  by planimeter, 307
  circle, 303
  co-ordinate methods, 309
  from drawings and plans, 304
  from survey notes, 308
  give and take lines, 304
  mean ordinate rule, 305
  ordinate overlay, 305

Areas – *continued*
parcels, on OS maps, 34
rectangle, 302
segment of circle, 303
Simpson's Rule, 307
squares or dots, 305
sub-division, 304, 310, 312
trapezium, 303
trapezoidal rule, 306
triangle, 303
Area excavation or fill, 271
Areas and parcels on OS maps, 34
Arithmetic mean, 12
Arrow, marking, 55
Atmospheric refraction, 94
Automatic level, 82, 84, 106
Automatic vertical indexing, 117
Auxiliary station calculations, 206

Back object/station, 125
Back sight, 93
Barometric heighting, 72
Base line, 8, 162, 196, 197
Basic survey methods, 5
Bearing and distance, radiation, 5, 221–39
Bearings:
grid, 18
magnetic, 18, 181
misclosure, 186
quadrant, 18
relative, 18
true, 18
whole circle, 18, 181
Beginning of curve, 276
Benchmarks:
density of OS, 74
lists of OS, 76
Ordnance Survey, 74, 102
TBMs, 74, 102
types, 75
Block plans, 256
Blunders, 11
Boning rod, 271
Booking observations and measurements:
angular measurement, 126–8, 174–9
building surveys, 248
EDM detail survey, 227–32
levelling, 98
linear control surveys, 165–7
linear measurements, 16
tape and offset detail surveys, 214–18
Boundary:
curvilinear, 304
OS administrative, 29, 31
rectilinear, 310
Bowditch's method, 191
Bracket bench mark, 74
Bubble:
adjustment position, 113
adjustments to level, 106
altitude, 117
circular, 83
plate, 113
staff, 90
Builder's level, 87
Builder's theodolite, 110
Building condition surveys, 42
check list, 42
equipment, 42
schedule of condition, 45
scope, 42
Building movement, 253
Building surveys, 241–55
accuracy requirements, 241
booking the measurements, 248
building movement, 253
bulging walls, 254
cracks in walls, 254
deformation measurements, 253
drawing equipment, 249
drawing material and size, 249
elevations, 1, 251
equipment, 242, 244
European projection, 249
field procedure, 244
first angle projection, 249
inaccessible distances, 248
layout on the drawing material, 249
leaning walls and buildings, 253
movement of buildings, 253
office procedure, 249
plans, 249
plot survey, 242
plotting, 250, 251
plotting scales, 241
preliminaries, 244
reconnaissance, 242, 244
recording measurements, 243
sections, 251
settlement and subsidence, 255
site plan, 241
sketching, 244
symbols, 252
system of measurement, 246
Bulking, 316

Cadastral survey, 3
Carrier wave (EDM), 137
Catenary taping, 58

Central meridian, 24
Centring the theodolite, 113
Chain:
  chainage, 52
  follower, 60
  Gunter's, 52
  land, 52
  leader, 60
  metric, 52
  surveyor's, 52
  surveys, 52
Check or proof line, 162
Circle locking lever, 113, 123
Circle microscope system, 119
Circle orienting drive, 113, 123
Circle reading eyepiece, 111, 118
Circular bubble, 83
Circular curves, *see* Curve ranging
Clinometer, 57
Closed traverse, 169
Closing co-ords./bearing, 187
Coding systems, 224
Coefficient of expansion of steel, 68
Coincidence optical micrometer, 120
Coincidence prism bubble reader, 83
Collimation adjustment, error, 104
Collimation height, 93
Collimation line, 79
Collin's Point resection, 207
Combining taping corrections, 70
Compass, prismatic, 168
Compensator unit, 85
Computers, 20
  software, 20
Construction level, 87
Contaminated land, 39
Contours, 7, 72
  direct contouring, 149
    by section levelling, 150
    by spot heights, 150
  horizontal equivalent (HE), 147
  indirect contouring, 150
    by grid levelling, 150
    by section levelling, 151
    by spot heights, 151
  interpolation and plotting of, 152
  location of, 149
  uses of, 148
  vertical interval (VI), 147, 148
Control, 4
  framework, 9, 161
  horizontal, in setting out, 259–65
  supplying plan, 8, 161–203
  vertical, 4, 267–71
Control surveys, 161–208
  co-ordinate problems, 203–8
  global positioning systems, 199–203
  introduction, 162
  linear, 162–7
  OS control information, 36
  traverse, 168–96
  triangulateration, 199
  triangulation, 196–8
  trilateration, 3, 8, 198
  tying to control, 205–8
County Series maps, 22
Co-ordinate problems:
  plane rectangular transformation, 203
    scale, 203
    shift, 203
    swing, 203
  linking to control:
    auxiliary station, 206
    intersection, 208
    resection, 207
    satellite station, 205
    trilateration, 205
Co-ordinates, 17
  areas from, 309
  Cartesian, 17
  closing, 185
  definitions, 18
  eastings, 19
  geographical, 17
  mathematical plane rectangular, 17
  northings, 19
  opening, 180
  OS National Grid, 17
  partial co-ordinates, 18
  polar (radiation) co-ordinates, 5, 17
  problems, 203
  reference latitude, 17
  reference meridian, 17
  setting out, 264
  surveying rectangular, 5, 17
  three-dimensional systems, 18
  transformation of, 203
  two-dimensional systems, 17
  x, y and z values, 18
Co-ordinatograph, 234
Copyright, 37
Correction factor, 65
Corrections to taped measurements, 64
  altitude (above MSL), 69
  combining, 70
  correction factor, 65
  local scale factor, 69
  sag, 68
  slope, 65
  standardization, 66

Corrections to taped measurements – *continued*
temperature variation, 67
tension variation, 67
Cosine rule, 20, 198, 205
Cross-hairs, 79
Cross-sections, 1, 155
Crown point of curve, 276
Cubic spline, 275
Cul-de-sac station, 226
Cumulative error, 11
Curvature and refraction error, 94
Curve ranging, 275–301
chords, 277
sub-chords, 277
cubic spline, 275
curve types, 275
degree of curve, 277
horizontal circular, 275
beginning of curve, 276
common problems, 289
inaccessible PI, 289
missing peg, 289, 291
obstructed sight, 289
crown point, 276
definitions, 275
deflection angle, 276
end of curve, 276
external distance, 278
general calculations, 277
intersection angle, 276
intersection point, 276
length of curve, 276
linear setting out methods:
deflection distances, 280
offsets from long chord, 283
offsets from tangent, 282
long chord, 278, 283
mid ordinate, 278
point of curvature, 276
radius, 275
rect./polar co-ords. method, 288
road centreline, 277, 281
specification 280
tangential deflection angles, 279
example calculation, 284
setting out by, 284, 286
tangents, 278
two theodolites method, 287
versine (versed sine), 289
through chainage, 277
transition curves:
angle consumed, 295
back angle, 295
cant, 290
centrifugal ratio, 291
and circular arc radius, 291
clothoid curve, or Euler spiral, 293
County Surveyors' Society, 296
cubic parabola, 293
deflection angle, 294
deviation angle, 294
introduction, 289
length, 293
radial force, 289
setting out spiral/circular arc, 297
shift, 296
spiral/circular arc geometry, 296
superelevation, 290
maximum allowable, 292
vertical curves, 298
concave, valley or sag, 298
convex, summit or crest, 298
grade angle, 299
gradient difference, 299
gradient sign convention, 299
gradients, 299
highest point, 300
K-values, 301
setting out, 301
sight distances and curve length, 301
simple parabola, 298, 299
Cut and fill, 273

Data recorder, 21
Datum, OS, 18, 74
Datum surface or line, 74
Declination, magnetic, 181
Deflection angles/ distances, *see* Curve ranging
Desk studies, 39
Detail drawings (setting out), 256
Detail line, 162
Detail surveys, 209–40
tape and offset, 209–21
radiation, 221–39
EDM radiation, 221–35
theodolite and tape, 239
ODM, 235–9
Detail type:
hard, 4
overhead, 4
relief, 4
soft, 4
underground, 4
Diagonal eyepiece, 269
Digital levels, 21, 82, 86, 106
Digital terrain models (DTM), 29, 144, 239
Direct distance measurement, 52–70
Abney level, 57
chains, 53

common tension applied, 54
fieldwork, 58
lifting a line, 59
line clearing and ranging, 59
line measurement, 60
optical square, 56
ranging rods, 55
standard temperature, 54
steel tapes, bands, 53
surface taping, 58
synthetic tape, 55
Documentation, setting out, 256
Double centre theodolite, 113, 122
Drop arrow, 66
Dumpy level, 82

Earth curvature, 94
Easements on land, 39
Easting, 19, 24
EDM radiation detail survey, 221–35
advantages/disadvantages, 222
basis, 221
equipment, 222
field procedure:
alternative measurements, 226
coding systems, 224
communication, 227
cul-de-sac station, 226
local reference object, 225
parent station, 226
setting up, 225
station selection, 223
station sketches, 224
observations:
control, 225
detail and heights, 225
recording data, 227–32
reflector, 226
office calculations, 233
office procedures, 232
preliminaries, 222
plotting/drawing detail, 234, 235
rectangular co-ordinatograph, 234
sight lengths, 222
suitability, 222
Electromagnetic distance measurement (EDM):
accuracy, 138, 140, 143
add-on distancer, 138
add-on keyboard, 138
applications, 144–5
background, 132, 136
carrier wave, 137
control programs, 144
orientation, 144
resection, 144
tie distance, 144
detail survey, 144
electronic tacheometer, 138
error sources and corrections, 142
atmospheric conditions, 142
instrumental errors, 143
NG scale factor, 143
prism constant, 142
slope, 143
frequency, 137
general field use, 138
of add-ons, 138
of total stations, 140
of timed-pulse distancers, 140
Geodimeter and Tellurometer, 136
infrared light, 138
levelling, 145
measurement principle, 137
modern development, 138
modulated infra-red light, 138
operational modes, 140
rapid tracking, 140
repeat, 140
stakeout, 144
standard, 140
tracking, 140
phase difference, 137
reflector prism, 137, 139
remote elevation measurement, 145
setting out, 144
tacheometry, 73, 221–35
targets and prism, 139
total station, 138
wavelength, 137
Electronic theodolite, 121
keyboard, 121
measurement systems, 121
Elevations, 1, 249
End of curve, 276
Engineer's level, 87
Engineer's theodolite, 110, 120
Errors:
accuracy, 10
alignment, 68
arithmetic mean, 12
blunders, 11
cumulative, 11
curvature and refraction, 94
Gaussian distribution, 11
gross, 11
in EDM measurements, 142
in levelling, 106
in stadia measurements, 136

Errors – *continued*
in taped measurements
alignment, 68
applied tension, 67
corrections, 64–70
end marking, 69
reading the tape, 69
mistakes, 11
normal distributions, 11
parallax, 80
permissible, in levelling, 102
precision, 10
principle of least squares, 12
probability distributions, 12
random, 11
residual, 12
standard, 13, 14
standard deviation, 13
systematic, 11
European projection, 249
Eyepiece, 79
Excavation control, 271
External distance (exsecant), 278

Face left, right, 116
False origin (NG), 24
Finder collimator, 116
Flying levels, 93
Focusing (telescope), 80
Footscrews, 111
Foresight, 93
Forward object/station, 125
Fundamental benchmarks, OS, 75

General purpose theodolite, 110, 120
Geodetic level, 84
Geodetic levelling, 76
Geodimeter, 136
Glass circle, theodolite, 111
Global positioning systems (GPS), 3, 9, 73, 199–202
applications, 202
artificial satellites, 199
binary timing codes, 199
control surveys, 199–203
controller, 201
co-ordinates, 201
differential, 199
heights, 201
measuring modes:
kinematic methods, 201
reoccupation surveying, 201
static positioning, 200
stop and go surveying, 200
phase measurement, 199
NAVSTAR, 9
real time, 201
reference station, 200
roving receiver, 200
surveys, 199–202
Gradienter screw, level, 84
Gradients, 7, 271
Graduation error, staff, 107
Grid bearing, 18
Grid levelling, 150
Grid North, 18
Grid references, OS, 24
Gross errors, 11

Height, 4, 7, 71, 74, 135
Height above MSL, correction for, 69
Height control, setting out, 265
Heighting, 4
barometric, 72
by levelling, 71–108
direct, 7
general, 7, 71
GPS, 73
hydrostatic (water), 72
indirect, 7
stadia, 72, 135
steel tape, 73
tacheometric, 72
trigonometric, 71
Horizontal circle, 110
Horizontal circular curves, *see* Curve ranging
Horizontal equivalent (HE), 147
Horizontal line, 73, 78
Horizontal plates, 110

Inaccessible distance, building surveys, 248
Inaccessible PI, 289
Infrared light, 138
Instrumental errors in EDM:
calibration procedure, 143
cyclic error, 143
scale error, 143
zero error, 143
Intermediate sight, 93
Internal focus telescope, 79, 116
Interpolation of contours, 152
Interpretation of OS maps, 31
Intersection, 6
Intersection angle, 276
Intersection calculation, co-ords., 208
Intersection point (PI), 276
Invar staff, 88

Invar tape, 55
Invert level, 274

Kepler, Johannes, 78

Land areas and parcels on OS maps, 34
Land-Form (OS), 29
Land-Line (OS), 29
Landplan Plots (OS), 27, 28, 31
Laser attachments, 270
Laser level, 82, 266, 272
Latitude, reference, 17
Layout of drawings, 249
Least squares, principle of, 12
Legal and statutory check list, 41
Legs, traverse, 168
Length of curve, 276
Level adjustments, permanent, 104–6
  automatic level, 106
  digital level, 106
  tilting level, 104
Level adjustments, temporary or station, 94
Level line, 73
Level, surveyors', 7, 71, 78
Level vial, 78
Levelling, 71–108
  accuracy, 102
  adjusting the level book, 100–3
  applications, 146
  backsight, foresight, 93
  benchmarks, 74
  booking and recording, 98
  change plate, 89
  changepoint (CP), 93
  checking reductions, 98–102
  classes of level, 87
  coincidence prism bubble reader, 83
  collimation adjustment, 104
  collimation error, 104
  collimation height, 93
  collimation height reduction, 100–1
  collimation line, 79
  compensator unit, 85
  computer application, 108
  contouring by, 149
  cross-sections, 155
  curvature and refraction error, 94
  datum, OS, 74
  datum surface or line, 74
  definitions, 73
  errors, sources, 106
  fieldwork, 92
  flying levels, 93
  fundamental benchmarks, OS, 75
  geodetic, 76
  gradienter screw, 84
  graduation error, 107
  grazing rays, 106
  grid, 150
  handles, steadying rods, 90
  hanging bracket, 90
  hydrostatic, 72
  intermediate sight, 93
  inverted staff, 102
  laser eyepiece, 92
  level line, 73
  level types, 79–106
    accuracy, 88
    automatic, 82, 84, 106
    builder's, 87
    construction, 87
    digital, 21, 82, 86, 106
    dumpy, 82
    engineer's, 87
    general purpose, 87
    geodetic, 84
    laser, 82, 266, 272
    precise, 84
    surveyor's, 7, 71, 78
    tilting, 79, 81–4, 104
  level vial, 78
  levelling up, 95
  lines of levels, 93
  Liverpool datum, OS, 74
  longitudinal sections, 1, 153
  mean sea level, 74
  misclosure, 100
  moving the level, 97
  Newlyn datum, OS, 74
  observing the staff, 95
  one-setup, 93
  ordinary or simple, 76
  parallax elimination, 95
  plane parallel plate micrometer, 90
  permanent adjustments, 83, 104–6
  permissible error, 102
  precise, 76, 156, 158
  principle of, 77
  reciprocal, 159
  reduced level, 18, 74
  reducing levels, 98–102
  refraction, 94
  rise and fall reduction, 98–100
  river crossing, 159
  series, 93
  setting up, 94
  simple levelling, 76
  spirit, 76, 78
  spirit level vial, 78

Levelling – *continued*
  spot heights, 72
  stadia hairs, using, 96
  staff-holder duties, 97
  staves or rods, 88–92
  sources of error, 106
  temporary bench mark (TBM), 74, 146
  terms used, 93
  tilting screw, 83
  two-peg test, 105
  using EDM, 145
  vertical axis, 83
  water level, 78
Levelling staves or rods, 7, 88–92
  change plate for, 89
  digital, 88
  handles and steadying rods, 90
  invar, 88
  ordinary, 88
  Philadelphia, 89
  precise, 88
  staff bubble, 90
  staff illuminator, GEB, 89, 92
Levelling up:
  level, 95
  theodolite, 112, 114
Lifting a line, 59
Line:
  bearing and length from co-ords., 19
  clearing, 59
  collimation, 59
  direct measurement, 60
  horizontal, 73
  level, 73
  lifting over a hill, 59
  ranging a, 59
  true length of, 66
Line ranger, 57
Linear control surveys:
  base line, 162
  booking, 165–7
  check or proof lines, 162
  detail lines, 162
  detail survey, 209–21
  field procedure, 163
  line:
    lifting over hill, 59
    measurement, 165
    ranging, 163
    selection, 163
  picking up detail, 165
  plotting, 167
  survey stations, 164, 166
  tie points, 164
  well conditioned triangles, 163
Linear measurement:
  direct, 52–70
  EDM, 136–44
  ODM, 132–6
Link traverse, 169
Liverpool datum, OS, 74
Local scale factor, 69
Location drawings, 256
Long chord, 278
Longitude, geographical, 23
Longitudinal sections, 153
Loop traverse, 169
Lower plate, 110

Magnetic bearing, 18, 181
Magnetic north, 18
Magnetic variation (declination), 181
Maps and plans, *see* OS maps
Marking materials, 58
Mean sea level (MSL), 74, 201
Measurement:
  angular, 124–8
  corrections to linear, 64
  direct, 52
  EDM, 55, 136–44
  errors in linear, 10, 64
  height, 71
  indirect, 52, 132–44
  line, 60
  notation conventions, 16
  obstacles to linear, 60–3
  optical distance, 132–6
  significant figures in, 16
  stadia, 72, 132
  tacheometric, 72, 132
  taping a line, 52
Minor control points, OS, 36
Misclosure, bearings, 186
Mistakes, 11
Most probable value (MPV), 12
Multiplication constant (stadia), 134

Nadir, 117
National Grid, 24
National maps, 22
NAVSTAR, 9
Newlyn datum, OS, 74
Normal distribution curve, 12
Northing, 19, 24

Objective pentaprism, 270
Objects of surveying, 1
ODM radiation detail survey, 235–9
  advantages/disadvantages, 235

ODM radiation detail survey – *continued*
basis, 235
field procedure, 236
observations:
control, 236
detail and heights, 236
recording data, 237
plotting, 239
reduction of observations:
axis height, 239
bearings, 239
co-ordinates, 239
height differences, 239
horizontal distance, 237
reduced levels, 239
staff intercept, 237
Offsetted base, 213
Offsets, 5, 57, 60
One set-up levelling, 93
Open traverse, 169
Opening co-ords. and bearing, 180, 184
Optical distance measurement (ODM):
accuracy, 136
direct reading tacheometer, 132
general, 132
horizontal wedge instruments, 132
radiation detail survey, 235–9
stadia measurement, 132
subtense bars, 132
vertical staff tache, 132–6
Optical square, 56
Optical theodolite, 109, 111, 237
Optical theodolite reading systems:
circle microscope, 119
coincidence optical micrometer, 120
optical micrometer, 120
optical scale, 119
Ordinary levelling, 76
Ordnance Survey (OS), 22–37
ADDRESS-POINT, 30
administrative boundaries, 29, 31
aerial photography, 36
benchmark information, 76
benchmarks, 34, 36, 74, 102
Boundary-Line, 30
contours, 28
control information, 36
co-ords. transformation service, 36
copyright, 37
datum, 18, 74
digital data, 29
digital map data revision, 30
general, 22
land parcels and areas, 34
Land-Form Panorama, 29
Land-Form Profile, 29
Land-Line, Land-Line Plus, 29
local scale factor, 69
map detail, 32
map interpretation, 31
map references, 24
maps, 22–4
Meridian Vector Data Product, 29
minor control points, 36
national GPS network, 36
National Grid (NG), 22–4
National Grid distance, 69
National Height Data set, 29
Network OSGB36, 22, 36
OSCAR, 30
parcels and Superplan, 36
raster data, 30
revision of digital map data, 30
revision points, 34, 36
Strategi, 30
symbols, 32
Transverse Mercator Projection, 23
traverse stations, 34, 36
triangulation network, 24, 162
triangulation stations, 34, 36
Orthomorphism, 23
OS maps, 22–37
basic map scales, 27
County Series, 22
detail, 32
Explorer, 28
interpretation, 31
land parcels and areas, 34
Landplan Plots, 27, 28, 31
Landranger Series, 28
large scale, 27
Old maps, 37
Pathfinder Series, 28
small scale, 27
Superplan Data Service, 31
Superplan Plots, 27, 31
Travelmaster, 37
1:1250, 28
1:2500, 27
1:5000, 28
1:10 000, 28
1:25 000, 28
1:50 000, 28
OS National Grid:
central meridian, 24
false origin, 24
general, 24
map references, 24
origin, 22
Overall measurements, 210

Parallax error, 80
Partial co-ordinates, 19
Pentaprism, objective, 270
Permissible error, levelling, 102
Phase difference (EDM), 137
Plan control (in setting out), 259–65
Plane parallel plate micrometer, 84
Plane surveying, 2
Planimeter, 307
Plot survey, 242
  angular measurement, 243
  booking the measurements, 243
  equipment, 242
  field procedure, 242
  linear measurement, 243
  preliminaries, 242
  reconnaissance, 243
Plotting:
  by polars, 234
  by rectangular co-ords., 234
Plus measurements, 213
Point of curvature, 276
Precise level, 84
Precise levelling:
  double scale staves, 158
  procedure, 156
  single scale staves, 157
  standard deviation, 88
Precision:
  of equipment, 16
  of measurement, 10
Precision optical plummet, 270
Precision theodolite, 109
Principal tangent to the level vial, 78
Principle of least squares, 12
Principles of survey:
  consistency, 9
  control, 9
  economy of accuracy, 9
  independent check, 10
  revision, 10
  safeguarding, 10
Prism and target, EDM, 139
Prism volume, 314
Prismatic compass, 168
Prismoidal rule, 315
Profiles, 258, 271
Projections, 249
Prolonging a straight line, 263, 271

Quadrant bearing, 18

Radiation detail surveys, 5
  EDM, 221–35
  ODM, 235–9
  theodolite and tape, 239
Radius of the Earth, 69
Radon, 39
Raising and measuring an offset, 211
Random errors, 11
Ranging a line, 163
Ranging rod or pole, 55
Reading systems, *see* Theodolite
Reciprocal levelling, 159
Rectangular co-ordinates, 17
Rectangular co-ordinatograph, 234
Reduced levels, 74, 103
Reference frame, 258
Reference latitude, 17
Reference meridian, 17
Reference object (RO), 125
Referencing a point, 171
Referencing station markers, 171
Reflector (EDM), 137, 139
Refraction, 94
Relative bearing, 18
Remote elevation measurement, 145
Representative fraction, 4
Resection, Collin's Point, 207
Restrictive covenants on land, 39
Revision point, OS, 34, 36
River crossing, 159
Roads:
  centreline, 277, 281
  chainage, 277
  cross-section, 1, 155
  longitudinal section, 1, 153
Running measurements, 210
Running offsets, 215

Sag correction, in taping, 68
Scale, scales, 4
Scale rule, 5
Schedule (setting out), 256
Sensor units, 272
Separate measurements, 210
Series levelling, 93
Setting out, 256–74
  accuracy, 257, 265, 270
  alignment laser, 270
  area excavation or fill, 271
  batter pegs, 273
  block plans, 256
  clarity, 257
  curved line in plan, 264
  cut and fill, 273
  data, documents, drawings, 256
  definitions, 256

detail drawings, 256
earth moving equipment, 272
equipment, 260, 266, 269
excavation control, 271
gradients, sloping line, 271
heights, height control, 265
horizontal angle, 260
horizontal distance, 263
invert level, 274
laser attachments, 270
laser levels, 266, 272
lining in between two marks, 261
location drawings, 256
offset pegs, 259
objective pentaprism, 270
optical plumbing, 268
optical plumbing instruments, 269
pegs, markers, 258, 259, 271
plan control, 259
  common tasks, 260
  methods available, 259
plumb-bob and string, 268
precision optical plummets, 270
profiles, 258, 271
prolonging a straight line, 263, 271
rectangular and polar co-ords., 264
reference frame, 258
rotating laser, 270
schedule, 256
sensor units, 272
sight rails, 271
site plans, 256
slope stakes, 273
small optical plummets, 269
specification, 256
stages in setting out:
  document inspection, 257
  preparing data, 258
  proving site drawings, 257
  site inspection, 257
  site procedure, 258
TBMs, 265, 267
telescope roof plummet, 269
theodolite diagonal eyepiece, 269
timing, 259
traveller, 271, 274
traverses, 259
trench excavation, 274
using EDM, 144
using polar/rectangular co-ordinates, 264
vertical alignment:
  control, 267
  control lines, 270
  laser transfer, 270
  optical plumbing 268
Setting up:
  level, 94
  theodolite 112–17
Settlement and subsidence, 255
Sexagesimal system, 109
Shift, co-ords. transform, 203
Shift, transition curve, 296
Sighting targets, angle measurement, 124
Significant figures, 16
Simple levelling, 76
Simple reversal, 125
Simpson's rule, 307, 315
Sine rule, 20
Site investigations, 35–51
  check lists, 40
  contaminated land, 39
  detail site survey, 41
  easements, 39
  existing building surveys, 42–5
  initial site appraisal, 38
  objectives, 38
  sub-surface investigations, 45–9
Site plans, 256
Slope correction, 65
Slope stakes, 273
Specification (setting out), 256
Spirit levelling, *see* Levelling
Spot height, 7, 149, 151
Spring balance, 58
Stadia measurement, *see* Vertical staff tacheometry
Stadia survey, *see* Vertical staff tacheometry
Staff holder duties, levelling, 97
Staff intercept, 237
Standard deviation, 13
Standard error:
  application of, 15
  of single observation, 13
  of arithmetic mean of a set, 13
  summary of expressions, 14–15
Standardization, tape, 66
Standards, theodolite, 110
Station markers, 171
Steel band, tape, 54
Step taping, 66
Straights, 216
Sub-surface investigations:
  general, 45
  ground water, 49
  harmful ground water, 49
  shrinkable soils, 49
  soil bearing capacity, 49
  soil identification, 47
  soil investigation methods, 45
  soil profiles, 47

Sub-surface investigations – *continued*
  test pits, 45
  width of strip founds., 50
Subtense bar, 132
Summit curve, 298
Superplan plots, 27
Supply of detail, 5
Supply of plan control, 5
Surface taping, 58
Survey drawings:
  cross-sections, 1
  elevations, 1
  longitudinal sections, 1
  maps, 1
  plans, 1
  sections, 1
Surveying:
  air, 1
  basic methods, 5
  cadastral, 3
  chain, 3, 52–64
  classifications of, 2
  control surveys, 161–203
  detail, 3, 209–39
  drawings, 1
  engineering, 3
  geodetic, 2
  GPS, 3, 9, 73, 199–202
  heighting, 7, 71–108
  hydrographic, 3
  introduction to, 1
  levelling, 71–108
  linear, 3, 52–64, 162, 209
  mining, 3
  objects of, 1
  operations, 1
  photogrammetry, 3
  plane, 2
  principles of, 9
  setting out, 2, 256–71
  tacheometer, 3, 132
  tasks, 3
  techniques, 3
  topographical, 3
  traverse, 3, 168–96
  triangulateration, 199
  triangulation, 3, 196–8
  trilateration, 3, 8, 198
Surveyor's level, 7, 71, 78
Surveys, building, 42
Systematic errors, 11

Tacheometry, *see* EDM and ODM survey
Tangent length, 275
Tangent point, 276
Tangential deflection angle, 279
Tape:
  carbon steel, 53
  catenary taping, 58
  coefficient of expansion of steel, 68
  field, 66
  grip, 58
  invar, 55
  spring balance, 58
  standard, 66
  steel band, 53
  step taping, 66
  surface taping, 58
  synthetic, 55
  taping a line, 52
  thermometer, 58
Tape and offset detail surveys, 209–21
  booking detail, 214–18
    common alignments, 215
    offsets, 214
    plus measurements, 216
    straights, 216
    tie points, 164
    ties, 216
  office procedure/plotting, 219–21
    completing the plan, 221
    drawing material, 219
    layout, 219
    North point, 221
    plotting detail, 220
    plotting framework, 219
    preliminaries, 219
    sheet size, 219
  offsets, 210
    raising and measuring, 211
    running, 215
    limitations, 210
  offsetted base, 213
  overall measurements, 210
  plus measurements, 213
  point of intercept, 211
  running measurements, 210
  separate measurements, 210
  step taping, 66
  straights, 211, 216
  ties, 211
Target and prism, EDM, 139
Targets, angle measurement, 124
Telescope:
  astronomical, 79
  collimation line, 78
  diaphragm, 78, 96
  eyepiece, 79
  eyepiece accessories, 92, 269

field of view, 80, 116
finder-collimator, 116
focusing, 80
internal focus, 79
Keplerian, 79
magnification, 116
objective lens, 79
parallax error, 80
resolving power, 116
reticule, 78, 96
stadia hairs, 79
surveying, 78
Temperature correction, taping, 67
Temporary bench marks (TBMs):
definition, 74
establishing, 146
in setting out, 265, 267
Tension correction, 67
Theodolite:
adjustments, 129
alidade, 110
altitude bubble, 117
angle of depression, 118
angle of elevation, 118
angular measurement, 109, 125–8
automatic vert. circle indexing, 117
basic construction, 110–12
booking observations, 126–8, 174–9
bubble adjustment position, 113
builder's, 110
centring, 113
centring rod, 112
centring thorn, 117
circle locking lever, 113, 123
circle microscope system, 119
circle orienting drive, 113, 123
circle reading eyepiece, 111, 118
classifications, 109
coincidence optical micrometer, 120
coincidence prism reader, 117
collimation line, 110
diagonal eyepiece, 269
double centre, 113, 122
EDM add-ons, 138
eyepiece, 116
face left or right, 116
finder-collimator, 116
focusing, 116
footscrews, 111–13
glass circles, 111
horizontal circle or plate, 110, 113
horizontal plate bubble, 113
internal focus telescope, 79, 116
levelling cams, head, screws, 112
levelling up, 112, 114, 115
lower plate, 110
motorized, 121
nadir, 117
observing angles, 116
observing procedure, 126
optical micrometer, 120
optical plummet, 114
optical reading systems, 118
optical scale system, 119
parallax, 80
pentaprism, 270
permanent adjustments, 129–30
plates, upper/lower, 110
plumb-bob, 112
precision, 109
reference object (RO), 125
repetition clamp, 113
repetition method, 128
setting on an angle, 122
setting up the instrument, 112–17
sighting targets, 124
single plate, 123
single second, 109
standards or 'A' frames, 110
swinging or turning, 125
telescope, 111, 116
transit or trunnion axis, 110
tribrach, 112
tripod, 112
types:
builder's, 110
electronic, 109, 121, 124
engineer's, 110, 120
general purpose, 110, 120
optical, 109, 111, 237
universal, 109
vernier, 111
upper/lower plates, 113
vertical angles, 118
vertical circle, 110, 116
wall plate or trivet, 112
zenith distance, 118
Theodolite adjustments, permanent, 129–30
collimation line, 129
optical plummet, 130
plate level, 129
transit axis, 129
vertical circle index, 129
vertical cross-hair orientation, 130
Theodolite and tape radiation survey, 239
Thermometer, 58
Three tripod traversing, 173
Through chainage (curves), 277, 301
Tie points, 164
Ties, 216

Tilting level, 79, 81–4, 104
Tilting screw, 83
Total station, 132
Topographical survey, 3
Topography, 3
Transformation of co-ords., 203
Transit axis, 110
Transit method, 193
Transition curves, *see* Curve ranging
Transverse Mercator Projection, 23
Traveller, 271
Traverse control surveys, 168–96
  abstract of observations, 179
  accuracy, 169, 172, 175, 186, 190
  adjustment methods:
    Bowditch's, 191
    equal adjustments, 193
    least squares, 194
    locating mistakes, 194
    transit, 193
  bearings:
    adjustment, 186
    closing, 185
    forward/back, 181
    misclosure, 186
    opening, 180
    whole circle, 181
  classification:
    closed/open, 169
    computed/plotted, 169
    link, 169
    loop or ring, 169, 183
    low accuracy, 169
    precise/semi-precise, 169
  closing co-ords/bearing, 187
  computation, 176–95
  data recording, 173–6
  legs (lines), 168
  misclosure and accuracy, 172, 190
  opening co-ords/bearing, 180, 184
  plotting, 195
  reconnaissance and layout, 170
  reduction of measured lengths, 178
  reference object (RO), 125
  referencing station markers, 171
  secondary traverses, 171
  station markers, 171
  three tripod method, 173
  traverse application:
    detail traverses, 173
    other traverses, 176
    setting out traverses, 171–3
Traverse survey, *see* Traverse control surveys
Trench excavation, 274
Triangulateration surveys, 199
Triangulation surveys, 8, 196–8
  adjustment:
    equal shifts, 198
    least squares, 198
  base line measurement, 198
  braced quadrilaterals, 196
  centre-point figures, 197
  check base line, 196
  computation, 198
  horizontal angle measurement, 198
  Ordnance Survey, 196
Tribrach, 112
  levelling screws or cams, 112
Trilateration surveys, 8, 198
Tripod, 82
  ball-and-socket head, 95
  three foot-screw type, 95
Trivet, 112
True bearing, 18
Two-peg test, 105
Tying traverse to control, 205–8

Universal theodolite, 109
Upper plate, 110

Variance, magnetic, 14
Vernier theodolite, 111
Versine (versed sine), 289
Vertical angle, 118
Vertical axis:
  level, 83
  theodolite, 110, 116
Vertical circle, 110, 116
Vertical control lines, 270
Vertical curves, *see* Curve ranging
Vertical interval (VI), 147
Vertical staff tacheometry (tache):
  accuracy, 136
  additive constant, 134
  distance measurement, 133
  error sources, 136
  heights, 135
  multiplication constant, 134
  stadia hairs, 133
  staff intercept, 133
  vertical distance, 135
  zenith distance, 135
Vertical transfer by laser, 270
Vertical transfer of points, 268
Volume calculations, 313–18
  changes in soil volume, 316
  cross-section areas, 316
  from contours, 316

from cross-sections:
mean areas rule, 314
prismoidal rule, 315
trapezoidal rule, 315
from spot heights, 317
prismoid, 314
simple solids, 313
Simpson's rule, 315

Walk over survey, 39
Wall plate, 112
Water level, 72
Weighted mean, 14
Weighted observations, 14
Well conditioned triangle, 163
Whole circle bearing, 18, 181
Wild, Heinrich, 111
Working through the bearings, 185

Zenith, zenith angle, zenith distance, 118, 134
Zero, zeros (angle measurement), 125